Nonlinear Optimization in $\mathbb{R}^n$

II. Transversality, Flows, Parametric Aspects

To

Steve Smale,

with deep respect.

Enschede, 8 okt 1986

Bert Jongen

F. Twilt

P. Jonker

Methoden und Verfahren der mathematischen Physik

Herausgegeben von B. Brosowski, Frankfurt
und E. Martensen, Karlsruhe

Band 32

Verlag Peter Lang
Frankfurt am Main · Bern · New York

H. Th. Jongen
P. Jonker
F. Twilt

Nonlinear Optimization in $\mathbb{R}^n$

II. Transversality, Flows, Parametric Aspects

Verlag Peter Lang
Frankfurt am Main · Bern · New York

CIP-Kurztitelaufnahme der Deutschen Bibliothek

Jongen, Hubertus Th.:

Nonlinear optimization in $\mathbb{R}^n$ [IRn] / H. Th. Jongen ;
P. Jonker ; F. Twilt. – Frankfurt am Main ;
Bern ; New York : Lang
NE: Jonker, Peter ; Twilt, Frank:
2. Transversality, flows, parametric aspects. – 1986.
(Methoden und Verfahren der mathematischen Physik ;
Bd. 32)
ISBN 3-8204-9798-6
NE: GT

ISSN 0170-9321
ISBN 3-8204-9798-6

Druck und Bindung: Weihert-Druck GmbH, Darmstadt

TABLE OF CONTENTS

PREFACE.

This book originates from the lecture "Nichtkonvexe Optimierung II", given by the first author during his stay at the University of Hamburg, in particular at the Institut für Angewandte Mathematik. It is preceded by the book (same authors):

NONLINEAR OPTIMIZATION IN $\mathbb{R}^n$, I. Morse Theory, Chebyshev Approximation. Vol. 29, Methoden und Verfahren der mathematischen Physik, Peter Lang Verlag (1983).

The latter book contains the chapters 1-5 (page 1-264).

The present volume contains the chapters 6-10 and it is organized as follows. In Chapter 6 we consider the dependence of critical points and critical values on the problem data. In particular, we deduce under a nondegeneracy assumption that critical points (resp. values) depend C^1- (resp. C^2-) Fréchet differentiable on the C^2-data. Then, we show that the concepts of stability and nondegeneracy coincide. Finally, a main result on global stability ("structural stability") is proved for compact, non-degenerate and separating optimization problems.
In Chapter 7 we treat transversality theory in finite dimensions. Many different points of view are brought together. Among the subjects we have (jet-) transversality theorems for manifolds and stratified sets, with applications to optimization. Particular attention is paid to openess-results ("stability").
In Chapter 8 we consider the gradient system of a differentiable function with respect to a variable (Riemannian) metric. The linearization of the resulting flow around regular and (hyperbolic) singular points is discussed, and (un)stable manifolds corresponding to singular points are introduced. Two famous results of S. Smale are explained: Firstly, it is shown how a given gradient system can be perturbed in order that the perturbed system is in general position (i.e. all stable and unstable manifolds intersect transversally). Secondly, we discuss possible variations of the critical values of the underlying function, which yields a so-called self-indexing function. For a gradient system (in general position) on a compact connected manifold we show, starting at some local minimum, how any other local minimum can be reached via a sequence of saddle points of index 1/local

minima (0-1-0 graph) on one hand, and via a sequence of local maxima/ minima on the other hand. Finally, we introduce the reflected gradient field F_k which has the property that, among the critical points of the underlying function, only those of index k are attractors for F_k.
In Chapter 9 we study the underlying differential equation for Newton's method for finding zeros of differentiable mappings from $\mathbb{R}^n$ to $\mathbb{R}^n$. An extension of this differential equation (or vector field) to the whole $\mathbb{R}^n$ by means of a suitable damping introduces additional singularities ("extraneous" singularities). These extraneous singularities will play an important role. Special attention will be paid to the case where the mapping on $\mathbb{R}^n$ is the derivative of a differentiable function, and a fairly complete study is done in the two-dimensional case. Finally, a small digression is made to two-dimensional systems arising from Newton's method for the search of zeros of meromorphic functions on the complex plane.
In Chapter 10 we consider optimization problems depending on parameters. After a general introduction into critical manifolds which arise in parametric problems, we specialize to the case where we have a one-dimensional parameter. Degeneracy- resp. bifurcation phenomena are considered from both a local and global point of view.

We would like to thank Bruno Brosowski for his invitation to publish these lecture notes and for his stimulation during the preparation of the manuscript.

The first author would like to thank all members of the Institut für Angewandte Mathematik (in particular Frau W. Bergmann, Klaus Glashoff and Wolf Hofmann) for their kind hospitality during his stay at the University of Hamburg. Moreover he is indebted to Dirk Siersma and Floris Takens for useful discussions.

Last but not least, we express our gratitude to Diny Ticheler for her careful and excellent preparation of the type-written version of this manuscript.

Bert Jongen, Peter Jonker, Frank Twilt
Enschede, The Netherlands, March 1986.

6. STABILITY OF OPTIMIZATION PROBLEMS.

6.1. On the dependence of critical points (-values) on the problem-data.

In Section 4.1 we studied the dependence of nondegenerate critical points and their values on a finite number of parameters. This was done in order to describe the change of the extremal set (-value) of the errorfunction in Chebyshev approximation problems.
In this section we will allow more general perturbations. In fact, these perturbations will describe -in a natural way- all possible perturbations. We start with a discussion on unconstrained problems and then, the ideas will be extended to constrained problems.

For fixed $\bar{x} \in \mathbb{R}^n$ and $r > 0$, we put, ($\|\cdot\|$ denoting the Euclidean norm):

$$B(\bar{x},r) = \{x \in \mathbb{R}^n \mid \|x-\bar{x}\| \leq r\}. \tag{6.1.1}$$

By $\mathring{B}(\bar{x},r)$ we denote the interior of $B(\bar{x},r)$. If no confusion is possible, we shortly write B, resp. $\mathring{B}$, instead of $B(\bar{x},r)$, resp. $\mathring{B}(\bar{x},r)$. Given the ball B, we introduce the following linear space in order to study some effects of <u>local</u> perturbations of C^2-functions:

$$C^2(B) = \{\phi \in C^2(\mathring{B},\mathbb{R}) \mid \text{all partial derivatives up to second order of } \phi \text{ are continuously extendable on the whole ball } B\}.$$

On $C^2(B)$ we define the norm $\|\cdot\|_B$, which turns $C^2(B)$ into a Banach space:

$$\|\phi\|_B = \max_{|\alpha|\leq 2} \sup_{x\in\mathring{B}} |\phi^\alpha(x)| , \tag{6.1.2}$$

where $\alpha = (\alpha_1,\ldots,\alpha_n) \in \mathbb{N}^n$, $|\alpha| = \sum_{i=1}^{n} \alpha_i$, $\phi^\alpha = \dfrac{\partial^{|\alpha|}}{\partial x_1^{\alpha_1}\ldots\partial x_n^{\alpha_n}}$

For sake of completeness we recall some details of Fréchet-differentiability in Banach spaces.
Let E, F be Banach spaces with norm $\|\cdot\|_E$, $\|\cdot\|_F$, and $\mathcal{O} \subset E$ be an open subset. Let $f\colon \mathcal{O} \to F$ be a map. Then, f is called Fréchet-differentiable at $\bar{x} \in \mathcal{O}$ if there exists a bounded linear map $L\colon E \to F$ such that

$$\|f(\bar{x}+y) - f(\bar{x}) - Ly\|_F = o(\|y\|_E). \qquad (*)$$

The map L is called the Fréchet-derivative of f at $\bar{x}$. Note, that a linear map L: $E \to F$ is bounded iff it is continuous.
Moreover, if f: $\mathcal{O} \to F$ is bounded in a neighborhood of $\bar{x}$ and if L is a linear map satisfying the above estimate (*), then L is bounded as well and hence, L is the Fréchet-derivative of f at $\bar{x}$. In fact, this follows from the following inequality:

$$\|Ly\|_F \leq \|f(\bar{x}+y) - f(\bar{x}) - Ly\|_F + \|f(\bar{x}+y) - f(\bar{x})\|_F.$$

Let f be Fréchet-differentiable at every $x \in \mathcal{O}$, the Fréchet-derivative at x being denoted by Df(x). Then, Df is a map from $\mathcal{O}$ to $L(E,F)$, where $L(E,F)$ is the Banach space consisting of all bounded linear maps from E to F, endowed with the usual operatornorm. Then, f is called C^1-Fréchet differentiable at $\bar{x}$ if Df: $\mathcal{O} \to L(E,F)$ is continuous at $\bar{x}$. If the map Df itself is (C^1-) Fréchet-differentiable, then we say that f is twice (C^2-) Fréchet-differentiable.
Let D^2f denote D(Df). Note that D^2f, evaluated at $\bar{x} \in \mathcal{O}$, is an element of $L(E,L(E,F))$. Let $L^2(E\times E,F)$ be the Banach space of bounded bilinear maps from $E \times E$ to F, with the norm

$$\|A\| = \sup_{\|e\|_E \leq 1, \|h\|_E \leq 1} \|A(e,h)\|_F .$$

Then, the map Φ: $L(E,L(E,F)) \to L^2(E\times E,F)$, defined by $[e \mapsto (h \mapsto L_e(h))] \mapsto [(e,h) \mapsto L_e(h)]$ is obviously a linear, bounded and bijective map between Banach spaces. Hence, Φ^{-1} is bounded as well. Thus Φ constitutes a topological isomorphism between $L(E,L(E,F))$ and $L^2(E\times E,F)$, (cf. also [53]).

<u>Lemma 6.1.1</u>. Let $f \in C^2(\mathbb{R}^n,\mathbb{R})$ and suppose that $\bar{x}$ is a nondegenerate critical point for f. Then we have:

a. There exists an $r > 0$ such that $\bar{x}$ is the only critical point for f in the ball $B := B(\bar{x},r)$.

Let f also denote the restriction of f to B.

b. Let $\mathcal{U} \subset \overset{\circ}{B}$ be a neighborhood of $\bar{x}$. Then, there exists an open neighborhood $\mathcal{O} \subset C^2(B)$ of f and a unique C^1-Fréchet differentiable map $\xi\colon \mathcal{O} \to \mathbb{R}^n$, $\xi(\mathcal{O}) \subset \mathcal{U}$, having the property: if $g \in \mathcal{O}$, then $\xi(g)$ is the unique zero in B of the derivative Dg and, moreover, $D^2g(\xi(g))$ is nonsingular.

c. The critical value-map ψ, defined by: $\psi\colon \mathcal{O} \to \mathbb{R}$, $\psi(g) = g(\xi(g))$, is C^2-Fréchet differentiable. □

Before proving Lemma 6.1.1 we consider the differentiability of some special maps T_α. These maps were used in [28], [38], in order to prove the local continuity of the Chebyshev-operator in Chebyshev approximation problems.

Note that $C^2(B) \times \mathbb{R}^n$ is a Banach space under the norm $\|\cdot\|$, where $\|(\phi,x)\| = \|\phi\|_B + \|x\|$. Of course, the set $C^2(B) \times \overset{\circ}{B}$ is an open subset of $C^2(B) \times \mathbb{R}^n$.

For $|\alpha| \le 2$, $\alpha = (\alpha_1,\dots,\alpha_n) \in \mathbb{N}^n$, the map T_α is well-defined:

$$T_\alpha\colon C^2(B) \times \overset{\circ}{B} \to \mathbb{R},\ T_\alpha(\phi,x) = \phi^\alpha(x). \tag{6.1.3}$$

<u>Lemma 6.1.2</u>. The map T_α is continuous for $|\alpha| \le 2$. Moreover, for $|\alpha| \le 1$, T_α is C^1-Fréchet differentiable.

<u>Proof</u>.

<u>Continuity</u>. $|T_\alpha(\phi,x) - T_\alpha(\psi,y)| = |\phi^\alpha(x) - \psi^\alpha(y)| =$

$= |\phi^\alpha(x) - \psi^\alpha(y) + \phi^\alpha(y) - \phi^\alpha(y)| \le |\phi^\alpha(y) - \psi^\alpha(y)| + |\phi^\alpha(x) - \phi^\alpha(y)| \le$

$\le \|\phi-\psi\|_B + |\phi^\alpha(x) - \phi^\alpha(y)|$. Since ϕ^α is continuous, it follows that T_α is continuous.

We prove the second assertion of the lemma for the case $\alpha = 0$, the proof of the case $|\alpha| = 1$ running along the same lines.

<u>Fréchet-differentiability</u>. Pick an arbitrary element (ϕ,x) from $C^2(B) \times \overset{\circ}{B}$. Let $y \in \mathbb{R}^n$ be such that $x+y \in \overset{\circ}{B}$. Then we have:

$$T_0(\phi+\psi, x+y) - T_0(\phi, x) = \phi(x+y) + \psi(x+y) - \phi(x) =$$

$$= D\phi(x)y + \psi(x) + D\psi(x)y + o(\|y\|) =$$

$$= D\phi(x)y + \psi(x) + o(\|(\psi,y)\|). \tag{6.1.4}$$

The map $DT_0(\phi,x)$: $C^2(B) \times \mathbb{R}^n \to \mathbb{R}$, $(\psi,y) \mapsto D\phi(x)y + \psi(x)$, is obviously linear, and since T_0 is continuous, it is bounded as well. Hence, in view of (6.1.4), $DT_0(\phi,x)$ is the Fréchet-derivative of T_0 at (ϕ,x).

<u>C^1-Fréchet differentiability</u>. By $[C^2(B) \times \mathbb{R}^n]^*$ we denote the space of bounded linear functionals on $C^2(B) \times \mathbb{R}^n$. So, we have to show that the map DT_0, defined by

$$DT_0\colon C^2(B) \times \mathring{B} \to [C^2(B) \times \mathbb{R}^n]^*,$$

is continuous w.r.t. $(\|\cdot\|, \|\cdot\|^*)$. Here, $\|\cdot\|^*$ is the norm on $[C^2(B) \times \mathbb{R}^n]^*$ induced by $\|\cdot\|$. We have:

$$\|DT_0(\phi,x) - DT_0(\tilde{\phi},\tilde{x})\|^* = \sup_{\|(\psi,y)\|\le 1} |(DT_0(\phi,x) - DT_0(\tilde{\phi},\tilde{x}))(\psi,y)| =$$

$$= \sup_{\|(\psi,y)\|\le 1} |(D\phi(x) - D\tilde{\phi}(\tilde{x}))y + \psi(x) - \psi(\tilde{x})| \le$$

$$\le \sup_{\|(\psi,y)\|\le 1} \{\|D\phi(x) - D\tilde{\phi}(\tilde{x})\| \cdot \|y\| + |\psi(x) - \psi(\tilde{x})|\}. \tag{6.1.5}$$

Since $\mathring{B}$ is convex, the line segment $[x,\tilde{x}]$ is contained in $\mathring{B}$. So, there exists a $\theta \in (0,1)$ such that:

$$|\psi(x) - \psi(\tilde{x})| = |D\psi(\tilde{x}+\theta(x-\tilde{x}))(x-\tilde{x})| \le$$

$$\le \|D\psi(\tilde{x}+\theta(x-\tilde{x}))\| \cdot \|x-\tilde{x}\|. \tag{6.1.6}$$

The inequality $\|(\psi,y)\| \le 1$ implies $\|\psi\|_B \le 1$, $\|y\| \le 1$, and from (6.1.5), (6.1.6), together with the continuity of T_α, $|\alpha| = 1$, we finally obtain the continuity of DT_0 at (ϕ,x). □

<u>Proof of Lemma 6.1.1.</u>

<u>Statement a</u> immediately follows from the fact that a nondegenerate critical point is an isolated critical point (cf. Section 1.1).

Statement b is proved by means of application of the Implicit Function Theorem for Banach spaces (cf. [56]). In fact, consider the map:

$$\mathcal{T}: \mathcal{C}^2(B) \times \mathring{B} \to \mathbb{R}^n, \ \mathcal{T}(\phi,x) = D^T\phi(x). \qquad (6.1.7)$$

From Lemma 6.1.2 we see that $\mathcal{T}$ is C^1-Fréchet differentiable. Moreover, we have $\mathcal{T}(f,\bar{x}) = 0$ and $D_x\mathcal{T}(f,\bar{x}) = D^2f(\bar{x})$. Since $\bar{x}$ is a nondegenerate critical point, the matrix $D^2f(\bar{x})$ is nonsingular. Consequently, we can apply the Implicit Function Theorem for Banach spaces, which gives us an open neighborhood $\mathcal{O}$ and a unique C^1-Fréchet differentiable map $\xi: \mathcal{O} \to \mathbb{R}^n$ such that $\mathcal{T}(g,\xi(g)) \equiv 0$. Now, Statement b follows, eventually by shrinking the open set $\mathcal{O}$ a little bit.

Statement c. From the fact that $g \in \mathcal{O}$ is a C^2-function on $\mathring{B}$ and that ξ is a C^1-map on $\mathcal{O}$ it follows that ψ is a C^1-Fréchet differentiable map. A short calculation shows:

$$\psi(g+\phi) - \psi(g) = Dg(\xi(g))\cdot D\xi(g)(\phi) + \phi(\xi(g)) + o(\|\phi\|_B). \qquad (6.1.8)$$

By construction we have $Dg(\xi(g)) \equiv 0$. The map $\phi \mapsto \phi(\xi(g))$ is linear, hence bounded since ψ is continuous. Consequently, we obtain, together with (6.1.8), that the derivative $D\psi(g)$ is represented by:

$$D\psi(g)(\phi) = \phi(\xi(g)) \text{ for all } \phi \in \mathcal{C}^2(B). \qquad (6.1.9)$$

Let us denote a dual space by an asterisk *. It remains to show that the map $D\psi(\cdot): \mathcal{O} \to \mathcal{C}^2(B)^*$ is C^1-Fréchet-differentiable. From (6.1.9) it is reasonable to expect that the derivative of $D\psi$ at g, say $\mathcal{D}_2\psi(g)$, will be represented by:

$$\mathcal{D}_2\psi(g)(\phi,h) = D\phi(\xi(g))\cdot D\xi(g)(h). \qquad (6.1.10)$$

At first sight, $\mathcal{D}_2\psi(g)$ in (6.1.10) seems not te be <u>symmetric</u> bilinear. However, since $\mathcal{T}(g,\xi(g)) \equiv 0$, we obtain by means of the chainrule and inserting the result for $D\xi(g)(h)$ in (6.1.10):

$$\mathcal{D}_2\psi(g)(\phi,h) = -D\phi(\xi(g))\cdot D^2g(\xi(g))^{-1}\cdot D^Th(\xi(g)). \qquad (6.1.11)$$

From (6.1.11) we see:

$$|\mathcal{D}_2\psi(g)(\phi,h)| \le \|D^2g(\xi(g))^{-1}\|\cdot\|\phi\|_B\cdot\|h\|_B,$$

the first in the righthand side norm standing for the matrix-norm induced by the Euclidean vectornorm. Hence, $\mathcal{D}_2\psi(g)$ is an element of $L^2(C^2(B) \times C^2(B), \mathbb{R})$. Consequently, the linear map

$$h \mapsto [\phi \mapsto D\phi(\xi(g)) \cdot D\xi(g)(h)]$$

is bounded (compare (6.1.10)).

In order to show that $\mathcal{D}_2\psi(g)$ <u>is</u> the second Fréchet-derivative of ψ at g, i.e. $\mathcal{D}_2\psi(g) = D^2\psi(g)$, it remains to establish the following formula:

$$\sup_{\|\phi\|_B \leq 1} |A(\phi,g,h)| = o(\|h\|_B), \tag{6.1.12}$$

where $A(\phi,g,h) = \phi(\xi(g+h)) - \phi(\xi(g)) - D\phi(\xi(g)) \cdot D\xi(g)(h)$.

Take a $\phi \in C^2(B)$ with $\|\phi\|_B \leq 1$.

The following equation holds for $g+h \in \mathcal{O}$:

$$\xi(g+h) = \xi(g) + D\xi(g) \cdot h + \varepsilon(h), \tag{6.1.13}$$

where $\varepsilon(h) \cdot \|h\|_B^{-1} \to 0$ as $\|h\|_B \to 0$.

Note that the linesegment $[\xi(g), \xi(g+h)]$ is contained in $\mathring{B}$. Then from (6.1.13) we obtain:

$$\phi(\xi(g+h)) = \phi(\xi(g)) + D\phi(\xi(g)) \cdot D\xi(g)(h) + D\phi(\xi(g)) \cdot \varepsilon(h) + \\ + \frac{1}{2} \alpha^T(h) \cdot D^2\phi(\xi(g) + \theta \cdot \alpha(h)) \cdot \alpha(h), \tag{6.1.14}$$

where $\alpha(h) = D\xi(g)(h) + \varepsilon(h)$ and θ some number in [0,1].

Note that $\|\alpha(h)\|^2 = o(\|h\|_B)$.

From (6.1.14) and the fact that $\|\phi\|_B \leq 1$, we obtain

$$|A(\phi,g,h)| \leq \|\varepsilon(h)\| \cdot \sqrt{n} + \frac{1}{2}n \|\alpha(h)\|^2, \tag{6.1.15}$$

(recall that $\|\cdot\|$ is the Euclidean norm).

Since the righthand side of (6.1.15) does not depend on ϕ and since $\|\varepsilon(h)\| = o(\|h\|_B)$, $\|\alpha(h)\|^2 = o(\|h\|_B)$, we see that (6.1.12) is established.

Finally, we have to show that $D^2\psi(\cdot)$ is continuous. In fact, from (6.1.11) it follows that $D^2\psi(\cdot)$ is continuous iff the map

$$\Phi: \mathcal{O} \to \text{(space of symmetric } n \times n\text{-matrices)},$$

given by $\Phi(\eta) = D^2\eta(\xi(\eta))^{-1}$ is continuous. But the latter fact is obvious. Hence, the proof of Lemma 6.1.1 is complete. □

<u>Remark 6.1.1</u>. The Euclidean ball $B := B(\bar{x},r)$ is closed and has a smooth boundary. Therefore, it follows from Corollary 3.1.4 that any $f \in C^2(B)$ is the restriction to B of some $\tilde{f} \in C^2(\mathbb{R}^n,\mathbb{R})$.

<u>Remark 6.1.2</u>. Although in Lemma 6.1.1 the most natural class of "data-perturbations" is used, its application to special parametrized families should be carried out carefully, as it will become clear in this remark.

Let $F \in C^k(\mathbb{R}^m \times \mathbb{R}^n,\mathbb{R})$, $(t,x) \mapsto F(t,x)$, where $t \in \mathbb{R}^m$, $x \in \mathbb{R}^n$ and where $k \geq 2$. We may consider F as an m-parameter family of functions on $\mathbb{R}^n$, t being the parameter.
Let $\bar{x} \in \mathbb{R}^n$ be a nondegenerate critical point for the function $F(\bar{t},\cdot)$, i.e. $D_xF(\bar{t},\bar{x}) = 0$ and $D_x^2F(\bar{t},\bar{x})$ is nonsingular. Choose a ball $B := B(\bar{x},r)$ such that $\bar{x}$ is the only critical point of $F(\bar{t},\cdot)$ in B.
Let $\mathcal{U} \subset \mathring{B}$ be a neighborhood of $\bar{x}$. Then, application of the Implicit Function Theorem yields an open neighborhood $\mathcal{V}$ of $\bar{t}$ and a unique C^1-map $\eta\colon \mathcal{V} \to \mathbb{R}^n$, $\eta(\mathcal{V}) \subset \mathcal{U}$, having the property: if $t \in \mathcal{V}$, then $\eta(t)$ is the unique zero in B of the derivative $D_xF(t,\cdot)$ and $D_x^2F(t,\eta(t))$ is nonsingular.
Furthermore, the critical value map $\phi\colon \mathcal{V} \to \mathbb{R}$, defined by $\phi(t) = F(t,\eta(t))$, is a C^2-function. In fact, a short calculation gives:

$$D_t\phi(t) = D_tF(t,\eta(t)), \tag{6.1.16}$$

$$D_t^2\phi(t) = D_t^2F - D_xD_t^TF\cdot(D_x^2F)^{-1}\cdot D_tD_x^TF\big|_{(t,\eta(t))}. \tag{6.1.17}$$

By $F(t,\cdot)$ we also denote the restriction of $F(t,\cdot)$ to B. If $k \geq 4$, the map $\mathcal{F}\colon \mathbb{R}^m \to C^2(B)$, $t \mapsto F(t,\cdot)$, is a C^2-Fréchet differentiable map. In that case, Formulae (6.1.16), (6.1.17) can be obtained from (6.1.9), (6.1.10) by means of the chain rule. Note that:

$$\phi(t) = \psi\circ\mathcal{F}(t) \quad \text{(also for } k = 2,3). \tag{6.1.18}$$

For the derivatives $D_t\mathcal{F}$, $D_t^2\mathcal{F}$ we obtain:

$$D_t\mathcal{F}(t) = D_tF(t,\cdot),\ D_t^2\mathcal{F}(t) = D_t^2F(t,\cdot). \tag{6.1.19}$$

We merely establish the formula for $D_t\mathcal{F}(t)$; the formula for $D_t^2\mathcal{F}(t)$ can be obtained in an analogous way.

Firstly, we have:

$$F(t+v,x) - F(t,x) = \int_0^1 \frac{d}{d\tau} F(t+\tau v,x)\,d\tau. \tag{6.1.20}$$

For $x \in B$ we obtain from (6.1.20):

$$|F(t+v,x) - F(t,x) - D_tF(t,x)v| \le (\sqrt{n})\|v\| \int_0^1 \|D_tF(t+\tau v,\cdot) - D_tF(t,\cdot)\|_\infty d\tau, \tag{6.1.21}$$

where $\|D_tF(\tilde{t},\cdot)\|_\infty := \max_{1\le i\le m} \{\max_{x\in B} |\frac{\partial}{\partial t_i} F(\tilde{t},x)|\}$.

From (6.1.21) we see that

$$\max_{x\in B} |F(t+v,x) - F(t,x) - D_tF(t,x)v| = o(\|v\|). \tag{6.1.22}$$

In an analogous way we obtain for $|\alpha| \le 2$, $\alpha = (\alpha_1,\dots,\alpha_n) \in \mathbb{N}^n$:

$$\max_{x\in B} |\frac{\partial^{|\alpha|}}{\partial x_1^{\alpha_1}\dots\partial x_n^{\alpha_n}} (F(t+v,x) - F(t,x) - D_tF(t,x)v)| = o(\|v\|),$$

and hence,

$$\|F(t+v,\cdot) - F(t,\cdot) - D_tF(t,\cdot)v\|_B = o(\|v\|),$$

which establishes the fact that $D_t\mathcal{F}(t) = D_tF(t,\cdot)$.

From (6.1.18) we obtain, by means of the chain rule, deleting the arguments:

$$D_t\phi\cdot v = D\psi\cdot D_t\mathcal{F}\cdot v\ , \tag{6.1.23}$$

$$v^T\cdot D_t^2\phi\cdot w = D^2\psi(D_t\mathcal{F}\cdot v, D_t\mathcal{F}\cdot w) + D\psi\cdot(v^T\cdot D_t^2\mathcal{F}\cdot w)\ , \tag{6.1.24}$$

where $v,w \in \mathbb{R}^m$.

A combination of (6.1.23) and (6.1.9) yields (6.1.16). Furhtermore, a combination of (6.1.24), (6.1.9) and (6.1.11) yields (6.1.17).

However, if k = 2, resp. 3, then $\mathcal{F}$ is merely a continuous, resp. C^1-Fréchet differentiable map. Hence, in the case k = 2,3, we cannot straightforwardly use the formal chain rule as above, in order to obtain (6.1.17), although the composite map, given by (6.1.18), is a C^2-map. This subtile "differentiability-gap" can be removed by noting that a function $F \in C^2(\mathbb{R}^m \times \mathbb{R}^n, \mathbb{R})$ can be approximated arbitrarily well, uniformly upto derivatives of second order with respect to a given compact subset of $\mathbb{R}^m \times \mathbb{R}^n$, by means of an $\tilde{F} \in C^4(\mathbb{R}^m \times \mathbb{R}^n, \mathbb{R})$. From this it follows that the result, obtained by the formal chain rule, is also true for an F of class C^2.

We will not dwell here on the technique of approximating an F of class C^2 by means of an F of class C^4, but refer for such constructions to [31]. □

By $C^2(B)^p$ we denote the finite product of p copies of $C^2(B)$. The space $C^2(B)^p$ becomes a Banach space under the norm:

$$\|(f_1,\ldots f_p)\|_B := \max_{1\leq i\leq p} \|f_i\|_B \,. \tag{6.1.25}$$

A constraint set M[h,g] is called regular in B if at every point $x \in M[h,g] \cap B$ the set of derivatives of the active constraints is linearly independent. If M[h,g] is regular in B, then it makes sense to speak about the critical points for $f_{|M[h,g]}$ restricted to B.

Some of the ideas in the proof of the following theorem are used independently by Fujiwara in [11].

Theorem 6.1.1. Let I, J be finite indexsets and $f, h_i, g_j \in C^2(\mathbb{R}^n, \mathbb{R})$, $i \in I$, $j \in J$. Suppose that the corresponding constraint set M[h,g] is regular, where $h = (h_i)_{i\in I}$ and $g = (g_j)_{j\in J}$. Moreover, let $\bar{x} \in M[h,g]$ be a nondegenerate critical point for $f_{|M[h,g]}$. Then, we have:

a. There exists an $r > 0$ such that $\bar{x}$ is the only critical point for $f_{|M[h,g]}$ in the ball $B := B(\bar{x}, r)$.

Let f, resp. h_i, g_j, also denote the restriction of f, resp. h_i, g_j, to B.

b. Let $\mathcal{U} \subset \mathring{B}$ be a neighborhood of $\bar{x}$. Then, there exists an open neighborhood $\mathcal{O} \subset C^2(B)^{1+|I|+|J|}$ of (f,h,g) and a unique C^1-Fréchet differentiable map $\xi\colon \mathcal{O} \to \mathbb{R}^n$, $\xi(\mathcal{O}) \subset \mathcal{U}$,

having the property:
if $(\tilde{f},\tilde{h},\tilde{g}) \in \mathcal{O}$, then $M[\tilde{h},\tilde{g}]$ is regular in B, and $\xi(\tilde{f},\tilde{h},\tilde{g})$ is the unique critical point for $\tilde{f}_{|M[\tilde{h},\tilde{g}]}$ restricted to B. Moreover, the critical point $\xi(\tilde{f},\tilde{h},\tilde{g})$ is nondegenerate and the index set of active inequality constraints $J_0(\xi(\tilde{f},\tilde{h},\tilde{g}))$ is constant.

c. The critical value map $\psi: \mathcal{O} \to \mathbb{R}$, defined by $\psi(\tilde{f},\tilde{h},\tilde{g}) = \tilde{f}(\xi(\tilde{f},\tilde{h},\tilde{g}))$ is C^2-Fréchet differentiable.

Proof. In view of the ideas in Section 4.1 and the fact that the proof of the theorem in the case $I = J = \emptyset$ is given in full detail in the proof of Lemma 6.1.1, we will restrict ourselves to the main idea and leave the details to the reader. From Corollary 3.2.1 we see that $\bar{x}$ is an isolated critical point. This implies Statement a. Now, we shrink the radius of the ball B such that the indexset of active inequality constraints, $J_0(x)$, is a subset of $J_0(\bar{x})$ for all $x \in B$.
Consequently, by continuity, the constraints $g_j, j \notin J_0(\bar{x})$, will play no role in the sequel. Let $\bar{\lambda}_i$, $\bar{\mu}_j$, $i \in I$, $j \in J_0(\bar{x})$ be the Lagrange parameters corresponding to the critical point $\bar{x}$. By λ, μ we denote the vectors $(\lambda_i)_{i\in I}$, $(\mu_j)_{j\in J_0(\bar{x})}$. Now, consider the following map T:

$$T: \mathring{B} \times \mathbb{R}^{|I|} \times \mathbb{R}^{|J_0(\bar{x})|} \times C^2(B)^{1+|I|+|J_0(\bar{x})|} \to \mathbb{R}^{n+|I|+|J_0(\bar{x})|},$$

$$T: \begin{pmatrix} x \\ \lambda \\ \mu \\ \tilde{f} \\ \tilde{h} \\ \tilde{g} \end{pmatrix} \longmapsto \begin{pmatrix} D^T\tilde{f}(x) - \sum_{i\in I} \lambda_i D^T\tilde{h}_i(x) - \sum_{j\in J_0(\bar{x})} \mu_j D^T\tilde{g}_j(x) \\ \\ -\tilde{h}_i(x),\ i \in I \\ \\ -\tilde{g}_j(x),\ j \in J_0(\bar{x}) \end{pmatrix} \tag{6.1.26}$$

Then, T is a C^1-Fréchet differentiable map, $T(\bar{x},\bar{\lambda},\bar{\mu},f,h,g) = 0$ and $D_{\begin{pmatrix} x \\ \lambda \\ \mu \end{pmatrix}} T(\bar{x},\bar{\lambda},\bar{\mu},f,h,g)$ is nonsingular (cf. Examples 3.2.6, 3.2.7).

Hence, by means of the Implicit Function Theorem for Banach spaces, Statement b is easily verified. Moreover, the Lagrange parameters $\lambda_i(\tilde{f},\tilde{h},\tilde{g})$, $\mu_j(\tilde{f},\tilde{h},\tilde{g})$, $i \in I$, $j \in J_o(\bar{x})$, corresponding to $\xi(\tilde{f},\tilde{h},\tilde{g})$, are C^1-Fréchet differentiable maps.

In order to establish Statement c, it is crucial to note the following:

$$\left.\begin{aligned} &\psi(\tilde{f}+\phi,\tilde{h}+\eta,\tilde{g}+\zeta) - \psi(\tilde{f},\tilde{h},\tilde{g}) = \\ &(\tilde{f}+\phi)(\xi(\tilde{f}+\phi,\tilde{h}+\eta,\tilde{g}+\zeta)) - \tilde{f}(\xi(\tilde{f},\tilde{h},\tilde{g})) + \\ &- \sum_{i\in I} \lambda_i(\tilde{f},\tilde{h},\tilde{g})\cdot(\tilde{h}_i+\eta_i)(\xi(\tilde{f}+\phi,\tilde{h}+\eta,\tilde{g}+\zeta)) + \\ &- \sum_{j\in J_o(\bar{x})} \mu_j(\tilde{f},\tilde{h},\tilde{g})\cdot(\tilde{g}_j+\zeta_j)(\xi(\tilde{f}+\phi,\tilde{h}+\eta,\tilde{g}+\zeta)) \end{aligned}\right\}. \tag{6.1.27}$$

In fact, just note that $\tilde{h}_i + \eta_i$, resp. $\tilde{g}_j + \zeta_j$ vanish at the point $\xi(\tilde{f}+\phi,\tilde{h}+\eta,\tilde{g}+\zeta)$, for $i \in I$, $j \in J_o(\bar{x})$.

Now, a few calculations, using (6.1.26), (6.1.27), show that the Fréchet-derivative $D\psi$ at the point $(\tilde{f},\tilde{h},\tilde{g})$ is represented by:

$$D\psi(*)(\phi,\eta,\zeta) = (\phi - \sum_{i\in I} \lambda_i(*)\eta_i - \sum_{j\in J_o(\bar{x})} \mu_j(*)\zeta_j)(\xi(*)), \tag{6.1.28}$$

for all $(\phi,\eta,\zeta) \in C^2(B) \times C^2(B)^{|I|} \times C^2(B)^{|J|}$, where $(*) = (\tilde{f},\tilde{h},\tilde{g})$. (Note that, locally, we may neglect the inequality constraints corresponding to the index set $J\backslash J_o(\bar{x})$).

The fact that $D\psi(\cdot)$ itself is a C^1-Fréchet differentiable map (and hence, ψ is C^2-Fréchet differentiable) runs along the same lines as in the proof of Lemma 6.1.1. □

<u>Remark 6.1.3</u>. We emphasize that Formula (6.1.28) describes the change of the critical point value up to first order. In this way, it can be interpreted as a "sensitivity result". A special perturbation is considered in detail in the following Example 6.1.1, the result of which can also be derived directly from (6.1.28). Furthermore, we note that the content of Remark 6.1.2 extends to the situation in Theorem 6.1.1 as well.

Example 6.1.1. Interpretation of Lagrange-parameters.
Let $f, h_i, g_j \in C^2(\mathbb{R}^n, \mathbb{R})$, $i = 1,\dots,m$, $j = 1,\dots,s$, assume that the corresponding constraint set $M := M[h,g]$ is regular and that the critical points for $f_{|M[h,g]}$ are nondegenerate.

For $a \in \mathbb{R}^m$, $b \in \mathbb{R}^s$, we put:

$$M(a,b) = \{x \in \mathbb{R}^n \mid h_i(x) = a_i,\ g_j(x) \geq b_j,\ i = 1,\dots,m,\ j = 1,\dots,s\}.$$

Let $\bar{x} \in M$ be a critical point for $f_{|M}$ with Lagrange parameters $\bar{\lambda}_i$, $i = 1,\dots,m$, $\bar{\mu}_j$, $j = 1,\dots,p$ (without loss of generality we assume that the first p inequality constraints are active at $\bar{x}$).
Choose $r > 0$ such that $\bar{x}$ is the only critical point for $f_{|M}$ in $B(\bar{x},r)$.
For $\|a\|$, $\|b\|$ sufficiently small, $M(a,b)$ is regular in the ball $B(\bar{x},r)$.

Let λ, resp. μ, be an m-, resp. p-vector and consider the map T:

$$T: \begin{pmatrix} x \\ \lambda \\ \mu \\ a \\ b \end{pmatrix} \longmapsto \begin{pmatrix} D_x^T[f(x) - \sum_{i=1}^{m} \lambda_i (h_i(x) - a_i) - \sum_{j=1}^{p} \mu_j (g_j(x) - b_j)] \\ h_i(x) - a_i,\ i = 1,\dots,m \\ g_j(x) - b_j,\ j = 1,\dots,p \end{pmatrix} \tag{6.1.29}$$

Then, T is a C^1-map with $T(\bar{x},\bar{\lambda},\bar{\mu},0,0) = 0$ and its partial derivative $D_{\begin{pmatrix} x \\ \lambda \\ \mu \end{pmatrix}} T(\bar{x},\bar{\lambda},\bar{\mu},0,0)$ is a nonsingular matrix.

In view of the Implicit Function Theorem, for sufficiently small $\|a\|$, $\|b\|$, we have C^1-maps $x(a,b)$, $\lambda(a,b)$, $\mu(a,b)$ such that

$$T(x(a,b), \lambda(a,b), \mu(a,b), a, b) \equiv 0,$$

where $x(a,b)$ is the only critical point for $f_{|M(a,b)}$ in $B(\bar{x},r)$ and where $x(a,b)$ is nondegenerate.
Now, the Lagrange-parameters $\bar{\lambda}_i$, $\bar{\mu}_j$ do have the following interpretation:

$$\boxed{\begin{aligned} &\frac{\partial}{\partial a_i} f(x(a,b))\Big|_{(0,0)} = \bar{\lambda}_i \ , \ i = 1,\ldots,m, \\ &\frac{\partial}{\partial b_j} f(x(a,b))\Big|_{(0,0)} = \bar{\mu}_j \ , \ j = 1,\ldots,p. \end{aligned}} \tag{6.1.30}$$

We merely verify the case i = 1 in (6.1.30). Firstly, we have:

$$\frac{\partial}{\partial a_1} f(x(a,b))\Big|_{(0,0)} = Df(\bar{x}) \cdot D_{a_1} x(0,0). \tag{6.1.31}$$

Since $h_i(x(a,b)) - a_i \equiv 0$, $g_j(x(a,b)) - b_j \equiv 0$, all i, j, we obtain:

$$\left.\begin{aligned} &Dh_1(\bar{x}) \cdot D_{a_1} x(0,0) = 1, \\ &Dh_i(\bar{x}) \cdot D_{a_1} x(0,0) = 0, \ i = 2,\ldots,m, \\ &Dg_j(\bar{x}) \cdot D_{a_1} x(0,0) = 0, \ j = 1,\ldots,p. \end{aligned}\right\} \tag{6.1.32}$$

Recall that

$$Df(\bar{x}) = \sum_{i=1}^{m} \bar{\lambda}_i Dh_i(\bar{x}) + \sum_{j=1}^{p} \bar{\mu}_j Dg_j(\bar{x}). \tag{6.1.33}$$

Multiplication of (6.1.33) with $D_{a_1} x(0,0)$ from the right and subsequent substitution of (6.1.32) in it yield the desired result. □

We proceed with the introduction of a suitable topology on the function-space $C^k(\mathbb{R}^n,\mathbb{R})$.

Let $\alpha = (\alpha_1,\ldots,\alpha_n) \in \mathbb{N}^n$, $|\alpha| = \sum_{i=1}^{n} \alpha_i$ and denote by $\partial^\alpha f$ the α-th partial derivative of f, where $f \in C^k(\mathbb{R}^n,\mathbb{R})$ and $|\alpha| \le k$.

Put $C_+(\mathbb{R}^n,\mathbb{R}) = \{\phi\colon \mathbb{R}^n \to \mathbb{R} \mid \phi \text{ continuous and } \phi(x) > 0 \text{ for all } x \in \mathbb{R}^n\}$.

We define the C^k-topology for $C^k(\mathbb{R}^n,\mathbb{R})$ by giving a basis for it (cf. [50], p. 47).

<u>Definition 6.1.1</u>. <u>(C^k-topology)</u>.

For fixed $k \in \mathbb{N}$, a basis for the C^k-topology for $C^k(\mathbb{R}^n,\mathbb{R})$ consists of all sets $V^k_{\phi,f}$:

$$V^k_{\phi,f} = \{g \in C^k(\mathbb{R}^n,\mathbb{R}) \mid |\partial^\alpha f(x)-\partial^\alpha g(x)| < \phi(x) \text{ for all } x \in \mathbb{R}^n, \text{ for all } \alpha \text{ with } |\alpha| \le k\}, \tag{6.1.34}$$

where $(\phi,f) \in C_+(\mathbb{R}^n,\mathbb{R}) \times C^k(\mathbb{R}^n,\mathbb{R})$.

In a straightforward way, the C^ℓ-topology for $C^k(\mathbb{R}^n,\mathbb{R})$, where $\ell \le k$, is defined, as well as the C^k-topology for $C^\infty(\mathbb{R}^n,\mathbb{R})$. Moreover, the C^∞-topology for $C^\infty(\mathbb{R}^n,\mathbb{R})$ has all sets of the type $V^k_{\phi,f}$, $k = 0,1,2,\ldots$, as a basis. The C^k-topology for a finite product of function spaces is defined to be the producttopology. □

Remark 6.1.4. The fact, that for fixed k the sets $V^k_{\phi,f}$ form a basis for a topology can be seen as follows. For simplicity we take $k = 0$.
Let $h \in V^o_{\phi,f} \cap V^o_{\psi,g}$. We have to show that there exists an $\xi \in C_+(\mathbb{R}^n,\mathbb{R})$ such that $V^o_{\xi,h} \subset V^o_{\phi,f} \cap V^o_{\psi,g}$ (cf. [50], p. 47).
In fact, choose $\xi \in C_+(\mathbb{R}^n,\mathbb{R})$ such that for all $x \in \mathbb{R}^n$:

$$\xi(x) < \min\{\phi(x) - |h(x) - f(x)|, \psi(x) - |h(x) - g(x)|\}.$$

Let $\ell \in V^o_{\xi,h}$. Then we have for all $x \in \mathbb{R}^n$:

$$|f(x)-\ell(x)| \le |f(x)-h(x)| + |h(x)-\ell(x)| < |f(x)-h(x)| + \xi(x) <$$

$$< |f(x)-h(x)| + \phi(x) - |h(x)-f(x)| = \phi(x).$$

Consequently, $V^o_{\xi,h} \subset V^o_{\phi,f}$ and analogously, $V^o_{\xi,h} \subset V^o_{\psi,g}$, and hence, $V^o_{\xi,h} \subset V^o_{\phi,f} \cap V^o_{\psi,g}$.

Note that the C^k-topology is strictly finer than the C^ℓ-topology for $k > \ell$. In order to show that all sets in $C^\infty(\mathbb{R}^n,\mathbb{R})$ of the type $V^k_{\phi,f}$, $k = 0,1,\ldots$, form a basis for a topology (in fact, the C^∞-topology), one has to note that $V^k_{\phi,f} \subset V^\ell_{\phi,f}$ if $k \ge \ell$. □

Remark 6.1.5. The addition $(f,g) \mapsto f+g$ is a continuous mapping in the C^k-topology, whereas the scalar multiplication $(\lambda,f) \mapsto \lambda f$ is not. In fact, put $f_n(x) \equiv \frac{1}{n}$ for $n = 1,2,\ldots$. Then, $f_n \not\to 0$. Hence, the space $C^\infty(\mathbb{R}^n,\mathbb{R})$ endowed with the C^k-topology, is a topological group, but not a topological vectorspace. Note: $f_i \to f$ as $i \to \infty$ (sequence) implies that $f_i = f$ outside some compact set $K \subset \mathbb{R}^n$ for i large enough.

Remark 6.1.6. The reason for introducing the C^k-topology (also called strong C^k-topology) comes from the fact that this topology takes the asymptotical behaviour of functions into account. For example: a function $f \in C^2(\mathbb{R}^n,\mathbb{R})$ is called nondegenerate if all its critical points are nondegenerate. Obviously, f is nondegenerate iff $\|Df(x)\| + |\det D^2f(x)| > 0$ for all $x \in \mathbb{R}^n$. From this observation it follows that the subset of $C^2(\mathbb{R}^n,\mathbb{R})$ consisting of all nondegenerate functions is C^2-open ("stability of nondegeneracy").

A weaker topology (the socalled weak C^k-topology) can be given by means of a metric. We will explain this w.r.t. the space $C^2(\mathbb{R}^n,\mathbb{R})$. Choose a countable number of points $\bar{x}_i \in \mathbb{R}^n$, $i = 1,2,\ldots$, such that the set of the balls $B(\bar{x}_i,1)$ cover $\mathbb{R}^n$. Each $B(x_i,1)$ generates a seminorm $|\cdot|_i$ on $C^2(\mathbb{R}^n,\mathbb{R})$:

$$|f|_i = \max_{x\in B(\bar{x}_{i,1})} \{|f(x)| + \sum_{i=1}^{n} |\frac{\partial}{\partial x_i} f(x)| + \sum_{i,j=1}^{n} |\frac{\partial^2}{\partial x_i \partial x_j} f(x)|\}. \tag{6.1.35}$$

The seminorms in (6.1.35) generate a metric $d(\cdot,\cdot)$ on $C^2(\mathbb{R}^n,\mathbb{R})$:

$$d(f,g) = \sum_{i=1}^{\infty} 2^{-i} \frac{|f-g|_i}{1+|f-g|_i}. \tag{6.1.36}$$

Note that $d(f_n,f) \to 0$ iff $|f_n-f|_i \to 0$ for all i. Thus, the convergence w.r.t. $d(\cdot,\cdot)$ is equivalent to the uniform convergence up to derivatives of second order on compact subsets of $\mathbb{R}^n$.
For $n = 1,2,\ldots$, we put $f_n(x) \equiv \frac{1}{n}$. Then, $d(f_n,0) \to 0$ (cf. also Remark 6.1.5). From this example we see that the metric $d(\cdot,\cdot)$ does not treat the behaviour "at infinity" very well. This comes from the damping factor 2^{-i} for i large. Furthermore, the set of nondegenerate functions is not open in the topology generated by $d(\cdot,\cdot)$. For example, take $n = 1$, put $\eta(x) = \frac{1}{\pi}\int_0^x \frac{\sin \pi t}{t}\,dt$ and define:

$$f(x) = \begin{cases} \eta(x), & \text{for } x \geq \frac{1}{2} \\ \eta(\frac{1}{2}) + (x-\frac{1}{2})\eta'(\frac{1}{2}) + \frac{1}{2}(x-\frac{1}{2})^2\eta''(\frac{1}{2}), & \text{for } x < \frac{1}{2}. \end{cases}$$

Then, $f \in C^2(\mathbb{R},\mathbb{R})$ and we have $Df(x) = 0$ iff $x = 1,2,\ldots$. Moreover, $D^2f(i) = \frac{1}{i}(-1)^i$, $i = 1,2,\ldots$. But then, given $\varepsilon > 0$, it is not difficult to construct an $f_\varepsilon \in C^2(\mathbb{R},\mathbb{R})$ such that f_ε has at least one <u>degenerate</u> critical point and $d(f,f_\varepsilon) < \varepsilon$.
We finally emphasize that $C^2(\mathbb{R}^n,\mathbb{R})$, endowed with the C^2-topology is not metrizable. In fact, note that there exists no <u>countable</u> base-neighborhood system for the zero function.

<u>Example 6.1.2</u>. For fixed m, $1 \le m \le n$, let $H \subset C^1(\mathbb{R}^n,\mathbb{R})^m$ be the following subset: $(h_1,\ldots,h_m) \in H$ iff $M[h] := \{x \in \mathbb{R}^n \mid h_i(x) = 0,\ i = 1,\ldots,m\}$ is regular. Then, H is C^1-open. To see this, note that $(h_1,\ldots,h_m) \in H$ iff $\sum_{i=1}^{m} |h_i(x)| + \sum_{\det} |\det(x)| > 0$ for all $x \in \mathbb{R}^n$, where $\det(x)$ stands for the determinant of an $(m\times m)$-submatrix of the $(n\times m)$-matrix $(Dh_1^T(x)|\ldots|Dh_m^T(x))$ and where the summation $\sum_{\det}$ runs over all those $(m\times m)$-submatrices.

<u>Example 6.1.3</u>. Let $m,s \in \mathbb{N}$ be given and define $F \subset C^2(\mathbb{R}^n,\mathbb{R})^{1+m+s}$ as follows: $(f,h_1,\ldots,h_m,g_1,\ldots,g_s) \in F$ iff $M[h,g]$ is regular and if all critical points for $f_{|M[h,g]}$ are nondegenerate. Then, F is C^2-open. The proof can be given in an analogous way as is done in Example 6.1.2. But, writing down an analogous "positive continuous function" is tedious in this case, and it will be omitted (cf. [34]); see, however, also Section 7.1.

<u>Definition 6.1.2</u>. Let $\Pi := \prod_{i\in P} \mathbb{R}^k$ be the Cartesian product of the spaces $\mathbb{R}^k$, where P is an index set (not necessarily finite). The <u>Box-topology</u> for Π is defined to be the topology generated by <u>all</u> products of open sets in $\mathbb{R}^k$ (cf. [50] , pp. 107, 108).

Note that, in case P is not finite, the Box-topology is strictly finer than the product topology for Π. In fact, if $P = \mathbb{N}$, $k = 1$, take the sequence $\{(\frac{1}{n},\frac{1}{n},\frac{1}{n},\ldots),\ n = 1,2,\ldots\}$. This sequence does not converge to $(0,0,0,\ldots)$. In a certain sense, the Box-topology is a discrete version of the C^k-topology for the space of smooth functions.

Theorem 6.1.2. Let $f,h_i,g_j \in C^2(\mathbb{R}^n,\mathbb{R})$, $i = 1,\ldots,m$, $j = 1,\ldots,s$ and assume that $M := M[h,g]$ is regular and that all critical points for $f_{|M}$ are nondegenerate. Let x^i, $i \in P$, denote the critical points for $f_{|M}$ with Lagrange parameters $\lambda_1^i,\ldots,\lambda_m^i,\mu_1^i,\ldots,\mu_s^i$, where μ_j^i is defined to be zero if $j \notin J_0(x^i)$, and let $\mathcal{B}$ denote the Box-topology for $\prod_{i\in P} \mathbb{R}^n \times \mathbb{R}^m \times \mathbb{R}^s$.

Then, there exist a C^2-neighborhood $\mathcal{O}$ of $(f,h_1,\ldots,h_m,g_1,\ldots,g_s)$ and a map T. T being continuous w.r.t. C^2 and $\mathcal{B}$,

$$T: \mathcal{O} \to \prod_{i\in P} \mathbb{R}^n \times \mathbb{R}^m \times \mathbb{R}^s,$$

where T maps an element $(\tilde{f},\tilde{h}_1,\ldots,\tilde{h}_m,\tilde{g}_1,\ldots,\tilde{g}_s)$ onto the point $\prod_{i\in P} (\tilde{x}^i,\tilde{\lambda}_1^i,\ldots,\tilde{\lambda}_m^i,\tilde{\mu}_1^i,\ldots,\tilde{\mu}_s^i)$ representing exactly the critical points and corresponding Lagrange-parameters for $\tilde{f}_{|M[\tilde{h},\tilde{g}]}$; moreover, every $\tilde{x}^i$ is a nondegenerate critical point and the indices (LI, LCI, QI, QCI) at $\tilde{x}^i$ are equal to those at x^i, $i \in P$. □

We will delete the proof of Theorem 6.1.2, since it is an immediate consequence if Theorem 6.1.1 and the nature of the topologies $\mathcal{B}$ and C^2.

Remark 6.1.7. The index set P in Theorem 6.1.2 is always countable, since nondegenerate critical points are isolated critical points.

6.2. Stability of critical points.

As in the preceding section, we start with a discussion on unconstrained problems.

Let $f \in C^2(\mathbb{R}^n,\mathbb{R})$ and let $\bar{x} \in \mathbb{R}^n$ be a critical point for f. If $D^2f(\bar{x})$ is nonsingular, then $\bar{x}$ is an isolated critical point and the Morse Lemma (cf. Theorem 2.8.2) tells us that there exist local C^1-coordinates in which f takes the form of a sum of squares. However, if rank $D^2f(\bar{x})$ equals k and $k < n$, then $\bar{x}$ need not to be an isolated critical point anymore. Nevertheless, in a neighborhood of $\bar{x}$ all possible critical points lie on a certain (n-k)-dimensional manifold through $\bar{x}$. The next theorem states that we can split up - in a nonlinear way - the coordinates locally into a "regular" part and a completely nonregular part. For sake of simplicity, the theorem is stated in a C^∞-version.

<u>Theorem 6.2.1.</u> (<u>Splitting-Theorem</u>).
Let $f \in C^\infty(\mathbb{R}^n, \mathbb{R})$, $f(0) = 0$, $Df(0) = 0$, rank $D^2f(0) = k$. Then, there exists a local C^∞-coordinate transformation $y = \Phi(x)$, sending the origin onto itself, such that:

(i) $$f \circ \Phi^{-1}(y) = \sum_{i=1}^{k} \pm y_i^2 + \phi(y_{k+1}, \ldots, y_n), \tag{6.2.1}$$

where the number of positive (negative) squares in (6.2.1) corresponds to the number of positive (negative) eigenvalues of $D^2f(0)$,

(ii) $\phi(0) = 0$, $D\phi(0) = 0$, $D^2\phi(0) = 0$ ($(n-k)\times(n-k)$ zero-matrix).

<u>Proof</u>. Firstly, we note that (ii) follows directly from (i), recalling Example 2.5.1 and Lemma 2.5.1.
By means of a linear coordinate transformation we may assume (cf. also the proof of Theorem 2.7.2) that $f(x) = x^T F(x) x$, where $F: \mathbb{R}^n \to \mathbb{R}^{n^2}$ is of class C^∞ and $F(0) = \operatorname{diag}(\lambda_1, \ldots, \lambda_k, \lambda_{k+1}, \ldots, \lambda_n)$, $\lambda_i \neq 0$, $i = 1, \ldots, k$, $\lambda_j = 0$, $j = k+1, \ldots, n$.
Consider the set $M := \{x \in \mathbb{R}^n \mid \frac{\partial f}{\partial x_i}(x) = 0,\ i = 1, \ldots, k\}$. Note that the matrix $\left(\frac{\partial^2 f}{\partial x_i \partial x_j}(0)\right)_{i,j=1,\ldots,k}$ is nonsingular.
Hence, in view of the Implicit Function Theorem, there exist an open neighborhood $\mathcal{O} \subset \mathbb{R}^{n-k}$ of $0 \in \mathbb{R}^{n-k}$ and C^∞-functions $u_i \in C^\infty(\mathcal{O}, \mathbb{R})$, $Du_i(0) = 0$, $i = 1, \ldots, k$, such that we have (locally): $x \in M$ iff $x_i = u_i(x_{k+1}, \ldots, x_n)$, $i = 1, \ldots, k$.
Put $z = \Psi(x)$, Ψ being defined by $z_i = x_i - u_i(x_{k+1}, \ldots, x_n)$, $i = 1, \ldots, k$, and $z_j = x_j$, $j = k+1, \ldots, n$. Then, Ψ is a local C^∞-coordinate transformation with $D\Psi(0) = I$ (identity).
With $g(z) := f \circ \Psi^{-1}(z)$, we obtain by the very construction:

$$\left.\begin{aligned} &\frac{\partial}{\partial z_i} g(0, \ldots, 0, z_{k+1}, \ldots, z_n) \equiv 0, \quad i = 1, \ldots, k, \\ &D^2 g(0) = 2F(0). \end{aligned}\right\} \tag{6.2.2}$$

Next, we put:

$$g(z) = \underbrace{g(z) - g(0,\dots,0,z_{k+1},\dots,z_n)}_{h(z)} + \underbrace{g(0,\dots,0,z_{k+1},\dots,z_n)}_{\phi(z_{k+1},\dots,z_n)}. \quad (6.2.3)$$

From (6.2.3) we see that $h(0,\dots,0,z_{k+1},\dots,z_n) = 0$. Hence,

$$\begin{aligned} h(z) &= \int_0^1 \frac{d}{d\tau} h(\tau z_1,\dots,\tau z_k,z_{k+1},\dots,z_n)\,d\tau = \\ &= \sum_{i=1}^{k} z_i \int_0^1 \frac{\partial}{\partial z_i} h(\tau z_1,\dots,\tau z_k,z_{k+1},\dots,z_n)\,d\tau = \\ &:= \sum_{i=1}^{k} z_i h_i(z). \end{aligned} \quad (6.2.4)$$

From (6.2.2), (6.2.3), (6.2.4) we see that $h_i(0,\dots,0,z_{k+1},\dots,z_n) \equiv 0$, $i = 1,\dots,k$. Analogously to the performance of (6.2.4) we may write

$h_i(z) = \sum_{j=1}^{k} z_j h_{ij}(z)$, $i = 1,\dots,k$, thus obtaining:

$$h(z) = \sum_{i,j=1}^{k} z_i z_j h_{ij}(z) = (z_1,\dots,z_k)H(z)(z_1,\dots,z_k)^\top. \quad (6.2.5)$$

We may assume that $H(z)$ in (6.2.5) is symmetric; otherwise we insert in (6.2.5) the matrix $\frac{1}{2}(H(z)^\top + H(z))$ instead of $H(z)$. A simple calculation shows that $H(0) = \mathrm{diag}(\lambda_1,\dots,\lambda_k)$.

In the proof of the Morse Lemma (Theorem 2.7.2, Step 3) we used an explicit "formula" for a local coordinate transformation, which was based on Lemma 2.5.3. Here we may use the same idea, treating $z_{k+1},\dots,z_k$ as additional parameters. This gives us a local C^∞-coordinate transformation $y = \tilde{\Phi}(z)$, where $y_j = z_j$, $j = k+1,\dots,n$, such that $h\circ\tilde{\Phi}^{-1}(y) = \sum_{i=1}^{k} \pm y_i^2$. Finally, we put $\Phi = \tilde{\Phi}\circ\Psi$ and the lemma is proved. □

We proceed with a concept of stability which is analogous to the one used by Kojima in [51]. W.r.t. ball $B := B(\bar{x},r)$ we use the following seminorm on $C^2(\mathbb{R}^n,\mathbb{R})$:

$$\| f \|_B = \max_{x\in B} \{|f(x)| + \sum_{i=1}^{n} |\frac{\partial}{\partial x_i} f(x)| + \sum_{i,j=1}^{n} |\frac{\partial^2}{\partial x_i \partial x_j} f(x)|\}. \quad (6.2.6)$$

Definition 6.2.1. Let $f \in C^2(\mathbb{R}^n,\mathbb{R})$ and $\bar{x} \in \mathbb{R}^n$ a critical point for f. The critical point $\bar{x}$ is called stable (for f) if for some $r > 0$ and each $\tilde{r} \in (0,r]$ there exists an $\alpha > 0$ such that for every $g \in C^2(\mathbb{R}^n,\mathbb{R})$ with $\|f-g\|_{B(\bar{x},r)} \le \alpha$, the ball $B(\bar{x},\tilde{r})$ contains a critical point for g which is unique in $B(\bar{x},r)$.

Lemma 6.2.1. Let $f \in C^2(\mathbb{R}^n,\mathbb{R})$ and let $\bar{x} \in \mathbb{R}^n$ be a critical point for f. Then, $\bar{x}$ is stable (for f) iff $\bar{x}$ is nondegenerate.

Remark 6.2.1. From Definition 6.2.1 it follows that a stable critical point is necessarily an isolated critical point. Moreover, from Lemma 6.2.1 we see: if $\bar{x}$ is stable, then for some $r > 0$ and each $\tilde{r} \in (0,r)$ there exists an $\alpha > 0$ such that $\|f-g\|_{B(\bar{x},r)} \le \alpha$ implies: $B(\bar{x},\tilde{r})$ contains a critical point $\tilde{x}$ for g, $\tilde{x}$ is stable (for g) and $\tilde{x}$ is the unique critical point for g in $B(\bar{x},r)$.

Proof of Lemma 6.2.1.
The "if"-part is an immediate consequence of Lemma 6.1.1.
The "only if"-part. Let $f \in C^2(\mathbb{R}^n,\mathbb{R})$ and let $\bar{x} \in \mathbb{R}^n$ be an isolated critical point for f. Choose an $r > 0$ such that $\bar{x}$ is the only critical point for f in $B(\bar{x},r)$. It suffices to show the following: if rank $D^2f(\bar{x}) < n$, then for every $\alpha > 0$ there exists a function $g_\alpha \in C^2(\mathbb{R}^n,\mathbb{R})$ with $\|f-g_\alpha\|_{B(\bar{x},r)} \le \alpha$, such that g_α has either no critical points in $B(\bar{x},r)$ or at least two critical points in $B(\bar{x},r)$. Of course, we may assume that $f(\bar{x}) = 0$ and $\bar{x} = 0$. We will treat three cases, and we start with the simplest case.

Case I. Let $n = 1$, $f \in C^\infty(\mathbb{R},\mathbb{R})$, $f(0) = 0$, $Df(0) = 0$, $D^2f(0) = 0$, and suppose that $\bar{x} = 0$ is the only critical point for f in the interval $[-r,r]$. Without loss of generality we may assume that $Df(x) > 0$ for $x \in (0,r]$. Then, there are two subcases: either $Df(x) > 0$ for $x \in [-r,0)$ (Subcase I.1), or $Df(x) < 0$ for $x \in [-r,0)$ (Subcase I.2).

Subcase I.1. For $c > 0$, the function $g_c(x) := f(x) + cx$ has no critical points in $B(0,r)$. Given $\alpha > 0$, c can be chosen so small that $\|f-g_c\|_{B(0,r)} \le \alpha$.

Subcase I.2. In Subcase I.1, we used linear perturbations. Now, we will use quadratic perturbations. For $k = 1,2,\ldots$, we put:

$$\phi_k(x) = \frac{k}{4}(Df(\tfrac{1}{k}) - Df(-\tfrac{1}{k}))x^2 + \frac{1}{2}(Df(\tfrac{1}{k}) + Df(-\tfrac{1}{k}))x.$$

Then, $D\phi_k(x)$ is the linear function satisfying:

$$D\phi_k(-\tfrac{1}{k}) = Df(-\tfrac{1}{k}) \ , \ D\phi_k(\tfrac{1}{k}) = Df(\tfrac{1}{k}).$$

By the very construction, $f(x) - \phi_k(x)$ has at least two critical points, namely at $x = \pm\frac{1}{k}$. Furthermore, we note that $\lim_{k\to\infty} k\,Df(\pm\frac{1}{k}) = 0$. Consequently, $\lim_{k\to\infty} \|\phi_k\|_{B(0,r)} = 0$. But then, the following is easily seen. Given $\alpha > 0$, there exists a k_α such that for all $k > k_\alpha$ the function $g_k(x) := f(x) - \phi_k(x)$ has at least two critical points in $B(0,r)$, and, moreover, $\|f-g_k\|_{B(0,r)} \le \alpha$.

Case II. Let $f \in C^\infty(\mathbb{R}^n,\mathbb{R})$, $f(0) = 0$, $Df(0) = 0$, rank $D^2f(0) = p$, $0 \le p < n$, and suppose that $\bar{x} = 0$ is the only critical point for f in the ball $B(0,r)$.

Let $\alpha > 0$ be given. In case $p < n-1$, we start with the following preparation step. Let V be an $n\times(n-p-1)$ matrix of rank $n-p-1$ whose columns are eigenvectors of $D^2f(0)$ corresponding to the zero eigenvalue. The matrix $V(V^TV)^{-1}V^T$ represents the orthogonal projection of $\mathbb{R}^n$ onto the linear subspace generated by the columns of V. For $\delta > 0$ we put $g_\delta(x) = \delta x^TV(V^TV)^{-1}V^Tx$. Then, $Df(0) + Dg_\delta(0) = 0$ and rank $(D^2f(0) + D^2g_\delta(0)) = n-1$. Choose δ so small that $\|g_\delta\|_{B(0,r)} \le \frac{\alpha}{2}$. If, besides $\bar{x} = 0$, the function $f + g_\delta$ has another critical point in $B(0,r)$, we are done. In case $p = n-1$, we put $g_\delta \equiv 0$.

So, now we assume that the origin is the only critical point for $f + g_\delta$ in $B(0,r)$. From Theorem 6.2.1 it follows that there exist open neighborhoods $\mathcal{O}$, $\mathcal{V}$, of the origin, $\mathcal{O} \subset B(0,r)$, and a C^∞-diffeomorphism $\Phi: \mathcal{O} \to \mathcal{V}$, such that $(f+g_\delta)\circ\Phi^{-1}(y) = \sum_{i=1}^{n-1} \pm y_i^2 + \psi(y_n)$, where $\psi(0) = D\psi(0) = D^2\psi(0) = 0$. Obviously, 0 is the only critical point for $(f+g_\delta)\circ\Phi^{-1}$ in $\mathcal{V}$. In particular, $y_n = 0$ is the only critical point for ψ. So, w.r.t. the function $\psi(y_n)$, we are in the situation of Case I.

Choose an $\varepsilon > 0$ such that $B(0,3\varepsilon)$ is contained in V and put

$$\zeta(y) = \phi(y_n)\cdot\eta_\varepsilon(\|y\|), \tag{6.2.7}$$

where $\phi(y_n)$ is either a linear or a quadratic function, in accordance with the idea in Subcase I.1, I.2. The function η_ε in (6.2.7) is of class C^∞ satisfying:

$$\begin{cases} \eta_\varepsilon(x) = 1, \text{ resp. } 0, \text{ for } x \le \varepsilon, \text{ resp. } x \ge 2\varepsilon, \\ 0 \le \eta_\varepsilon(x) \le 1, \text{ for all } x. \end{cases}$$

Note that the derivatives of $\eta_\varepsilon(x)$ up to order two are bounded on $\mathbb{R}$. By a precise inspection of first derivatives it follows that we can choose $\phi(y_n)$ in such a way that $f + g_\delta + \zeta\circ\Phi$ has either no critical points in $\mathcal{O}$ or at least two critical points in $\mathcal{O}$. Note that ζ vanishes identically outside $\Phi^{-1}(B(0,2\varepsilon))$. So, we can extend ζ by means of the zero function on the whole $\mathbb{R}^n$; denote this function by $\tilde{\zeta}$. An appropriate choice of $\phi(y_n)$ in (6.2.7) will yield the inequality $\|\tilde{\zeta}\|_{B(0,r)} \le \frac{\alpha}{2}$. Altogether, with such a choice, the function $g_\delta + \tilde{\zeta}$ serves as the desired perturbation of the original function f.

Case III. Let f satisfy the assumptions as in Case II, however under the restriction: $f \in C^2(\mathbb{R}^n,\mathbb{R})$.
For $h,g \in C^2(\mathbb{R}^n,\mathbb{R})$, we put

$$\mu_{h,g}(x) = (h-g)(0) + D(h-g)(0)x + \tfrac{1}{2}x^T D^2(h-g)(0)x.$$

The space $C^\infty(\mathbb{R}^n,\mathbb{R})$ is C^2-dense in $C^2(\mathbb{R}^n,\mathbb{R})$. (We will not dwell on this type of approximations, but refer to [31] for details). From this and the fact that $B(0,r)$ is compact, it is easily seen that, given $\alpha > 0$, there exists a function $g \in C^\infty(\mathbb{R}^n,\mathbb{R})$ such that the following two inequalities hold simultaneously:

$$\|f-g\|_{B(0,r)} \le \frac{\alpha}{4}, \quad \|\mu_{f,g}\|_{B(0,r)} \le \frac{\alpha}{4}.$$

Now, $g + \mu_{f,g}$ is of class C^∞ and its partial derivatives up to order two at the origin coincide with those of f. Moreover $\|f-g-\mu_{f,g}\|_{B(0,r)} \le \frac{\alpha}{2}$. If 0 is not the only critical point for $g + \mu_{f,g}$ in $B(0,r)$, we are done.

Otherwise, with the function $g + \mu_{f,g}$, we are in the situation of Case II and we can perturb it suitably in the $\|\cdot\|_{B(0,r)}$ sense; so we are done again. This, finally, completes the proof of Lemma 6.2.1. □

We proceed with a discussion on stability of critical points in constrained problems. Let B be a ball in $\mathbb{R}^n$ and let $C^2(\mathbb{R}^n,\mathbb{R})^p$ be the product of p copies of $C^2(\mathbb{R}^n,\mathbb{R})$. Then, we use the following seminorm on $C^2(\mathbb{R}^n,\mathbb{R})^p$:

$$\|(f_1,\ldots,f_p)\|_B = \max_{1\le i\le p} \|f_i\|_B . \tag{6.2.8}$$

<u>Definition 6.2.2</u>. Let $f,h_i,g_j \in C^2(\mathbb{R}^n,\mathbb{R})$, $i \in I$, $j \in J$ (I, J finite) and suppose that the corresponding constraint set $M[h,g]$ is regular.
A critical point $\bar{x}$ for $f_{|M[h,g]}$ is called <u>stable</u> if for some $r > 0$ and each $\tilde{r} \in (0,r]$ there exists an $\alpha > 0$ such that for every $(\tilde{f},\tilde{h},\tilde{g}) \in C^2(\mathbb{R}^n,\mathbb{R})^{1+|I|+|J|}$ with $\|(f,h,g) - (\tilde{f},\tilde{h},\tilde{g})\|_{B(\bar{x},r)} \le \alpha$ and such that $M[\tilde{h},\tilde{g}]$ is regular in $B(\bar{x},r)$: the ball $B(\bar{x},\tilde{r})$ contains a critical point for $\tilde{f}_{|M[\tilde{h},\tilde{g}]}$ which is unique in $B(\bar{x},r)$.

<u>Theorem 6.2.2</u>. Let $f,h_i,g_j \in C^2(\mathbb{R}^n,\mathbb{R})$, $i \in I$, $j \in J$ (I, J finite) and suppose that the corresponding constraint set $M[h,g]$ is regular.
Then, a critical point $\bar{x}$ for $f_{|M[h,g]}$ is stable iff $\bar{x}$ is nondegenerate.

<u>Proof</u>.
<u>The "if"-part</u> is an immediate consequence of Theorem 6.1.1.
<u>The "only if"-part</u>. First of all, we may assume that $\bar{x}$ is an isolated critical point. We will show: if $\bar{x}$ is degenerate, then $\bar{x}$ is not stable. This will be done in the following way: we keep the constraints h_i, g_j, $i \in I$, $j \in J$ fixed and define suitable perturbations of the object function f. But then, we may restrict ourselves to a discussion in local C^2-coordinates for the constraint set $M[h,g]$.
Hence, we discuss the following situation. Let $f \in C^2(\mathbb{R}^n,\mathbb{R})$, let $\bar{x} = 0$ be the only critical point for $f_{|\mathbb{R}^k\times\mathbb{H}^{n-k}}$ in the ball $B(0,r)$ and let $\bar{x}$ be degenerate. The latter fact means, besides that $\frac{\partial f}{\partial x_i}(0) = 0$, $i = 1,\ldots,k$:

$$\begin{cases} \text{either } A \text{ is singular and } \frac{\partial f}{\partial x_j}(0) = 0 \text{ for some } j \in \{k+1,\dots,n\}, & \underline{\text{(Case I)}} \\ \text{or } A \text{ is singular and } \frac{\partial f}{\partial x_j}(0) \neq 0 \text{ for all } j \in \{k+1,\dots,n\}, & \underline{\text{(Case II)}} \\ \text{or } A \text{ is nonsingular and } \frac{\partial f}{\partial x_j}(0) = 0 \text{ for some } j \in \{k+1,\dots,n\}, & \underline{\text{(Case III)}} \end{cases}$$

$$\text{where } A = \left(\frac{\partial^2 f}{\partial x_i \partial x_j}\right)(0)_{i,j=1,\dots,k}. \tag{6.2.9}$$

Case I. Without loss of generality, $\frac{\partial f}{\partial x_j}(0) = 0$ for $j \in \{k+1,\dots,k+s\}$, and $\frac{\partial f}{\partial x_j}(0) \neq 0$ for $j \in \{k+s+1,\dots,n\}$. For $\varepsilon > 0$, consider the function $g_\varepsilon(x) := f(x) + \varepsilon(\sum_{j=k+1}^{k+s} x_j)$. Then, the origin is a critical point for $g_{\varepsilon|\mathbb{R}^k \times \mathbb{H}^{n-k}}$. Obviously, we have: $\| f-g_\varepsilon \|_{B(0,r)} \to 0$ as $\varepsilon \to 0$. If, for some sufficiently small ε, there is a critical point for $g_{\varepsilon|\mathbb{R}^k \times \mathbb{H}^{n-k}}$ in $B(0,r)$, different from the origin, we are done. Otherwise, replacing f by g_ε, we are in the situation of Case II.

Case II. Consider perturbations of $f(x_1,\dots,x_n)$ of the form $\psi(x_1,\dots,x_k)$ and exploit the ideas used in the proof of Lemma 6.2.1.

Case III. Without loss of generality we assume that $\frac{\partial f}{\partial x_{k+1}}(0) = 0$. Since the matrix A (cf. (6.2.9)) is nonsingular, it follows that 0 is a stable critical point for $f_{|\mathbb{R}^k \times \{0_{n-k}\}}$, where 0_{n-k} is the origin in $\mathbb{R}^{n-k}$. So, it suffices to create at least one (additional) critical point in $\mathbb{R}^k \times \mathbb{H}^1 \times \{0_{n-k-1}\}$ with $(k+1)$-coordinate greater than zero, by means of a (locally) arbitrarily small C^2-perturbation of f. To this aim we consider the following two-parameter family of functions:

$$g(x,\beta,\gamma) = f(x) + \tfrac{1}{2}\beta(x_{k+1}-\gamma)^2. \tag{6.2.10}$$

If we choose β,γ sufficiently small, then, $\| f-g \|_{B(0,r)}$ is small.

Since the coordinates $x_{k+2},\dots,x_n$ will play no role anymore in the sequel, we may assume that $\underline{x \in \mathbb{R}^{k+1}}$.

Let b denote the column vector $\left(\frac{\partial^2 f(0)}{\partial x_i \partial x_{k+1}}\right)_{i=1,\dots,k}$ and put $c = \frac{\partial^2 f(0)}{\partial^2 x_{k+1}}$.

Then we have:

$$D_x^2 g(0,\beta,\gamma) = \left(\begin{array}{c|c} A & b \\ \hline b^T & c+\beta \end{array}\right) \tag{6.2.11}$$

A short calculation shows that $D_x^2 g(0,\beta,\gamma)$ is nonsingular iff $b^T A^{-1} b \neq c+\beta$. Now, choose $\beta \neq 0$ such that $b^T A^{-1} b \neq c+\beta$. Keeping this β fixed, we see that $0 \in \mathbb{R}^{k+1}$ is a nondegenerate critical point for $g(\cdot,\beta,0)$. By means of the Implicit Function Theorem we obtain (locally) a C^1-map ξ from $\mathbb{R}$ to $\mathbb{R}^{k+1}$ such that $D_x g(\xi(\gamma),\beta,\gamma) \equiv 0$. If the (k+1)-component of $\frac{d\xi(0)}{d\gamma}$ is unequal to zero, then for all small γ, γ either negative or positive, we have created a critical point for $g(\cdot,\beta,\gamma)_{|\mathbb{R}^k \times \mathbb{H}^1}$ with a (k+1)-component greater than zero, and this completes our proof. Well, a straightforward calculation shows $(\xi(\gamma) = (\xi_1(\gamma),\ldots,\xi_{k+1}(\gamma)))$:

$$\frac{d\xi_{k+1}}{d\gamma}(0) = \beta(-b^T A^{-1} b + c + \beta)^{-1}. \tag{6.2.12}$$

Since $\beta \neq 0$ and $b^T A^{-1} b \neq c+\beta$, we see from (6.2.12) that $\frac{d\xi_{k+1}}{d\gamma}(0) \neq 0$. □

<u>Remark 6.2.2</u>. If we would replace in Definition 6.2.2 the word critical point by (+) Kuhn-Tucker point (i.e. all Lagrange parameters corresponding to active inequality constraints are nonnegative), then certain critical points which are (+) Kuhn-Tucker points, but degenerate as critical points, become stable.

For example, consider $f(x) = x^2$ on the set $\mathbb{H}^1$. Then, $0 \in \mathbb{H}^1$ is a degenerate critical point, but a (+) Kuhn-Tucker point.

For $|\beta|$, $|\gamma|$ small, put $g(x,\beta,\gamma) = f(x) + \beta(x-\gamma)^2$.

If $\beta\cdot\gamma < 0$, then the origin is the only critical point for $g(\cdot,\beta,\gamma)_{|\mathbb{H}^1}$, and, in fact, a (+) Kuhn-Tucker point. However, if $\beta\cdot\gamma > 0$, then there are two critical points for $g(\cdot,\beta,\gamma)_{|\mathbb{H}^1}$, only one of them being a (+) Kuhn-Tucker point!

So the restriction to the set of (+) Kuhn-Tucker points gives rise to a concept of stability which also allows some degeneracies. In particular, this concept of stability is called "strong stability" by M. Kojima. Moreover, within that context, a slight relaxation is also allowed w.r.t. the regularity of the constraint set M[h,g]. These considerations are carried out very carefully in the interesting exposé [51].

<u>Remark 6.2.3</u>. For an extensive bibliography on optimization problems depending on parameters, we refer to the recent book [4].

<u>6.3. Structural stability of optimization problems</u>.

In the preceding section we studied stability of critical points, which was a local consideration. Now, we turn over to a global study of stability, called "structural" stability. For sake of simplicity, we present a C^∞-version of the idea.

Let $m,s \in \mathbb{N}$ and $f,h_i,g_j \in C^\infty(\mathbb{R}^n,\mathbb{R})$, $i \in \{1,\ldots,m\}$, $j \in \{1,\ldots,s\}$. Furthermore, let the constraint set $M = M[h,g]$ be regular. We recall that $f_{|M}$ is separating if distinct critical points for $f_{|M}$ have distinct functional values.

<u>Theorem 6.3.1</u>. (<u>Theorem on Structural Stability</u>).
Let $M[h,g]$ be a <u>compact</u>, regular constraint set and $f_{|M}$ nondegenerate, separating. Then, there exists a C^2-neighborhood $\mathcal{O}$ of $(f,h_1,\ldots,h_m,g_1,\ldots,g_s)$ such that: for every $(\tilde{f},\tilde{h}_1,\ldots,\tilde{h}_m,\tilde{g}_1,\ldots,\tilde{g}_s) \in \mathcal{O}$ there exist C^∞-diffeomorphisms $\tilde{\phi}: \mathbb{R}^n \to \mathbb{R}^n$, $\tilde{\psi}: \mathbb{R} \to \mathbb{R}$, satisying:

(i) $\tilde{\phi}$, resp. $\tilde{\psi}$, equals the identity outside a compact set in $\mathbb{R}^n$, resp. $\mathbb{R}$,

(ii) $\tilde{\phi}$ maps $M[h,g]$ onto $\tilde{M} = M[\tilde{h},\tilde{g}]$,

(iii) $\tilde{f}_{|\tilde{M}} = \tilde{\psi}\circ f\circ\tilde{\phi}^{-1}{}_{|\tilde{M}}$. □

The following commutative diagram illustrates Theorem 6.3.1:

$$\begin{array}{ccc} \mathbb{R}^n \supset M & \xrightarrow{\ f\ } & \mathbb{R} \\ \Big\downarrow \tilde{\phi} & & \Big\downarrow \tilde{\psi} \\ \mathbb{R}^n \supset \tilde{M} & \xrightarrow{\ \tilde{f}\ } & \mathbb{R} \end{array}$$

Before proving Theorem 6.3.1, we discuss some aspects of it. From (i) $\tilde{\psi}$ preserves the ordering structure of $\mathbb{R}$, i.e. $x < y$ implies $\tilde{\psi}(x) < \tilde{\psi}(y)$. In fact, note that $D\tilde{\psi}$ is equal to 1 outside a compact subset of $\mathbb{R}$. Hence, $D\tilde{\psi}$ is positive on the whole $\mathbb{R}$ since $D\tilde{\psi}$ does not vanish. Consequently,

$\tilde{\psi}$ is strictly monotone increasing. From (ii) it follows that the topological structures of $M[h,g]$ and $\tilde{M}[\tilde{h},\tilde{g}]$ are equal, since, in particular, $\tilde{\phi}$ is a homeomorphism. Moreover, $\tilde{\phi}$ is stratification-preserving, i.e. $\tilde{\phi}$ maps k-dim. strata of $M[h,g]$ onto k-dim. strata of $\tilde{M}[\tilde{h},\tilde{g}]$ (cf. also Corollary 3.1.3). See Fig. 6.3.1.

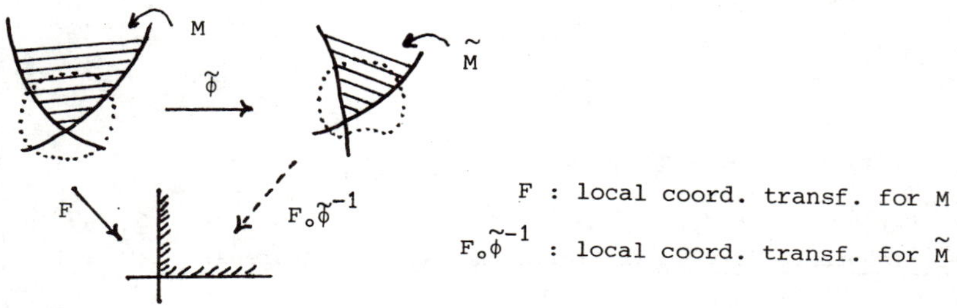

Fig. 6.3.1

From (iii) it follows that $\tilde{\phi}$ maps critical points for $f_{|M}$ onto critical points for $\tilde{f}_{|\tilde{M}}$ and preserves the corresponding indices (LI,LCI,QI,QCI). Since $\tilde{\psi}$ is strictly monotone increasing, lower level sets are mapped onto lower level sets. In fact, let $\tilde{x} \in \tilde{M}$ with $\tilde{f}(\tilde{x}) \le \alpha$. Then, $\tilde{\psi}\circ f\circ\tilde{\phi}^{-1}(\tilde{x}) \le \alpha$ and thus, $f\circ\tilde{\phi}^{-1}(\tilde{x}) \le \tilde{\psi}^{-1}(\alpha)$. Conversely, let $y \in M$ with $f(y) \le \tilde{\psi}^{-1}(\alpha)$. Then, $f\circ\tilde{\phi}^{-1}\circ\tilde{\phi}(y) \le \tilde{\psi}^{-1}(\alpha)$ and thus, $\tilde{\psi}\circ f\circ\tilde{\phi}^{-1}\circ\tilde{\phi}(y) \le \alpha$, i.e. $\tilde{f}(\tilde{\phi}(y)) \le \alpha$.

Finally, in Fig. 6.3.2, we sketched some level lines of a nondegenerate function $f_1 \in C^\infty(\mathbb{R}^2,\mathbb{R})$ which is <u>not</u> separating. By means of an (arbitrarily small) C^2-perturbation of f_1 we may obtain the functions f_2, f_3, as sketched in Fig. 6.3.2.b,c. Note that the functional values of f_1 coincide at the critical points p_1^1, p_2^1. The function f_2, resp. f_3, is obtained from f_1 by increasing the functional value of p_1^1, resp. p_2^1, thereby leaving the value at p_2^1, resp. p_1^1, unchanged. We emphasize that the critical point p_1^1 for f_2 has the same indices as the critical point p_1^1 for f_3. However, the "topological type" of p_1^1 for f_2 is different from the topological type of p_1^1 for f_3. To see this, just look at the behaviour of lower level sets for increasing functional values.

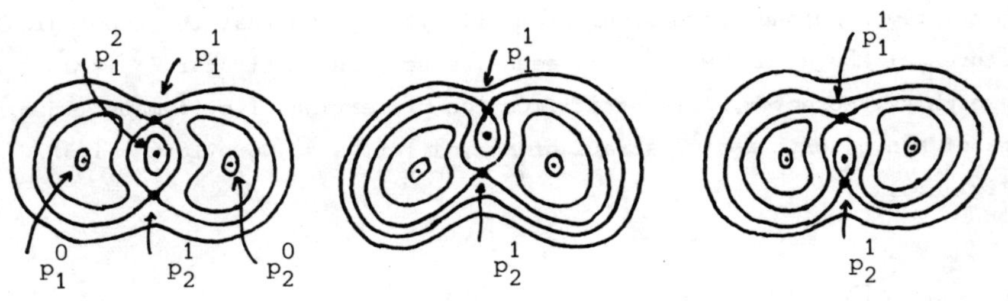

p_j^i : critical point with QI = i.

Fig. 6.3.2

Proof of Theorem 6.3.1. We will give the proof in 4 steps.

Step 1. We remark that $M[h,g] = \emptyset$ iff the following associated function is positive for all $x \in \mathbb{R}^n$: $\sum_{i=1}^{m} |h_i(x)| + \sum_{j=1}^{s} |\min\{0, g_j(x)\}|$.
Hence, if $M[h,g] = \emptyset$, then $M[\tilde{h},\tilde{g}] = \emptyset$ for all $(\tilde{h}_1,\ldots,\tilde{h}_m,\tilde{g}_1,\ldots,\tilde{g}_s)$ in a C^2-neighborhood $\mathcal{U}$ of $(h_1,\ldots,h_m,g_1,\ldots,g_s)$. Therefore, we assume in the sequel that $M[h,g] \neq \emptyset$.

Step 2. In Formula (6.1.34), the set $V^k_{\phi,f}$ is defined as a subset of $C^k(\mathbb{R}^n,\mathbb{R})$. Since we are dealing now with C^∞-functions, we denote $V^k_{\phi,f} \cap C^\infty(\mathbb{R}^n,\mathbb{R})$ again by $V^k_{\phi,f}$.
For any given compact neighborhood K of M (i.e. the interior of the compact set K contains M) we may choose a continuous function $\phi: \mathbb{R}^n \to \mathbb{R}$, $\phi(x) > 0$ for all $x \in \mathbb{R}^n$, such that for $\tilde{f} \in V^2_{\phi,f}$, $\tilde{h}_i \in V^2_{\phi,h_i}$, $\tilde{g}_j \in V^2_{\phi,g_j}$, we have:

2(i) $M[\tilde{h},\tilde{g}]$ is regular and contained in K (cf. Step 1),

2(ii) $\tilde{f}_{|M[\tilde{h},\tilde{g}]}$ is nondegenerate (cf. Example 6.1.3).

So, we start by choosing a compact neighborhood K of M, and a function ϕ as above such that 2(i), (ii) hold, and we put

$$\mathcal{O} = V^2_{\phi,f} \times \mathop{X}_i V^2_{\phi,h_i} \times \mathop{X}_j V^2_{\phi,g_j}.$$

Note that $\mathcal{O}$ is an open, <u>convex</u> set (by the very definition of $V^2_{\phi,f},\ldots$).

Step 3. Take an element $(\tilde{f},\tilde{h}_1,\ldots,\tilde{h}_m,\tilde{g}_1,\ldots,\tilde{g}_s) \in \mathcal{O}$. We will construct a C^∞-diffeomorphism $\tilde{\phi}$: $\mathbb{R}^n \to \mathbb{R}^n$ having the following properties:

3(i) $\tilde{\phi}$ equals the identity outside a compact subset of $\mathbb{R}^n$,

3(ii) $\tilde{\phi}$ maps $M[h,g]$ onto $\tilde{M} = M[\tilde{h},\tilde{g}]$,

3(iii) $\tilde{\phi}$ maps the critical pointset of $f_{|M}$ onto the critical pointset of $\tilde{f}_{|\tilde{M}}$.

The intuitive idea of constructing the diffeomorphism $\tilde{\phi}$ is clarified by considering a certain related problem in $\mathbb{R}^{n+1} = \mathbb{R}^n \times \mathbb{R}$. Let $z = (x,t)$ denote a point in $\mathbb{R}^{n+1}$. The last coordinate t will play a special role. We embed M, $\tilde{M} \subset \mathbb{R}^n$ into $\mathbb{R}^{n+1}$ in the following simple way: $M \to M \times \{0\}$, $\tilde{M} \to \tilde{M} \times \{1\}$. Now, we consider $M \times \{0\}$ and $\tilde{M} \times \{1\}$ as levelsets of the special function $\eta(x,t) = t$, restricted to a cylinder-like manifold $\mathcal{M}$ "connecting" $M \times \{0\}$ and $\tilde{M} \times \{1\}$ (cf. Fig. 6.3.3). Then, we construct a special smooth 1-parameter group $\Phi(u,x,t)$ of diffeomorphisms of $\mathbb{R}^{n+1}$ onto $\mathbb{R}^{n+1}$, u being the parameter, such that $(\tilde{\phi}(x),1) := \Phi(1,x,0)$.

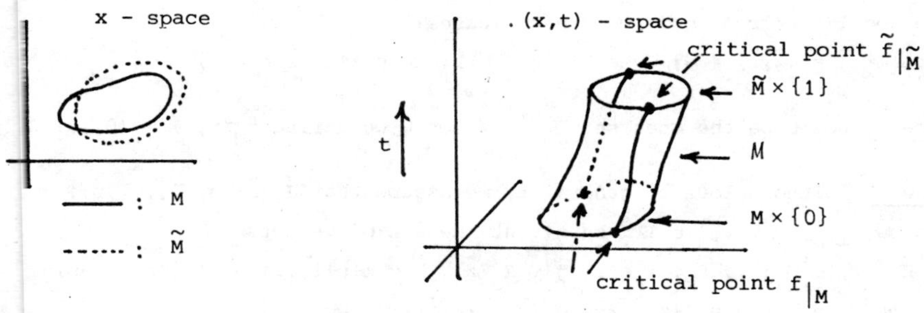

Fig. 6.3.3

Put $F(x,t) = (1-t)f(x) + t\tilde{f}(x)$, $H_i(x,t) = (1-t)h_i(x) + t\tilde{h}_i(x)$, $G_j(x,t) = (1-t)g_j(x) + t\tilde{g}_j(x)$, $i \in I = \{1,\ldots,m\}$, $j \in J = \{1,\ldots,s\}$ and

$$\mathcal{M} = \{(x,t) \in \mathbb{R}^{n+1} \mid H_i(x,t) = 0, G_j(x,t) \geq 0,\ i \in I,\ j \in J,\ 0 \leq t \leq 1\}.$$

We may look at $\mathcal{M}$ as a new constraint set, where H_i, G_j and "$t \geq 0$", "$1-t \geq 0$", are the constraints. Note that, as $t \in [0,1]$, F is a convex combination of f and $\tilde{f}$, the same being true for H_i, G_j. Since $\mathcal{O}$ is convex, $(F(\cdot,t),\ldots,H_i(\cdot,t),\ldots,G_j(\cdot,t),\ldots)$ stays in $\mathcal{O}$ for $t \in [0,1]$. From 2(i) it follows that $\mathcal{M}$ is a compact ($\mathcal{M} \subset K \times [0,1]$!), regular constraint set. In fact, to show the regularity of $\mathcal{M}$, let $\bar{z} = (\bar{x},\bar{t}) \in \mathcal{M}$ and let $J_o(\bar{z})$ denote the index set of the active inequality constraints G_j at $\bar{z}$. In the "worst case" we have to show that the set of vectors (in $\mathbb{R}^{n+1}$)

$$\{D^T H_i(\bar{z}),\ D^T G_j(\bar{z}),\ i \in I,\ j \in J_o(\bar{z})\} \cup \{(0,\ldots,0,1)^T\}$$

is linearly independent. But this clearly follows from the fact that the vectors (in $\mathbb{R}^n$) $D_x^T H_i(\bar{z})$, $D_x^T G_j(\bar{z})$, $i \in I$, $j \in J_o(\bar{z})$ are linearly independent.

Put $\Sigma = \{(\bar{x},\bar{t}) \in \mathcal{M} \mid \bar{x}$ critical point for $F(\cdot,\bar{t})_{|M[H(\cdot,\bar{t}),G(\cdot,\bar{t})]}\}$.
We proceed with the construction of a special vectorfield $\mathcal{F}$ on $\mathbb{R}^{n+1}$, being tangent to $\mathcal{M}$ (cf. Section 3.2), except for the "top and bottom" (i.e. $t = 0$ or 1). To this aim we construct locally defined vectorfields which will be glued together by means of a C^∞-partition of unity of $\mathbb{R}^{n+1}$ (cf. Section 2.2), thus giving us the desired vectorfield $\mathcal{F}$. Concerning the local constructions, we treat 3 cases:
Case I: $\bar{z} = (\bar{x},\bar{t}) \notin \mathcal{M}$; Case II: $\bar{z} \in \mathcal{M}\setminus\Sigma$; Case III: $\bar{z} \in \Sigma$.

<u>Case I</u>. Let U be the open set $\mathbb{R}^{n+1}\setminus\mathcal{M}$. On U we define $\mathcal{F}_U(x,t) = (0,\ldots,0,1)^T$.

<u>Case II</u>. Without loss of generality we assume that $J_o(\bar{z}) = \{1,\ldots,p\}$.
Choose $\xi_{m+p+1},\ldots,\xi_n \in \mathbb{R}^n$ such that the set of vectors $\{D_x^T H_i(\bar{z}),\ D_x^T G_j(\bar{z}),\ \xi_k,\ i \in I,\ j \in J_o(\bar{z}),\ k = m+p+1,\ldots,n\}$ forms a basis for $\mathbb{R}^n$ and put $y = \Psi(z)$, where $(\bar{z} = (\bar{x},\bar{t}))$:
$y_i = H_i(z)$, $i \in I$, $y_j = G_{j-m}(z)$, $j = m+1,\ldots,m+p$, $y_k = \langle\xi_k, x-\bar{x}\rangle$, $k = m+p+1,\ldots,n$ and $y_{n+1} = t-\bar{t}$.
Then, Ψ is a local coordinate system (in the sense of Def. 3.1.1), say: $\Psi\colon V \to W$, V and W open in $\mathbb{R}^{n+1}$, and $0 = \Psi(\bar{z})$, $J_o(z) \subset J_o(\bar{z})$ for all $z \in V$.
On V we define the following vectorfield $\mathcal{F}_V$:

$$\mathcal{F}_V(z) = (D\Psi^{-1})\circ\Psi(z)\cdot(0,\ldots,0,1)^T.$$

Note that the last component of $\mathcal{F}_V$ is identically equal to 1. Moreover,

F_V is tangent to the stratum of M through $\bar{z}$ (cf. Lemma 3.3.2 (3) and Fig. 3.3.1.b). Finally, we take an open neighborhood U of $\bar{z}$, the closure $\bar{U}$ being compact, $\bar{U} \subset V$, such that $U \cap \Sigma = \emptyset$ (note that Σ is a closed subset of $\mathbb{R}^{n+1}$), and define the vectorfield F_U to be the restriction of F_V to U.

Case III. Let $\bar{z} \in \Sigma$. Again, we assume that $J_0(\bar{z}) = \{1,\dots,p\}$. Let $\bar{\lambda} = (\bar{\lambda}_1,\dots,\bar{\lambda}_m)$, $\bar{\mu} = (\bar{\mu}_1,\dots,\bar{\mu}_p)$ be the corresponding Lagrange parameters at the critical point $\bar{x}$ ($\bar{z} = (\bar{x},\bar{t})$) for $F(\cdot,\bar{t})_{|M[H(\cdot,\bar{t}),G(\cdot,\bar{t})]}$. Consider the C^∞-mapping $T\colon \mathbb{R}^{n+m+p+1} \to \mathbb{R}^{n+m+p}$

$$T\colon \begin{pmatrix} x \\ \lambda \\ \mu \\ t \end{pmatrix} \mapsto \begin{pmatrix} D_x^T[F - \sum\limits_{i\in I} \lambda_i H_i - \sum\limits_{j\in J_0(\bar{z})} \mu_j G_j] \\ H_i,\ i \in I \\ G_j,\ j \in J_0(\bar{z}) \end{pmatrix} .$$

Then, $T(\bar{x},\bar{\lambda},\bar{\mu},\bar{t}) = 0$, and from Step 2, 2(i), (ii) it follows that the matrix of partial derivatives $D_{\begin{pmatrix} x \\ \lambda \\ \mu \end{pmatrix}} T(\bar{x},\bar{\lambda},\bar{\mu},\bar{t})$ is nonsingular (cf. Example 3.2.6).

Hence, in view of the Implicit Function Theorem, there exist C^∞-mappings $x(t)$, $\lambda(t)$, $\mu(t)$, such that $T(x(t),\lambda(t),\mu(t),t) \equiv 0$ and $(x(\bar{t}),\lambda(\bar{t}),\lambda(\bar{t})) = (\bar{x},\bar{\lambda},\bar{\mu})$. Note that $z(t) := (x(t),t)$ serves as a local parametrization of the 1-dim. manifold Σ.

Let U be a sufficiently small open neighborhood of $\bar{z}$. Analogously as in the Case II, on U we define the vectorfield F_U:

$$F_U(z) = (D\tilde{\Psi}^{-1})\circ\tilde{\Psi}(z)\cdot(0,\dots,0,1)^T,$$

where $\tilde{\Psi}(x,t) = \Psi(x,t) - A\Psi(x(t),t)$, A being the (n+1)-matrix diag(1,...,1,0) and Ψ as in Case II. By the very construction, it is easily seen that F_U is tangent to both Σ and the stratum of M to which $\bar{z}$ belongs. Moreover, the last component of F_U is identically equal to 1.

The sets $\mathbb{R}^{n+1}\backslash M$, $M\backslash\Sigma$ and Σ form a partition of $\mathbb{R}^{n+1}$, so the open sets U as taken in Case I, II, III, cover the whole $\mathbb{R}^{n+1}$. Then, the locally defined vectorfields F_U can be glued together by means of a suitable C^∞-partition of unity of $\mathbb{R}^{n+1}$. In this way we obtain a C^∞-vectorfield F on

$\mathbb{R}^{n+1}$. Note that in gluing locally defined fields together, at each fixed point we take convex combinations. Consequently, the vectorfield F has the following properties:

(a) the last component of F is identically equal to 1,

(b) outside a compact set, F is identically equal to $(0,\dots,0,1)^T$, thus bounded,

(c) F is tangent to the strata of M (except at "top and bottom") and F is tangent to Σ.

From (b) it follows that F generates a (C^∞-) one-parameter group $\Phi(u,x,t)$ of diffeomorphisms of $\mathbb{R}^{n+1}$ onto $\mathbb{R}^{n+1}$, u being the parameter (cf. Theorem 2.3.3). From (a), (c) we see that $\Phi(u,x,0) \in M$ whenever $(x,0) \in M$ and $u \in [0,1]$. Note that we can write $\Phi(1,x,t)$ as $(\phi(x,t),1+t)$, where $\phi \in C^\infty(\mathbb{R}^{n+1},\mathbb{R}^n)$ and $\phi(x,t) = x$ outside a compact subset of $\mathbb{R}^{n+1}$. Because of the fact that $\Phi(1,\cdot,\cdot)$ is a diffeomorphism, we obtain:

$$\begin{pmatrix} D_x\phi(x,t) & D_t\phi(x,t) \\ 0 & 1 \end{pmatrix} \text{ is nonsingular.}$$

Consequently, $D_x\phi(x,t)$ is nonsingular.

Finally, we put $\tilde{\phi}(x) := \phi(x,0)$. We have to show that $\tilde{\phi}$ is a diffeomorphism from $\mathbb{R}^n$ onto $\mathbb{R}^n$. Since $D\tilde{\phi}(x)$ is nonsingular for all x, it suffices to show that $\tilde{\phi}$ is bijective.

The injectivity of $\tilde{\phi}$ follows from the fact that $D\tilde{\phi}$ is nonsingular and $\tilde{\phi}$ is defined on the whole (convex) $\mathbb{R}^n$.

The surjectivity of $\tilde{\phi}$. We show that $\tilde{\phi}[\mathbb{R}^n]$ is an open and closed subset of $\mathbb{R}^n$. Since $\mathbb{R}^n$ is connected, it then follows that $\tilde{\phi}[\mathbb{R}^n] = \mathbb{R}^n$. The map $\tilde{\phi}$ is a local diffeomorphism, hence an open mapping. This shows the openess of $\tilde{\phi}[\mathbb{R}^n]$. In order to show that $\tilde{\phi}[\mathbb{R}^n]$ is closed, we note that $\tilde{\phi}$ is equal to the identity map outside a compact subset of $\mathbb{R}^n$. So, we may choose a radius $r > 0$ such that $\tilde{\phi}$ equals the identity map outside the closed ball $B(0,r)$ with center at 0 and radius r. Let $\mathring{B}(0,r)$ be the interior of $B(0,r)$. Then, $\mathbb{R}^n = (\mathbb{R}^n \backslash \mathring{B}(0,2r)) \cup B(0,2r)$. The set $\tilde{\phi}[B(0,2r)]$ is compact, since $\tilde{\phi}$ is continuous and $B(0,2r)$ compact. Furthermore, $\tilde{\phi}[\mathbb{R}^n \backslash \mathring{B}(0,2r)] = \mathbb{R}^n \backslash \mathring{B}(0,2r)$ is a closed set. Consequently, $\phi[\mathbb{R}^n]$ -being equal to the union of the two closed sets $\tilde{\phi}[\mathbb{R}^n \setminus \mathring{B}(0,2r)]$ and $\tilde{\phi}[B(0,2r)]$- is closed.

So, $\tilde{\phi}$ is a diffeomorphism. The fact that $\tilde{\phi}$ maps $M[h,g]$ onto $M[\tilde{h},\tilde{g}]$ is easily checked, as well as the fact that $\tilde{\phi}$ maps the critical pointset of $f_{|M}$ onto the critical pointset of $\tilde{f}_{|\tilde{M}}$. This proves Step 3.

Step 4. In view of the foregoing steps, we may proceed with the following assumptions (eventually after reducing the open set $\mathcal{O}$ in Step 2 and noting that the critical values of $f_{|M[h,g]}$ depend continuously on f, h_i, g_j (cf. Section 6.1).

4(i) $M = M[h,g]$ is nonempty, compact, regular,

4(ii) $F(\cdot,t)_{|M}$ is nondegenerate for all $t \in [0,1]$, where $F(x,t) = (1-t)f(x) + t\tilde{f}(x)$,

4(iii) the critical pointset of $f_{|M}$ and $\tilde{f}_{|M}$ coincide,

4(iv) let x^i, $i = 1,\ldots,k$, be the critical points for $f_{|M}$. Put $\alpha_i = f(x^i)$, $\beta_i = \tilde{f}(x^i)$ and for $\varepsilon_i > 0$ let $I(\varepsilon_i) = (\frac{1}{2}(\alpha_i+\beta_i) - \varepsilon_i, \frac{1}{2}(\alpha_i+\beta_i) + \varepsilon_i)$. Then, there exist $\varepsilon_i > 0$, $i = 1,\ldots,k$, such that: $\alpha_i, \beta_i \in I(\varepsilon_i)$, $i = 1,\ldots,k$ and $I(2\varepsilon_i) \cap I(2\varepsilon_j) = \emptyset$ for $i,j = 1,\ldots,k$, $i \neq j$.

Let η be a C^∞-vectorfield (1-dimensional) on $\mathbb{R}$ having the following properties: $0 \le \eta(x) \le 1$ for all $x \in \mathbb{R}$, $\eta(x) = 1$ on $\bigcup_{i=1}^{k} I(\varepsilon_i)$ and $\eta(x) = 0$ on $\mathbb{R} \setminus \bigcup_{i=1}^{k} I(2\varepsilon_i)$.

For each $t \in \mathbb{R}$ let ξ_t be the following C^∞-vectorfield on $\mathbb{R}$:

$$\xi_t(x) = \begin{cases} t(\beta_i - \alpha_i)\eta(x) \text{ on } I(2\varepsilon_i),\ i = 1,\ldots,k \\ 0, \text{ elswhere}. \end{cases}$$

For each $t \in \mathbb{R}$, ξ_t generates a (unique) C^∞-one-parameter group of diffeomorphisms ψ_t: $\mathbb{R} \times \mathbb{R} \to \mathbb{R}$, $(u,x) \mapsto \psi_t(u,x)$. We put

$$\tilde{\psi}_t(x) = \psi_t(1,x). \tag{6.3.1}$$

It may be verified that $\tilde{\psi}_t(x)$ -viewed at as a function of the variables

(t,x)- is of class C^∞. Note that $\tilde{\psi}_t$ is equal to the identity outside a compact subset of $\mathbb{R}$. In fact, $\tilde{\psi}_1$ will serve as the diffeomorphism $\tilde{\psi}$ of our theorem. Note furthermore, that $\tilde{\psi}_0$ equals the identity on the whole $\mathbb{R}$. Next, we put:

$$H(x,t) = \tilde{\psi}_t^{-1} \circ [(1-t)f(x) + t\tilde{f}(x)]. \tag{6.3.2}$$

<u>Remark 6.3.1</u>. From 4(ii) and the fact that $\tilde{\psi}_t$, and thus $\tilde{\psi}_t^{-1}$, is a diffeomorphism, it follows that $H(\cdot,t)_{|M}$ is nondegenerate for all $t \in [0,1]$.

<u>Remark 6.3.2</u>. From the very construction of $\tilde{\psi}_t$ it follows:

$$H(x^i,t) = f(x^i), \; i = 1,\dots,k, \text{ and } t \in [0,1]. \tag{6.3.3}$$

As in Step 3 we define local vectorfields F_U which will be glued together, thus resulting in a special smooth vectorfield F on $\mathbb{R}^{n+1}$.

Put $\mathcal{M} = M \times [0,1]$, $\Sigma = \bigcup_{i=1}^{k} \{x^i\} \times [0,1]$. Note that the compact set Σ represents all critical points for $H(\cdot,t)_{|M}$, $t \in [0,1]$. Let $z = (x,t)$ denote again a point in $\mathbb{R}^{n+1}$, $x \in \mathbb{R}^n$, $t \in \mathbb{R}$. Concerning the local constructions of F_U, we treat 3 cases:

Case I: $\bar{z} = (\bar{x},\bar{t}) \notin \mathcal{M}$; Case II: $\bar{z} \in \mathcal{M}\backslash\Sigma$; Case III: $\bar{z} \in \Sigma$.

<u>Case I</u>. Let U be the open set $\mathbb{R}^{n+1}\backslash\mathcal{M}$. On U we define $F_U(x,t) = (0,\dots,0,1)^T$.

<u>Case II</u>. Let $(\bar{x},\bar{t}) \in \mathcal{M}\backslash\Sigma$. Then, $\bar{x}$ is not a critical point for $H(\cdot,\bar{t})_{|M}$. Without loss of generality we assume that $J_0(\bar{x}) = \{1,\dots,p\}$ (the index set of active inequality constraints g_j). As in Step 3, Case II, choose $\xi_{m+p+1},\dots,\xi_n \in \mathbb{R}^n$ such that the set

$$\{D^T h_i(\bar{x}),\; D^T g_j(\bar{x}), \xi_k, i \in I, j \in J_0(\bar{x}), k = m+p+1,\dots,n\}.$$

forms a basis for $\mathbb{R}^n$.

Put $y = \Psi(z)$, where $y_i = h_i(x)$, $i \in I = \{1,\dots,m\}$, $y_j = g_{j-m}(x)$, $j = m+1,\dots,m+p$, $y_k = \langle\xi_k, x-\bar{x}\rangle$, $k = m+p+1,\dots,n$, $y_{n+1} = t-\bar{t}$ $(\bar{z} = (\bar{x},\bar{t}))$. Then, Ψ is a local coordinate system, say Ψ: $V \to W$, V, W open in $\mathbb{R}^{n+1}$ and $0 = \Psi(\bar{z})$, $J_0(x) \subset J_0(\bar{x})$ for all $(x,t) \in V$ (eventually, by shrinking V).

Since $\bar{x}$ is not a critical point for $H(\cdot,\bar{t})_{|M}$, hence not a corner-point of M, we have: $m+p < n$. Moreover, $0 \in \mathbb{R}^{n-m-p}$ is not a critical point for $H\circ\Psi^{-1}(0,\ldots,0,y_{m+p+1},\ldots,y_n,0)$. Consequently, for some $i \in \{m+p+1,\ldots,n\}$ and some open neighborhood $\hat{W}$ of $0 \in \mathbb{R}^{n+1}$, $\hat{W} \subset W$, we have: $\left|\frac{\partial}{\partial y_i} H\circ\Psi^{-1}(y)\right| \geq \alpha > 0$ for all $y \in \hat{W}$. On $\hat{W}$ we put

$$\mathcal{H}(y_1,\ldots,y_{n+1}) = (0,\ldots,0,\underset{\uparrow i\text{-th coordinate}}{1},0,\ldots,0)^T.$$

Let U be an open neighborhood of $\bar{z}$, $\bar{U}$ compact, $\bar{U} \subset \Psi^{-1}(\hat{W})$. On U we define F_U as follows:

$$F_U(z) = \begin{pmatrix} -\dfrac{D_tH(z)[D\Psi^{-1}(v)\cdot\mathcal{H}(v)]_n}{D_xH(z)[D\Psi^{-1}(v)\cdot\mathcal{H}(v)]_n} \\ 1 \end{pmatrix}\begin{matrix}\}n \\ \}1\end{matrix}, \tag{6.3.4}$$

where $[(\alpha_1,\ldots,\alpha_n,\alpha_{n+1})^T]_n := (\alpha_1,\ldots,\alpha_n)^T$ and $v = \Psi(z)$.

Note that the denominator in (6.3.4) is bounded away from zero on U. Obviously, we have $U \cap \Sigma = \emptyset$. A crucial fact is that we have on U:

$$DH(z)\cdot F_U(z) \equiv 0. \tag{6.3.5}$$

<u>Remark 6.3.3</u>. Formula (6.3.5) has the following important consequence: if we (locally) integrate the vectorfield F_U, then <u>H is constant on the integralcurves</u>.

<u>Case III</u>. Let $\bar{z} = (\bar{x},\bar{t}) \in \Sigma$ and let $\Psi\colon V \to W$, $y = \Psi(z)$, $0 = \Psi(\bar{z})$, be a coordinate transformation as defined in Case II. Then, for all y_{n+1}, the origin $0 \in \mathbb{R}^n$ is a nondegenerate critical point for $H\circ\Psi^{-1}(\cdot,y_{n+1})_{|\hat{M}}$, where $\hat{M} = \{(y_1,\ldots,y_n) \in \mathbb{R}^n \mid y_i = 0,\ y_j \geq 0,\ i = 1,\ldots,m,\ j = m+1,\ldots,m+p\}$. Note that $H\circ\Psi^{-1}(0,y_{n+1})$ is <u>constant</u> and equal to $f(\bar{x})$. For each fixed y_{n+1} there exists a local C^∞-coordinate transformation $\Phi_{y_{n+1}}$ on $\mathbb{R}^n$, $\Phi_{y_{n+1}}(0) = 0$, such that (cf. Theorem 3.2.1, Remark 3.2.5):

$$H\circ\Psi^{-1}(\cdot,y_{n+1})\circ\Phi^{-1}_{y_{n+1}}(0,\ldots,0,\xi_{m+1},\ldots,\xi_n) = f(\bar{x}) + \sum_{i=m+1}^{m+p} \pm\,\xi_i + \sum_{j=m+p+1}^{n} \pm\,\xi_j^2. \tag{$*$}$$

An inspection of the proof of Theorem 3.2.1 shows that, at least for small values of y_{n+1}, we may choose $\Phi_{y_{n+1}}(y_1,\ldots,y_n)$ such that $\Phi(y_1,\ldots,y_n,y_{n+1}) := \Phi_{y_{n+1}}(y_1,\ldots,y_n)$ is of class C^∞. (This shows in particular, that the number of $\pm$ signs in the linear and quadratic part of (*) is constant).

So, let $\tilde{W}$ be an open neighborhood of $0 \in \mathbb{R}^{n+1}$, $\tilde{W} \subset W$, such that $\tilde{\Phi}: \tilde{W} \to \tilde{\Phi}(\tilde{W})$ is a local coordinate transformation, where

$$\xi = \tilde{\Phi}(y), \quad \begin{pmatrix} \xi_1 \\ \vdots \\ \xi_n \\ \xi_{n+1} \end{pmatrix} = \begin{pmatrix} \Phi(y_1,\ldots,y_n,y_{n+1}) \\ \\ y_{n+1} \end{pmatrix}. \tag{6.3.6}$$

Let U be an open neighborhood of $\bar{z}$, $\bar{U}$ compact $\bar{U} \subset \Psi^{-1}(\tilde{W})$. Finally, on U we put:

$$F_U(z) = D\Psi^{-1}(v)\cdot(D\tilde{\Phi}^{-1} \circ \tilde{\Phi}(v))\cdot(0,\ldots,0,1)^T, \text{ where } v = \Psi(z).$$

<u>Remark 6.3.4.</u> In view of the very construction of F_U it follows that the last component of F_U equals one and, moreover, H is constant on the integral-curves of F_U.

By means of a suitable C^∞-partition of unity of $\mathbb{R}^{n+1}$, we glue the locally defined vectorfields F_U together, thus obtaining a <u>bounded</u> vectorfield F. Consequently, F generates a (unique) C^∞-one-parameter group $\hat{\Phi}(u,x,t)$ of diffeomorphisms of $\mathbb{R}^{n+1}$ onto $\mathbb{R}^{n+1}$, u being the parameter (cf. Theorem 2.3.3). Note that the last component of F is identically equal to 1. Analogously as in Step 3 we may write

$$(\phi(x,t),1+t) := \hat{\Phi}(1,x,t) \text{ and } \tilde{\phi}(x) := \phi(x,0),$$

and we obtain again that $\tilde{\phi}$ is a C^∞-diffeomorphism, mapping M onto M and being equal to the identity outside a compact subset of $\mathbb{R}^n$. From Remark 6.3.3, 6.3.4, it follows:

$$H\circ\hat{\Phi}(u,x,0) = H\circ\hat{\Phi}(0,x,0), \text{ for all } u \in [0,1],\ x \in M. \tag{6.3.7}$$

Consequently, substituting u = 1 in (6.3.7), we obtain:

$$H(\tilde{\phi}(x),1) = H(x,0), \text{ for all } x \in M. \tag{6.3.8}$$

Substitution of (6.3.2) in (6.3.8) yields:

$$\tilde{\psi}_1^{-1} \circ \tilde{f} \circ \tilde{\phi}(x) = \tilde{\psi}_0^{-1} \circ f(x), \text{ for all } x \in M.$$

Recalling that $\tilde{\psi}_0$ is equal to the identity, and putting $\tilde{\psi} = \tilde{\psi}_1$, we finally obtain:

$$\tilde{\psi}^{-1} \circ \tilde{f} \circ \tilde{\phi}_{|M} = f_{|M}, \text{ or, since } \tilde{\phi} \text{ maps M onto M,}$$

$$\tilde{f}_{|M} = \tilde{\psi} \circ f \circ \tilde{\phi}^{-1}_{|M}.$$

This completes the proof of Theorem 6.3.1. □

7. TRANSVERSALITY.

7.1. Introduction, Sard's Theorem, Regular Optimization Problems.

Let us consider in $\mathbb{R}^2$ two smoothly embedded circles C_1, C_2 (i.e. C_1 and C_2 are manifolds in $\mathbb{R}^2$ which are diffeomorphic with the circle S^1, cf. Fig. 7.1.1.a.).

C_1 and C_2 are called <u>transversal at an intersectionpoint</u> ($\bar{x}$) if at this point they intersect under an angle unequal to 0 or π; more formally, this means: the tangentspaces of C_1, C_2 at $\bar{x}$ span the whole $\mathbb{R}^2$ (= space of embedding), cf. Fig. 7.1.1.b. In the case of non-transversal intersection, the tangentspaces of C_1 and C_2 at the intersectionpoint coincide, thus spanning a one-dimensional subspace of $\mathbb{R}^2$ (Fig. 7.1.1.c).

C_1 and C_2 are called <u>transversal</u> if either they are transversal at each intersectionpoint or they do not intersect at all.

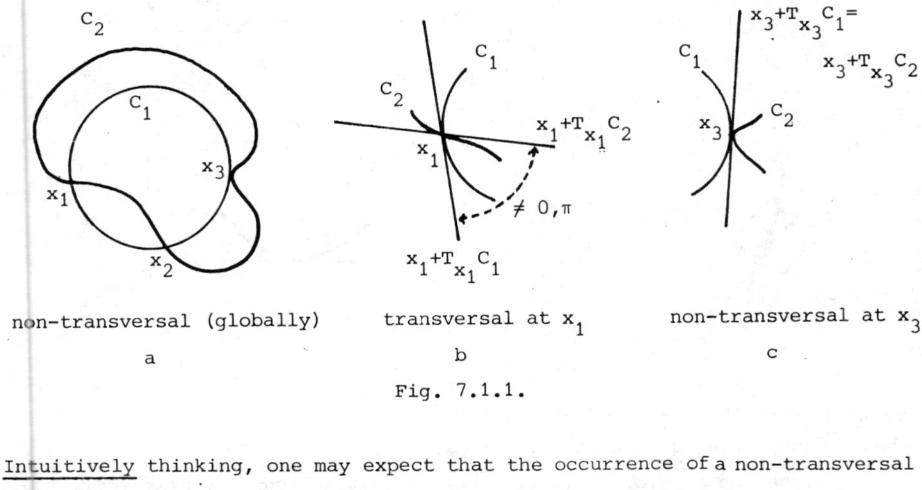

non-transversal (globally) — a; transversal at x_1 — b; non-transversal at x_3 — c

Fig. 7.1.1.

<u>Intuitively</u> thinking, one may expect that the occurrence of a non-transversal intersectionpoint, such as the point x_3, is very exceptional w.r.t. (unrestricted) smooth perturbations of C_1, C_2. Now we translate this intuition into mathematical ideas. Let us parametrize an open <u>$\mathbb{R}^2$-neighborhood</u> $\mathcal{O}$ of C_1 smoothly with one parameter α, such that C_1 is represented by $C_1 \times \{0\}$ (Fig. 7.1.2.a). (If C_1 would lie in $\mathbb{R}^n$, $n > 2$, we should need (n-1)-parameters to do so). Next we cut out a suitable open subset $\mathcal{O}_1 \subset \mathcal{O}$, $\mathcal{O}_1$ containing the point x_3 (Fig. 7.1.2.b). To each point $x \in \mathcal{O}_1 \cap C_2$ we assign the value of the parameter α, say $f(x)$, having the property that

$x \in C_1 \times \{f(x)\} \cap C_2$. In this way we obtain a function f (in fact a parameter-distribution) on $\mathcal{O}_1 \cap C_2$. We may proceed by sending $\mathcal{O}_1$ diffeomorphically onto an open subset $\mathcal{O}_2 \subset \mathbb{R}^2$, thereby leaving x_3 fixed, such that C_2 is represented by a coordinate axis. In this coordinate-frame, the function f has the property that $f(x_3) = 0$, $Df(x_3) = 0$. So, $f^{-1}(0)$ contains a critical point, namely the point x_3 (Fig. 7.1.3).

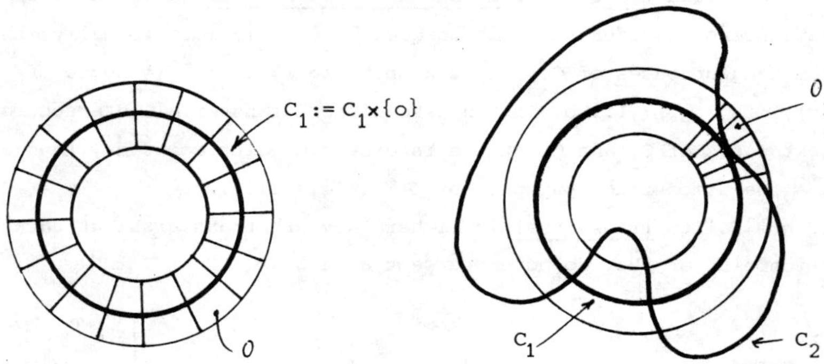

Fig. 7.1.2.

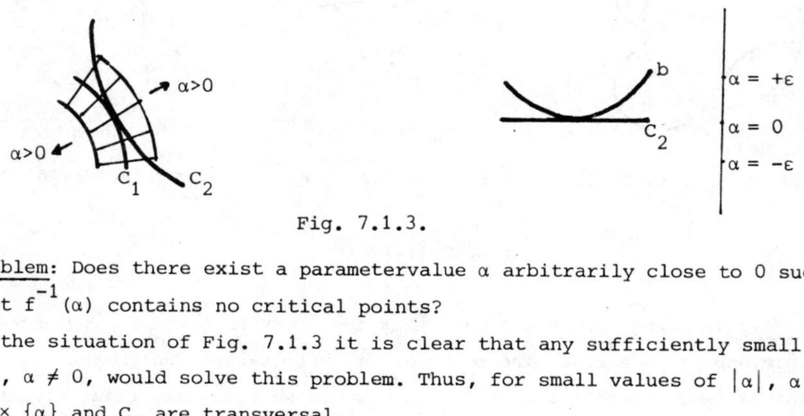

Fig. 7.1.3.

Problem: Does there exist a parametervalue α arbitrarily close to 0 such that $f^{-1}(\alpha)$ contains no critical points?

In the situation of Fig. 7.1.3 it is clear that any sufficiently small $|\alpha|$, $\alpha \neq 0$, would solve this problem. Thus, for small values of $|\alpha|$, $\alpha \neq 0$, $C_1 \times \{\alpha\}$ and C_2 are transversal.

Note that - in the situation as suggested in the above example - another phenomenon is used which is crucial for the concept of transversality: if C_1 and C_2 are transversal (at an intersectionpoint), then this remains true after sufficiently small smooth perturbations of C_1 and C_2. This fact basically follows from the Implicit Function Theorem.

Remark 7.1.1. It is very important to note that we parametrized a whole $\mathbb{R}^2$-neighborhood of the point x_3 with the circles $C_1 \times \{\alpha\}$. Any concept of "transversality" will always be related to a certain class of admissible perturbations. For example, let us consider the set G of all one-dimensional manifolds in $\mathbb{R}^2$ of the form graph(f), where $f \in C^\infty(\mathbb{R},\mathbb{R})$ is such that $f(0) = 0$ and $Df(0) = 0$.

Obviously, all manifolds in G are non-transversal at 0. Thus, under arbitrary smooth perturbations within G, the non-transversality is conserved (cf. Fig. 7.1.4.a). However if we change our concept of transversality as follows: graph (f_1) and graph (f_2), both in G, are "transversal at 0" iff graph (Df_1) and graph (Df_2) are transversal at 0 in the sense of Fig. 7.1.1.b, then smooth perturbations of the form $f(x) + \alpha x^2$, $\alpha \in \mathbb{R}$ arbitrary, will yield analogous phenomenae as discussed before (cf. Fig. 7.1.4.b). Note that a whole $\mathbb{R}^2$-neighborhood of 0 is parametrized by the manifolds graph$(Df+2\alpha)$, where graph$(f) \in G$.

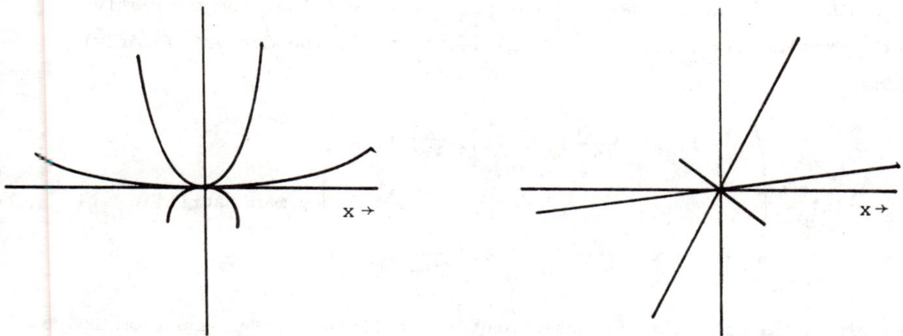

graph$(f(x)+\alpha x^2)$; $f \equiv 0$
(a)

graph$(Df(x)+2\alpha x)$; $f \equiv 0$
(b)

Fig. 7.1.4.

Remark 7.1.2. If we choose $\mathbb{R}^3$ as the embedding space for the circles C_1, C_2 and if we may parametrize an open $\mathbb{R}^3$-neighborhood of C_1 by means of 2 parameters α_1, α_2, where $C_1 := C_1 \times \{(0,0)\}$, then we may expect that for almost all values of (α_1,α_2) the circle $C_1 \times \{(\alpha_1,\alpha_2)\}$ does not intersect C_2 at all, thus being transversal to C_2.

Note that the tangentspaces of C_1, C_2 are 1-dimensional, so they never span the whole embedding space $\mathbb{R}^3$, being 3-dimensional. Another way of looking at this is, that in general, one needs only three "independent" equations for the determination of a point in $\mathbb{R}^3$, whereas C_1, C_2 are <u>both</u> definable - locally - by means of 2 equations.

Although in Fig. 7.1.3 it is obvious how to choose α such that $f^{-1}(\alpha)$ does not contain a critical point, in general such choices are not obvious at all (Fig. 7.1.5). In fact, the possibility of many such choices is based on the famous Theorem of Sard which is crucial in transversality theory.

Fig. 7.1.5.

In order to formulate Sard's Theorem, we need some preliminaries. Let $f \in C^1(\mathbb{R}^n, \mathbb{R}^m)$ and $\bar{x} \in \mathbb{R}^n$. At the point $\bar{x}$ we consider the derivative $Df(\bar{x})$. Putting $f(x_1,\ldots,x_n) = (f_1(x),\ldots,f_m(x))^T$, the derivative $Df(\bar{x})$ becomes:

$$Df(\bar{x}) = \begin{pmatrix} \frac{\partial}{\partial x_1} f_1(\bar{x}) & \frac{\partial}{\partial x_2} f_1(\bar{x}) & \cdots & \frac{\partial}{\partial x_n} f_1(\bar{x}) \\ \vdots & \vdots & & \vdots \\ \frac{\partial}{\partial x_1} f_m(\bar{x}) & \frac{\partial}{\partial x_2} f_m(\bar{x}) & \cdots & \frac{\partial}{\partial x_n} f_m(\bar{x}) \end{pmatrix}, \quad m\times n\text{-matrix.}$$

The point $\bar{x}$ is called a <u>critical</u> point for f if the linear map from $\mathbb{R}^n$ to $\mathbb{R}^m$ given by $\xi \mapsto Df(\bar{x})\xi$ is <u>not surjective</u>. In other words, $\bar{x}$ is <u>regular</u> (i.e. not critical) for f iff the set $\{\frac{\partial}{\partial x_i} f(\bar{x}),\ i = 1,\ldots,n\}$ spans the whole $\mathbb{R}^m$ (i.e. $Df(\bar{x})\,[\mathbb{R}^n] = \mathbb{R}^m$). Consequently, in case $m > n$, every $\bar{x} \in \mathbb{R}^n$ is critical for f.
A $y \in \mathbb{R}^m$ is called a <u>regular</u>, resp. <u>critical value</u> for f if $f^{-1}(y)$ contains no critical points, resp. contains at least one critical point.

<u>Remark 7.1.3.</u> In case m = 1 we have $f \in C^1(\mathbb{R}^n, \mathbb{R})$. Then $\bar{x}$ is critical iff the set $\{\frac{\partial}{\partial x_i} f(\bar{x}),\ i = 1,\ldots,n\}$ does not span $\mathbb{R}$, thus $\bar{x}$ is critical iff $\frac{\partial}{\partial x_i} f(\bar{x}) = 0$, $i = 1,\ldots,n$. So, in case m = 1 the above definition of a

critical point coincides with the definition of a critical point as given in Section 1.1.

Note that apart form the concept of critical point, in literature we also have the so called singular points: Let $g \in C^1(\mathbb{R}^n,\mathbb{R}^m)$, then $\bar{x} \in \mathbb{R}^n$ is called a singular point for g if rank $Dg(\bar{x})$ is not maximal, i.e. rank $Dg(\bar{x}) < \min(n,m)$. Obviously, a singular point is critical and both concepts coincide in the case $m = 1$.

Definition 7.1.1. A subset $A \subset \mathbb{R}^n$ is said to be of Lebesgue-measure zero (or shortly: thin), if for every $\varepsilon > 0$ a sequence of cubes $W_i \subset \mathbb{R}^n$ exists with $A \subset \bigcup_{i=1}^{\infty} W_i$ and $\sum_{i=1}^{\infty} |W_i| < \varepsilon$. Here, $|W|$ stands for the volume of the cube W, i.e. $|W| = (2a)^n$, $W = \{(x_1,\dots,x_n) \in \mathbb{R}^n \mid |x_i-\bar{x}_i| \le a,\ i = 1,\dots,n\}$.

Example 7.1.1. A countable subset of $\mathbb{R}^n$ is thin. The union of a countable number of thin subsets of $\mathbb{R}^n$ is thin again. As an application, the set of rational numbers ($\mathbb{Q}$) is thin in $\mathbb{R}$.

Remark 7.1.4. Let $A \subset \mathbb{R}^n$ be thin. It is easily seen (exercise) that $\mathbb{R}^n \setminus A$ is dense in $\mathbb{R}^n$. However, the converse is not true (take for A the subset of $\mathbb{R}$ consisting of the irrational numbers). Moreover, it can be shown that $\mathbb{R}^n \setminus B$ need not be thin, even if B is open and dense.

Theorem 7.1.1. (Sard's Theorem, C^∞-case).
Let $f \in C^\infty(\mathbb{R}^n,\mathbb{R}^m)$. Then the set of critical values of f has Lebesgue-measure zero.

For a nice proof of Theorem 7.1.1 we refer to [7]. From this theorem it follows immediately that, in case $m > n$, a smooth mapping from $\mathbb{R}^n$ to $\mathbb{R}^m$ is never surjective.

Remark 7.1.5. Let X be a k-dimensional smooth manifold in $\mathbb{R}^n$, $x \in X$ and let $g \in C^\infty(X,\mathbb{R}^m)$. Using local coordinates, the concept of the derivative $Dg(x)$ - as a linear map from T_xX ($\cong \mathbb{R}^k$) to $\mathbb{R}^m$ - can be introduced (cf. Exercise 3.2.2). Since the concept of critical point - as defined above - is of local nature, this concept can be transferred to this more general situation: x is critical point for g if $Dg(x)[T_xX] \neq \mathbb{R}^m$.

Since the Euclidean space $\mathbb{R}^n$ (and thus also X) satisfies the second axiom of countability, it is possible to select a countable set of local coordinate systems $\phi_i: U_i \to \mathbb{R}^k$, $i = 1,2,\ldots$ such that $\{U_i\}$ constitutes an open covering of X. Moreover, we may assume that the ϕ_i's are diffeomorphisms. Note that the set of critical values for g just equals the union of the sets of critical values for all the mappings $g \circ \phi_i^{-1}$. From Theorem 7.1.1 together with Example 7.1.1 it will be clear that the proposition of Sard's Theorem is also true for a smooth mapping from X to $\mathbb{R}^m$.

From now on we suppose that X is a C^∞-Manifold with Generalized Boundary (MGB) in $\mathbb{R}^n$, cf. Section 3.1. We may introduce the concept of a critical point for the mapping $g \in C^\infty(X,\mathbb{R}^m)$ in a similar way as in the case that X is a smooth manifold (compare again Exercise 3.2.2). The proposition of Sard's Theorem remains also true. This is a consequence of Sard's Theorem for smooth manifolds as well as the following two observations:

- X is the countable union of its strata Σ_α, $\alpha = 1,2,\ldots$,
- the set of critical values for g is contained in the union of the sets of critical values for the mappings $g_{|\Sigma_\alpha}$.

Finally, let us suppose that dim X < n. Then, the inclusion map $i: X \to \mathbb{R}^n$ is obviously of the class C^∞ and the set of critical values for i is just X. Hence, X is a thin subset of $\mathbb{R}^n$ (and thus $\mathbb{R}^n \backslash X$ is dense in $\mathbb{R}^n$). Note that X itself is not a dense subset of $\mathbb{R}^n$. This follows from the fact that X - as an MGB - is locally closed in $\mathbb{R}^n$ (cf. Exercise 3.1.4).

In the sequel we say that a property holds <u>for almost all</u> $x \in \mathbb{R}^n$ if it holds outside a thin subset of $\mathbb{R}^n$.

<u>Remark 7.1.6.</u> Let X be a k-dimensional MGB in $\mathbb{R}^n$ of the class C^∞ and let $f \in C^\infty(X,\mathbb{R})$. Then for almost all $\alpha \in \mathbb{R}$ the sets $f^{-1}((-\infty,\alpha])$ and $f^{-1}(\alpha)$ are MGB's of class C^∞ of dimension k resp. k-1 (if not empty).
The proof of this statement follows immediately from Lemma 3.2.7 and Remark 7.1.5.

<u>Remark 7.1.7.</u> Sard's Theorem is in general not true for mappings $f \in C^k(\mathbb{R}^n,\mathbb{R}^m)$, $1 \le k < \infty$. In fact, in [76], Whitney has constructed a counterexample, where $f \in C^1(\mathbb{R}^2,\mathbb{R})$ is not constant on a <u>connected</u> set A

consisting of critical points (In that case $f(A)$ is connected and contains an interval, say (α,β), $\beta > \alpha$; thus $f(A)$ cannot be thin!).
However, Sard's Theorem remains true for $f \in C^k(\mathbb{R}^n,\mathbb{R}^m)$, provided that $k > \max(n-m,0)$, cf. [74, pp. 45-55]. Thus a mapping $f \in C^1(\mathbb{R}^n,\mathbb{R}^m)$, $m > n$ can never be surjective. However, there exist surjective <u>continuous</u> mappings from $\mathbb{R}^n$ to $\mathbb{R}^m$, $m > n$! Compare for this phenomenon the "Peano space-filling curve" in [33, pp. 122,123].

We recall that a smooth function defined on an MGB is called nondegenerate if all its critical points are nondegenerate (cf. Chapter 3). As an application of Sard's Theorem we obtain:

<u>Theorem 7.1.2.</u> For $0 \le m \le n$, let $M = \mathbb{R}^m \times \mathbb{H}^{n-m}$ and $f \in C^\infty(\mathbb{R}^n,\mathbb{R})$. For $u \in \mathbb{R}^n$, put $F_u(x) = f(x) - u^T x$. Then, for almost all $u \in \mathbb{R}^n$, $F_{u|M}$ is nondegenerate.

<u>Proof</u>. We give the proof in two steps:
<u>Step 1</u>. Firstly we deal with the case $m = n$ (> 0), i.e. $M = \mathbb{R}^n$. Let $\bar{x}$ be a critical point for F_u, thus $Df(\bar{x}) = u^T$. Note that $\bar{x}$ is degenerate (as a critical point for F_u) iff $\bar{x}$ is critical for the mapping $G: \mathbb{R}^n \to \mathbb{R}^n$: $x \mapsto D^T f(x)$. Hence, F_u is nondegenerate iff u is a regular value for G. Now, application of Sard's Theorem yields the result.

<u>Step 2</u>. Next, we consider the case $m < n$. In the sequel, it will be convenient to introduce the following terminology: $F_{u|M}$ is called nice on a subset of M if all critical points for $F_{u|M}$ which are situated in this subset are nondegenerate.
Let Σ be an arbitrary stratum of M. Without loss of generality we may assume that Σ is of the form:

$$\Sigma = \{x = (y,z) \in \mathbb{R}^n \mid y = (y_1,\ldots,y_{m+\ell}),\ y_{m+1} > 0,\ldots,y_{m+\ell} > 0;\quad z = (z_1,\ldots,z_{n-m-\ell}) = 0\}$$

where $0 \le \ell \le n-m$ (see Fig. 7.1.6).

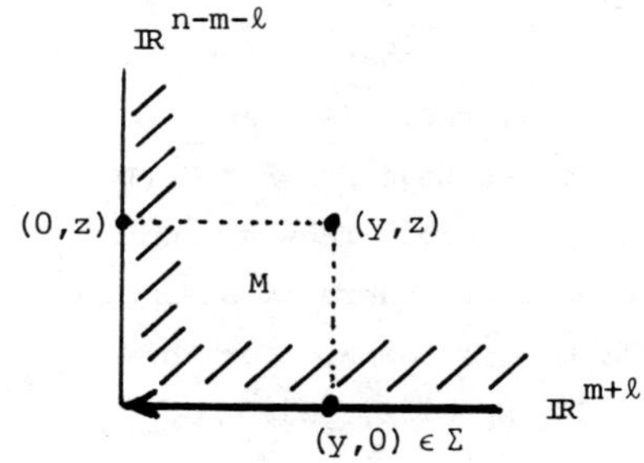

Fig. 7.1.6.

Let $\bar{x} = (\bar{y},0) \in \Sigma$ be a nondegenerate critical point for $F_{u|M}$. It is not difficult to show (cf. Definition 3.2.2, Lemma 3.2.6) that this means:

$$\begin{cases} D_y F_u(\bar{x}) = 0 \\ \pi(\bar{x}) = \left|\frac{\partial F_u}{\partial z_1}(\bar{x})\cdot\frac{\partial F_u}{\partial z_2}(\bar{x}) \dots \frac{\partial F_u}{\partial z_{n-m-\ell}}(\bar{x})\right| > 0 & (ND_1) \\ \det(D_y^2 F_u(\bar{x})) \neq 0. & (ND_2) \end{cases}$$

Thus $F_{u|M}$ is nice on Σ iff

$$\|D_y F_u(\cdot)\| + |\det(D_y^2 F_u(\cdot))|\pi(\cdot) \text{ is strictly positive on } \Sigma. \qquad (*)$$

We consider a countable covering of Σ consisting of the compact sets C_r (⊂ Σ), $r = 1,2,\dots$. In view of the Condition (*) we have: the subset U_{C_r} of points $u \in \mathbb{R}^n$ for which $F_{u|M}$ is nice on C_r, is <u>open and thus measurable</u> in $\mathbb{R}^n$. It follows that $U := \bigcap_{r=1}^{\infty} U_{C_r}$ (= subset of $\mathbb{R}^n$ for which $F_{u|M}$ is nice on Σ) is measurable in $\mathbb{R}^n$.

We proceed by proving that the complement (say U^c) of U in $\mathbb{R}^n$ has measure zero. To this aim, we apply some version of Fubini's Theorem (cf. for example [23], [63]) which states:

> Let S be a measurable set in $\mathbb{R}^p = \mathbb{R}^s \times \mathbb{R}^{p-s}$ $(0 < s < p)$.
> We denote a point in $\mathbb{R}^p$ by (a,b), $a \in \mathbb{R}^s$, $b \in \mathbb{R}^{p-s}$.
> For $c \in \mathbb{R}^s$, let
>
> $$S_c = \{b \in \mathbb{R}^{p-s} \mid (c,b) \in S\}$$
>
> Then S is thin in $\mathbb{R}^p$ iff S_c is thin in $\mathbb{R}^{p-s}$ for almost all $c \in \mathbb{R}^s$.

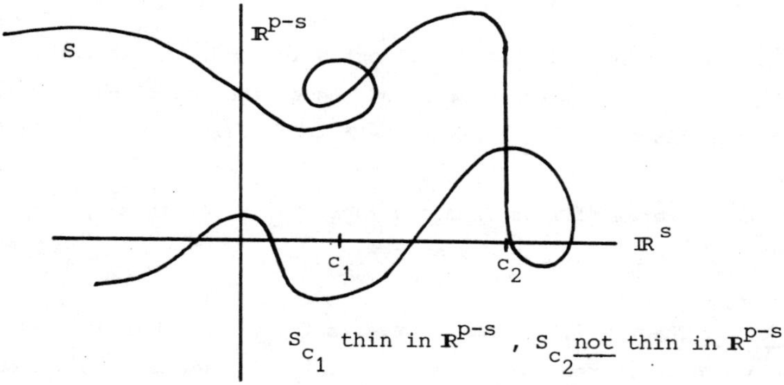

Fig. 7.1.7.

Consider an arbitrary point $\bar{v} = (\bar{v}_1,\ldots,\bar{v}_{m+\ell})$ of $\mathbb{R}^{m+\ell}$ and let $\bar{u} = (\bar{v},0) \in \mathbb{R}^n$. Note that (for $\bar{x} = (\bar{y},0) \in \Sigma$) we have:

$$D_y F_{(\bar{v},0)}(\bar{x}) = D_y F_{(\bar{v},w)}(\bar{x}), \text{ where } w \in \mathbb{R}^{n-m-\ell} \text{ is arbitrary} \qquad (**)$$

We distinguish between the following alternatives:

α. In Σ there exists a degenerate critical point $(\bar{x})$ for $F_{\bar{u}|\Sigma}$;

β. $F_{\bar{u}|\Sigma}$ is nondegenerate.

<u>In case α</u> it is obvious that for all $u = (\bar{v},w)$, $w \in \mathbb{R}^{n-m-\ell}$, $\bar{x} = (\bar{y},0) \in \mathbb{R}^n$ is a degenerate critical point for $F_{u|M}$, i.e. $u \in U^c$. It follows that the whole $(n-m-\ell)$-dimensional linear variety $\bar{U} = \{(\bar{v},w)\,|\,w \in \mathbb{R}^{n-m-\ell}\}$ is contained in U^c and hence $\bar{U} \cap U^c$ is not thin in $\mathbb{R}^{n-m-\ell}$.

As a consequence of the result of Step 1 - use that Σ is open in $\mathbb{R}^{m+\ell}$ - we have that for almost all $\bar{v} \in \mathbb{R}^{m+\ell}$ the alternative β is valid.

<u>In case β</u>, the set K of critical points for $F_{\bar{u}|\Sigma}$ is countable (finite or infinite) since each critical point is nondegenerate and thus isolated. One easily verifies that for $\bar{x} \notin K$ we have: $\bar{x}$ is non-critical for $F_{(\bar{v},w)|M}$ in Σ, where $w \in \mathbb{R}^{n-m-\ell}$ is arbitrary (use **). On the other hand for $\bar{x} \in K$ we have: $\{\bar{v}\} \times \mathbb{R}^{n-m-\ell} \cap U^c = \{\bar{v}\} \times \bigcup_{\bar{x}\in K} W_{\bar{x}}$ where $W_{\bar{x}}$ is the set of $w \in \mathbb{R}^{n-m-\ell}$ such that $F_{(\bar{v},w)}$ is degenerate in $\bar{x}$. One easily verifies, by using explicitely the nondegeneracy conditions ND1 and ND2, that $W_{\bar{x}}$ is the union of finitely many hyperplanes in $\mathbb{R}^{n-m-\ell}$ and hence $\{\bar{v}\} \times \mathbb{R}^{n-m-\ell} \cap U^c$ has measure zero in $\mathbb{R}^{n-m-\ell}$.

Altogether we have proved that for almost all $v \in \mathbb{R}^{m+\ell}$ the intersection $\{v\} \times \mathbb{R}^{n-m-\ell} \cap U^c$ has measure zero and hence, the set of parameters $u \in \mathbb{R}^n$ for which the function $F_{u|M}$ fails to be nice on the stratum Σ, is thin.

Finally, the theorem follows from the fact that M consists of finitely many strata. □

Remark 7.1.8. Theorem 7.1.2 remains true if $f \in C^2(\mathbb{R}^n, \mathbb{R})$. In that case, the function G (cf. proof of Theorem 7.1.2) is of the class C^1, compare also Remark 7.1.7.

Theorem 7.1.3. Let X ($\subset \mathbb{R}^n$) be a closed Manifold with Generalized Boundary (MGB) of the class C^∞, and let $\mathcal{M}(X)$ denote the subset of mappings in $C^\infty(\mathbb{R}^n, \mathbb{R})$ for which the restriction to X is nondegenerate. Then $\mathcal{M}(X)$ is is C^k-open and C^k-dense in $C^\infty(\mathbb{R}^n, \mathbb{R})$, for all $k \geq 2$. (Note that this automatically means that $\mathcal{M}(X)$ is dense for $k = 0,1$).

Before we give a proof of this theorem, we present an example which clarifies that the condition "X is a closed MGB in $\mathbb{R}^n$" is necessary.

Example 7.1.2. Let X be the open interval (0,1) of $\mathbb{R}$. Consider a function $f \in C^\infty(\mathbb{R}, \mathbb{R})$ such that the derivative Df vanishes outside X but has no zeros on X. As far as the existence of such a function f is concerned, we refer to Section 2.6 (take for f(x) the function $\phi_3(x+1)$, see Fig. 2.6.2). Clearly, the restriction $f_{|X}$ is nondegenerate. For $\varepsilon > 0$, we define $f_\varepsilon(x) = f(x+\varepsilon x)$. Obviously, for any $\varepsilon > 0$, the function $f_{\varepsilon|X}$ fails to be nondegenerate. Now, for $k \geq 2$ and for $\phi \in C^0(\mathbb{R}, \mathbb{R})$ arbitrary and strictly positive, let $V^k_{\phi,f}$ be a base-neighborhood of f w.r.t. the C^k-topology on $C^\infty(\mathbb{R}, \mathbb{R})$, cf. Definition 6.1.1. Since the function ϕ has a strictly positive minimum on [0,1] and moreover the functions f and f_ε as well as their derivatives vanish outside X, one easily shows that, for $\varepsilon > 0$ sufficiently small, we have: $f_\varepsilon \in V^k_{\phi,f}$. Consequently, the set $\{f \in C^\infty(\mathbb{R}, \mathbb{R}) \mid f_{|X}$ is nondegenerate$\}$ is not C^k-open in $C^\infty(\mathbb{R}, \mathbb{R})$. Note that if we take for X the closed interval [0,1], then $f_{|X}$ fails to be non-degenerate.

Remark 7.1.9. The condition "X is closed in $\mathbb{R}^n$" may be skipped if we consider $M(X)$ as a subset of $C^\infty(X,\mathbb{R})$, where the latter space is endowed with a special topology which is analogous to the C^k-topology on $C^\infty(\mathbb{R}^n,\mathbb{R})$. However, we shall not dwell on the details here.

Now, we give the proof of the preceding theorem:

Proof of Theorem 7.1.3.

Open Part. Since X is a smooth MGB in $\mathbb{R}^n$, around each $x \in X$, a local C^∞-coordinate system of $\mathbb{R}^n$, say $\psi: U \to V$, exists such that

$$\psi(U \cap X) = \{(y_1,\dots,y_m,y_{m+1},\dots,y_{m+p},\dots,y_n) \in V \mid y_1=\dots=y_m=0, y_{m+1}\geq 0,\dots y_{m+p}\geq 0\} \tag{*}$$

Now, it is easily seen that we may select a countable set of such coordinate-systems, say $\psi_i: U_i \to V_i$, $i = 1,2,\dots$, with the additonal property that the closure $(\bar{U}_i)$ of U_i in $\mathbb{R}^n$ is compact for all i.

Let $\{\theta_\alpha\}_{\alpha=0,1,2,\dots}$ be a partition of unity (cf. Section 2.1) subordinate to the open covering $\{U_\alpha\}_{\alpha=0,1,2,\dots}$ of $\mathbb{R}^n$, where $U_0 = \mathbb{R}^n\backslash X$.

Consider a function $f \in M(X)$. This means that, adopting the terminology as used in the proof of Theorem 7.1.2, for all $i = 1,2,\dots$, the restriction $f_{|X}$ is nice on supp θ_i, or - equivalently - $f\circ\psi_i^{-1}{}_{|\psi_i(U_i\cap X)}$ is nice on any of the sets $\Sigma_i := \psi_i(X \cap \text{supp}\,\theta_i) \cap \Sigma$ where Σ is any stratum for $\{o\} \times \mathbb{H}^p \times \mathbb{R}^{n-m-p}$, $o \in \mathbb{R}^m$. (**)

Note that, the number p depends on i and that the sets Σ_i may not be assumed to be compact.

Let $\bar{y}$ be an arbitrary point in Σ_i. Then, we may assume that, without loss of generality, $\bar{y}$ is of the form

$$\bar{y} = (\underbrace{0,\dots,0}_{m \text{ times}},\underbrace{0,\dots,0}_{k \text{ times}},\bar{y}_{m+k+1},\dots,\bar{y}_{m+p},\bar{y}_{m+p+1},\dots,\bar{y}_n),$$

where the "semi-free" coordinates $\bar{y}_{m+k+1},\dots,\bar{y}_{m+p}$ are strictly positive and the "free coordinates" $\bar{y}_{m+p+1},\dots,\bar{y}_n$ are arbitrary $(0 \leq k \leq p)$. Compare Fig. 7.1.8. (Note that in particular this means that Σ is of dimension $n-m-k$).

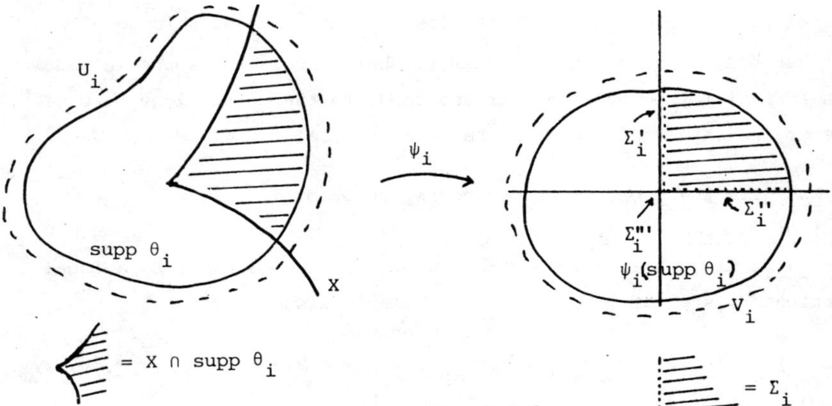

Fig. 7.1.8.

We put $f_i = f \circ \psi_i^{-1}$ and define, for $y \in V_i$,

$$\sigma_i(y) = \sum_{\ell=m+k+1}^{n} \left|\frac{\partial}{\partial y_\ell} f_i(y)\right| + \prod_{\ell=m+1}^{m+k} \left|\frac{\partial}{\partial y_\ell} f_i(y)\right| \cdot \Delta(y),$$

$$\text{where } \Delta(y) := \left|\det\left(\frac{\partial^2 f_i}{\partial y_r \partial y_s}(y)\right)\right|, \quad r,s = m+k+1,\ldots,n.$$

Note that in σ_i the dependence on the stratum Σ is a consequence of the fact that the definition of $\sigma_i(\cdot)$ depends on the (number of) semi-free and free coordinates in Σ.
One easily checks (cf. Section 3.2) that (**) is equivalent with:

$\sigma_i(\cdot)$ is strictly positive on Σ_i.

If $\bar{\Sigma}$ stands for the closure of Σ we even have

$\sigma_i(\cdot)$ is strictly positive on the <u>compact</u> set $\bar{\Sigma}_i$.

In fact, note that the complement $\bar{\Sigma}\backslash\Sigma$ is the union of strata with dimension strictly lower than $\dim(\Sigma)$. Hence, for an arbitrary $\tilde{y} \in \bar{\Sigma}_i\backslash\Sigma_i$ we have either $\tilde{y}$ is non-critical for $f_{i|\psi_i(U_i\cap X)}$ and thus at least one of the partial derivatives of f_i w.r.t. the (semi-)free coordinates of the stratum to which $\tilde{y}$ belongs does not vanish, or
$\tilde{y}$ is critical for $f_{i|\psi_i(U_i\cap X)}$ and thus at least one of the partial derivatives of f_i w.r.t. the coordinates $y_{m+k+1},\ldots,y_{m+p}$ just equals a Lagrange parameter of $\tilde{y}$ (w.r.t. $f_{i|\psi_i(U_i\cap X)}$) and thus does not vanish as well (use nondegeneracy).

Altogether we conclude that the functions $\sigma_i(\cdot)$, which are strictly positive on Σ_i do attain a strictly positive minimum on $\bar{\Sigma}_i$. From this one easily derives that after a C^2-perturbation (cf. Section 6.1) of $f_{i|\Sigma_i}$ which is sufficiently small, the resulting function remains nice on Σ_i. Consequently, for any $i = 1,2,\ldots$ a strictly positive real, say ε_i, exists such that any function $g \in C^\infty(\mathbb{R}^n,\mathbb{R})$ for which

$$|f-g|_{X \cap \operatorname{supp} \theta_i} < \varepsilon_i$$

we have: $g_{|X}$ is nice on supp θ_i. Here $|\cdot|$ stands for the C^k-seminorm on $C^\infty(\mathbb{R}^n,\mathbb{R})$ w.r.t. $X \cap \operatorname{supp} \theta_i$, $k \geq 2$. (Note that at this point we use the facts that ψ_i is a diffeomorphism and that there are only finitely many strata Σ). Since the sets supp θ_i, $i = 1,2,\ldots$, constitute a <u>locally finite</u> covering (of X), for any index $i_0(\geq 1)$, the set $I(i_0)$ of indices i, with supp $\theta_i \cap \operatorname{supp} \theta_{i_0} \neq \emptyset$ is <u>finite</u>. We define $\bar{\varepsilon}_{i_0} := \min_{i \in I(i_0)} \varepsilon_i$ and moreover $\bar{\varepsilon}_0 := 1$. Now we introduce the function ρ on $\mathbb{R}^n$ as follows

$$\rho(x) = \sum_{\alpha=0}^{\infty} \bar{\varepsilon}_\alpha \cdot \theta_\alpha(x).$$

Note that ρ is continuous and strictly positive!

Consider an arbitrary point $x_0 \in X \cap \operatorname{supp} \theta_{i_0}$. Then,

$$\rho(x_0) = \sum_{i \in I(i_0)} \bar{\varepsilon}_i \cdot \theta_i(x_0) \leq \sum_{i \in I(i_0)} \varepsilon_{i_0} \cdot \theta_i(x_0) \leq \varepsilon_{i_0}.$$

By $V^k_{\rho,f}$ we denote the C^k-neighborhood of f, determined by ρ (cf. Definition 6.1.1). Then, for any $g \in V^k_{\rho,f}$ we conclude that $g_{|X}$ is nondegenerate.

<u>Dense Part</u>. Let $g \in C^\infty(\mathbb{R}^n,\mathbb{R})$ be an arbitrary function and let $V^k_{\phi,g}$ be an arbitrary C^k-baseneighborhood of g. In order to prove the density of $M(X)$ we have to show that $V^k_{\phi,g} \cap M(X)$ is non-empty. To this aim we consider the open covering $\{U_\alpha\}_{\alpha=0,1,\ldots}$ of $\mathbb{R}^n$ and the partition of unity $\{\theta_\alpha\}$ subordinate to this covering which was introduced in the Open Part of this proof. It is easily seen that the collection $\{\operatorname{int}(\operatorname{supp} \theta_\alpha)\}_{\alpha=0,1,2\ldots}$ constitutes an open covering of $\mathbb{R}^n$ as well. Hence, a partition of unity - say $\{\chi_\alpha\}$ - exists subordinate to this latter covering. Note that - since the U_i's, $i = 1,2,\ldots$, are chosen such that their closures are compact - we have $\operatorname{int}(\operatorname{supp} \theta_i) \subsetneq U_i$. In view of Lemma 2.2.1, for any i, a function

$\xi_i \in C^\infty(\mathbb{R}^n,\mathbb{R})$ exists with the following properties:

$$0 \le \xi_i(x) \le 1, \quad \text{all } x \in \mathbb{R}^n$$
$$\xi_i(x) = 1, \quad \text{on a neighborhood of supp}(\chi_i)$$
$$\text{supp}(\xi_i) \subset \text{int(supp } \theta_i).$$

Now, we focus our attention to the coordinate system $\psi_1: U_1 \to V_1$, with $\psi_1(X \cap U_1) = M_1 \cap V_1$, where $M_1 = \{o\} \times \mathbb{H}^p \times \mathbb{R}^{n-m-p}$, $o \in \mathbb{R}^m$. Let:

$$\varepsilon_1 := \min_{x \in \text{supp}(\theta_1)} (\phi(x)),$$

where ϕ is the continuous, strictly positive function which determines $V^k_{\phi,g}$ (thus $\varepsilon_1 > 0!$). Moreover, we define

$$g_1 := g \circ \psi_1^{-1}.$$

In view of Lemma 2.2.1(b) we may extend the function g_1 to a function $h_1 \in C^\infty(\mathbb{R}^n,\mathbb{R})$ in such a way that h_1 and g_1 coincide on some open neighborhood of $\psi_1(\text{supp } \theta_1)$ which is contained in V_1. Now, we apply (a slight generalization of) Theorem 7.1.2 to the function h_1: for almost all $u \in \mathbb{R}^n$, the restriction to M_1 of the functions $\bar{h}_1(y) := h_2(y) - u^T y$ is nondegenerate. From this it follows (since $\psi_1: U_1 \to V_1$ is a diffeomorphism) that the restriction of $h_1^* := h_1 \circ \psi_1$ to $X \cap U_1$ is nice on supp χ_1. By choosing $\|u\|$ sufficiently small, we may assume that

$$|\xi_1 \cdot (g - h_1^*)|_{\text{supp } \theta_1} < \varepsilon_1,$$

where, as in the Open Part of this proof, $|\cdot|$ stands for the C^k-seminorm on $C^\infty(\mathbb{R}^n,\mathbb{R})$ w.r.t. supp θ_1 (= compact subset of $\mathbb{R}^n$). (Note that for the latter estimate we need the fact that on a neighborhood of $\psi_1(\text{supp } \theta_1)$ the functions g_1 and h_1 coincide). The locally defined function $\bar{h}_1 \circ \psi_1$ is extended to a global one in the following way:

$$g_1^*(x) := g(x) + \xi_1(x).(h_1^*(x) - g(x)), \quad x \in U_1$$
$$g_1^*(x) := g(x) \quad , \quad x \in \mathbb{R}^n \setminus U_1.$$

It is easily shown that $g_1^* \in C^\infty(\mathbb{R}^n,\mathbb{R})$, $g^*_{1|X}$ is nice on supp χ_1 and $g_1^* \in V^k_{\phi,g}$.

Now, we perturb the function g_1^* into a function g_2^* by changing its values on supp θ_2 a little bit (and keeping them constant outside supp θ_2) in such a way that: $g_2^* \in C^\infty(\mathbb{R}^n,\mathbb{R})$, $g^*_{2|X}$ is nice on supp χ_2, $g_2^* \in V^k_{\phi,g}$ and moreover, $g^*_{2|X}$ is also nice on supp χ_1. To this aim we extend $g_1^* \circ \psi_2^{-1}$ to a function $h_2 \in C^\infty(\mathbb{R}^n,\mathbb{R})$ in a similar way as we defined the function h_1 above. Now we apply Theorem 7.1.2 to h_2 and obtain functions $\bar{h}_2(y) = h_2(y) - w^T y$ with the property that for almost all $w \in \mathbb{R}^n$ the restriction $\bar{h}_{2|M_2}$ is nondegenerate, where M_2 is of the form $\{o\} \times \mathbb{H}^{p'} \times \mathbb{R}^{n-m-p'}$, $o \in \mathbb{R}^m$, and fulfils the condition $\psi_2(X \cap U_2) = M_2 \cap V_2$. The function $\bar{h}_2 \circ \psi_2$ can be extended to a function $g_2^* \in C^\infty(\mathbb{R}^n,\mathbb{R})$ in a similar way (using ξ_2) as we extended $\bar{h}_1 \circ \psi_1$ to g_1^* (see above). By a suitable choice of the parameter w (namely, w such that $\bar{h}_{2|M_2}$ is nondegenerate and $\|w\|$ sufficiently small) we can assure that: $g_2^* \in V^k_{\phi,g}$ and $g^*_{2|X}$ is nice on supp $\chi_1 \cup$ supp χ_2. Note that the nicety of $g^*_{2|X}$ on supp χ_2 follows in the same way as the nicety of $g^*_{1|X}$ on supp χ_1, whereas the fact that $g^*_{2|X}$ is nice on supp χ_1 is based on the following "openess property": if $g^*_{1|X}$ is nice on supp χ_1 then this remains true after a sufficiently small C^k-perturbation of g_1^*. This property can be proved in a similar way as the openess of $M(X)$ (see above). Now, we proceed inductively by constructing a sequence of functions $\{g_i^*\}$ such that $g_i^* \in C^\infty(\mathbb{R}^n,\mathbb{R}) \cap V^k_{\phi,g}$ and $g^*_{i|X}$ is nice on $\bigcup_{j=1}^{i}$ supp χ_j.

Since the covering $\{\text{supp } \theta_i\}_{i=1,2,\dots}$ of X is locally finite we may conclude that for any $\bar{x} \in \mathbb{R}^n$ there exists a number $N(\bar{x})$ such that for $j_1, j_2 > N(\bar{x})$ we have $g^*_{j_1}(\bar{x}) = g^*_{j_2}(\bar{x})$. From this it follows that $g^* := \lim_{j\to\infty} g_j^*$ is well-defined and fulfils the following conditions:

> $g^* \in C^\infty(\mathbb{R}^n,\mathbb{R})$, $g^* \in V^k_{\phi,g}$ and $g^*_{|X}$ is nondegenerate (for the latter condition we need the fact that $\{\text{supp } \chi_i\}$ constitutes a covering of X).

□

<u>Remark 7.1.10</u>. The technique we used in the proof of Theorem 7.1.3 is a typical example of a so called "local-global" construction.

We recall that a separating function is a smooth function with the property that at its critical points it attains different values. The following generalization of Theorem 7.1.3 has already been announced - in a preliminary form - in Chapter 5 (cf. Remark 5.2.2).

Theorem 7.1.4. Let X be a compact MGB in $\mathbb{R}^n$. Then, the set of all nondegenerate, separating functions $f_{|X}$, $f \in C^\infty(\mathbb{R}^n,\mathbb{R})$, is C^k-open and C^k-dense in $C^\infty(\mathbb{R}^n,\mathbb{R})$, $k \geq 2$, where the density remains true also in the case $k = 0,1$.

Proof. We begin by stipulating that a nondegenerate critical point for the restriction to X of a smooth function on $\mathbb{R}^n$ is isolated (in X). This follows immediately from the so called generalized Morse lemma (cf. Theorem 3.2.1). Consequently, since X is compact, the number of critical points for a nondegenerate function on X is finite. Firstly, we pay attention to the "open part" of the theorem. So, let $f_{|X}$, $f \in C^\infty(\mathbb{R}^n,\mathbb{R})$, be nondegenerate and separating. We must show that a C^k-open neighborhood, say $V^k_{\phi,f}$, of f exists such that all $g \in V^k_{\phi,f}$ are nondegenerate and separating. The critical points for $f_{|X}$ are denoted by $\bar{x}_1,\ldots,\bar{x}_N$. Around each of these critical points $\bar{x}_i$, we choose a coordinate system $\psi_i\colon U_i \to V_i$, according to Definition 3.1.1, with the additional property that $U_i \cap U_j = \emptyset$ if $i \neq j$. Around each $\bar{x}_i$ we choose a closed ball B_i which is contained in U_i. Moreover, by choosing the balls B_i sufficiently small, we assure that

$$\max_{x \in B_i} |f(x) - f(\bar{x}_i)| < \frac{1}{2} \cdot \min_{\substack{i,j=1,\ldots,N \\ i \neq j}} |f(\bar{x}_i) - f(\bar{x}_j)| \qquad (*)$$

Application of Theorem 6.1.1 yields the following property: after a sufficiently small perturbation of $f_{|B_i}$ w.r.t. the C^2-norm, the resulting function restricted to X has only one critical point in $\mathring{B}_i$, which is nondegenerate. Note that as a matter of fact we need a slight modification of Theorem 6.1.1.b, since the latter theorem deals with a globally defined RCS, whereas in the present situation only $X \cap U_i$ can be described as a (local) RCS.
On $C := X \setminus \bigcup_{i=1}^{N} \mathring{B}_i$ the function $f_{|X}$ has no critical points. We contend that after a sufficiently small C^2-perturbation of f the restriction to X of the

resulting function does not have critical points on C as well. This can be proved by using similar techniques as used in the "open part" of the proof of Theorem 7.1.3. In fact, instead of considering the functions σ_i (see the proof of Theorem 7.1.3) we have to introduce for each Σ as an associated function:

$$\tilde{\sigma}_i(y) = \sum_{\ell=m+k+1}^{n} \left|\frac{\partial}{\partial y_\ell} f_i(y)\right|,$$

which turns out to be strictly positive on $\bar{\Sigma}_i$. (We emphasize that since a corner of X always gives rise to a critical point for $f_{|X}$, corners are <u>not</u> contained in C; hence the definition of $\tilde{\sigma}_i$ does always make sense).
Altogether we conclude that after a sufficiently small C^2-perturbation of $f_{|X}$, the resulting function remains nondegenerate and has exactly N critical points, say $\tilde{x}_1,\ldots,\tilde{x}_N$, such that for all $i = 1,\ldots,N$, both $\bar{x}_i$ and $\tilde{x}_i$ are contained in $\mathring{B}_i$ and moreover (use Condition (*) above) all these perturbed functions remain separating.
We proceed by proving the "<u>dense part</u>" of our assertion. So, let us consider the arbitrary, nondegenerate function $f_{|X}$, as introduced in the "open part" of this proof, which however is <u>not</u> necessarily separating in this case. Since the set $\mathcal{M}(X)$ of smooth functions with nondegenerate restriction to X is C^k-dense in $C^\infty(\mathbb{R}^n,\mathbb{R})$, it suffices to prove that $f_{|X}$ can be approximated (in the C^k-sense) arbitrary well by a nondegenerate separating function. To this aim we consider around each critical point $\bar{x}_i$, $i = 1,\ldots,N$, a closed ball, having all the properties of the ball B_i as introduced in the "open part", with the exception of Condition (*). We denote these balls again by B_i, whereas by B_i' we denote a ball with center $\bar{x}_i$ and radius which is strictly smaller than the radius of B_i. Obviously, the collection of sets $\{\mathbb{R}^n \setminus \bigcup_{i=1}^{N} B_i', \mathring{B}_i;\ i = 1,\ldots,N\}$ constitutes an open covering of $\mathbb{R}^n$. This covering admits a C^∞-partition of unity. The function of this partition for which the support is contained in $\mathring{B}_i$ is denoted by ρ_i. Now, we consider the function

$$g_\tau := f + \sum_{i=1}^{N} i\tau\rho_i,\ \tau > 0,$$

where - for the moment - the positive number τ is still arbitrary.

Obviously, outside $\bigcup_{i=1}^{N} B_i$, the functions g_τ and f coincide, whereas, on the balls B_i', i = 1,...,N, the derivatives of g_τ and f coincide. We contend that if we choose τ sufficiently small, then $g_{\tau|X}$ does not have critical points on the set $R := \bigcup_{i=1}^{N} (B_i \backslash \mathring{B}_i')$. This follows from the facts that on R, the restriction $f_{|X}$ has no critical points and that $g_{\tau|X}$ is arbitrarily C^k-close to $f_{|X}$ if τ is chosen sufficiently small (compare the proof of the "open part"). Altogether we may conclude that there exists a strictly positive real - say τ_1 - such that for all $\tau \in (0,\tau_1)$ the functions $g_{\tau|X}$ are non-degenerate with $\bar{x}_1,\dots,\bar{x}_N$ as critical points.
Finally, it is not difficult to show that - for τ sufficiently small - the functions $g_{\tau|X}$ are always separating. This completes the proof of the Dense Part of the theorem. □

<u>Remark 7.1.11</u>. In the stituation of Theorem 7.1.4 (proof of the "open part") the following property holds: For all i = 1,...,N, the critical points $\bar{x}_i$, $\tilde{x}_i$ (for $f_{|X}$ and a small C^2-perturbation of $f_{|X}$ respectively) belong to the same X-stratum; moreover, they have the same linear and quadratic (co-) indices. We will not give a proof of this statement but merely refer to the special case, we dealt with in Example 3.2.7.

<u>Remark 7.1.12</u>. To a certain extent, the result of Theorem 7.1.4 remains true if we replace the condition "X is compact" by the weaker condition that X must be closed:
Let X be a <u>closed</u> MGB in $\mathbb{R}^n$ and assume that the restriction to X of the smooth function f has the following properties:

1. Nondegenerate with finitely many critical points.
2. Separating.

Then, a C^k-open neighborhood ($k \geq 2$) of f in $C^\infty(\mathbb{R}^n,\mathbb{R})$ - say $\mathcal{V}$ - exists such that for all $g \in \mathcal{V}$ the properties 1. and 2. above hold. It can be proved that - in general - this is not true anymore if $f_{|X}$ has infinitely many critical points.
The density property (cf. Theorem 7.1.4) remains always true.

We conclude this section by presenting the so called Main Theorem on Regular Optimization Problems. To this aim we extend our terminology and introduce one more lemma.
Let X be a <u>closed</u> MGB in $\mathbb{R}^n$ and let A be a subset of X. We define:

$$E(X;A) := \{f \in C^\infty(\mathbb{R}^n,\mathbb{R}) \mid f^{-1}(0) \cap A \text{ does not contain critical points for } f_{|X}\}.$$

<u>Lemma 7.1.1.</u> Let X be a closed MGB. Then, for $k \geq 1$ the set $E(X,X)$ is C^k-open and dense in $C^\infty(\mathbb{R}^n,\mathbb{R})$.

<u>Proof.</u> We only give a rough sketch of the proof since it runs exactly along the same lines as the proof of Theorem 7.1.3.
<u>Open Part.</u> In the situation of the proof of Theorem 7.1.3 (open part) we consider - instead of the functions σ_i - functions $\tilde{\sigma}_i$ given by

$$\tilde{\sigma}_i(y) = |f_i(y)| + \sum_{\ell=m+k+1}^{n} \left| \frac{\partial}{\partial y_\ell} f_i(y) \right| \; ; \quad y \in V_i.$$

Note that $f \in E(X;\, \text{supp}\, \theta_i)$ if and only if $\tilde{\sigma}_i$ is strictly positive on the corresponding set Σ_i, whereas $f \in E(X;\, \text{supp}\, \theta_i)$ even implies that $\tilde{\sigma}_i$ is strictly positive on $\bar{\Sigma}_i$. From these facts the openess of $E(X;X)$ may be derived in an analogous way as the openess property in Theorem 7.1.3.
<u>Dense Part.</u> Let $g \in C^\infty(\mathbb{R}^n,\mathbb{R})$ be arbitrary and consider an arbitrary C^k-base neighborhood $V^k_{\phi,g}$ of g. For almost all $\alpha \in \mathbb{R}$, we have: $(g-\alpha) \in E(X;X)$, and thus $g-\alpha\xi_1 \in E(X,\, \text{supp}\, \chi_1)$. This follows from Remarks 7.1.5, 6. Therefore it is always possible to choose α in such a way that $(g-\alpha\xi_1) \in V^k_{\phi,g}$ and moreover $(g-\alpha\xi_1) \in E(X,\, \text{supp}\, \chi_1)$. For such a choice of α we put $g_1^* := g-\alpha\xi_1$. Now, we consider functions of the form $g_1^* - \beta$, $\beta \in \mathbb{R}$, and note that for almost all β these functions are contained in $E(X;X)$. Consequently , for almost all β, we have: $(g_1^*-\beta\xi_2) \in E(X;\, \text{supp}\, \chi_2)$. Therefore we can always choose β in such a way that
$(g_1^*-\beta\xi_2) \in E(X;\, \text{supp}\, \chi_2) \cap V^k_{\phi,g}$ and moreover $(g_1^*-\beta\xi_2) \in E(X;\, \text{supp}\, \chi_1)$. Note that the latter condition can be fulfilled in view of the openess of $E(X;\, \text{supp}\, \chi_1)$ which may be proved in a similar way as the openess of $E(X;X)$. For such a choice of β, we put $g_2^* := g_1^*-\beta\xi_2$. Now we proceed by considering functions $g_2^* - \gamma\xi_3$ etc. and construct a function g^* with the property that

$g^* \in E(X;X) \cap V^k_{\phi,g}$. □

Theorem 7.1.5. (Main Theorem on Regular Optimization Problems).
Let $m,s \in \mathbb{N}$ be given and let $F \subset C^\infty(\mathbb{R}^n,\mathbb{R})^{1+m+s}$ be defined as follows: $(f,h_1,\ldots,h_m,g_1,\ldots,g_s) \in F$ iff $M[h,g]$ is regular and $f_{|M[h,g]}$ nondegenerate, where h,g stand for $(h_1,\ldots,h_m)$, $(g_1,\ldots,g_s)$ respectively.
Then: F is C^k-open for $k \geq 2$; F is C^k-dense for all k.

Proof. The C^k-open part of this theorem was already stated in Example 6.1.3. Here we consider the C^k-dense part. Again, it suffices to show that $V^k_{\phi,f} \times \mathop{X}\limits_{i=1}^{m} V^k_{\phi,h_i} \times \mathop{X}\limits_{j=1}^{s} V^k_{\phi,g_j}) \cap F \neq \emptyset$ for an arbitrary, but fixed $k \geq 0$.
We give the proof in several steps:

Step 1. Let h_1 be an arbitrary smooth function on $\mathbb{R}^n$. In view of Lemma 7.1.1 a function $\tilde{h}_1 \in V^k_{\phi,h_1}$ exists which is contained in $E(\mathbb{R}^n;\mathbb{R}^n)$. As a consequence of the latter property we have: $M[\tilde{h}_1]$ is an RCS (and thus, in particular a closed MGB). If $M[\tilde{h}_1] = \emptyset$, then we are done since in that case also $M[\tilde{h}_1,h_2,\ldots,h_m,g_1,\ldots,g_s]$ is empty. If $M[\tilde{h}_1] \neq \emptyset$, goto Step 2.

Step 2. Suppose that $M[\tilde{h}_1] \neq \emptyset$ and regular. As a consequence of Lemma 7.1.1 a function $\tilde{h}_2 \in V^k_{\phi,h_2}$ exists which is contained in $E(M[\tilde{h}_1]; M[\tilde{h}_1])$; thus $M[\tilde{h}_1,\tilde{h}_2]$ is an RCS. If $M[\tilde{h}_1,\tilde{h}_2] = \emptyset$, then we are done; otherwise we may proceed and look at a function $\tilde{h}_3 \in V^k_{\phi,h_3} \cap E(M[\tilde{h}_1,\tilde{h}_2]; M[\tilde{h}_1,\tilde{h}_2])$ etc.

Step 3. Suppose that $M[\tilde{h}_1,\ldots,\tilde{h}_m] \neq \emptyset$ and regular, whereas $(\tilde{h}_1,\ldots,\tilde{h}_m) \in \mathop{X}\limits_{i=1}^{m} V^k_{\phi,h_i}$. Again, Lemma 7.1.1 garantees the existence of a function $\tilde{g}_1 \in V^k_{\phi,g_1}$ which is contained in $E(M[\tilde{h}_1,\ldots,\tilde{h}_m]; M[\tilde{h}_1,\ldots,\tilde{h}_m])$.
This latter property implies (cf. Lemma 3.2.7 for the basic idea) that $M[\tilde{h}_1,\ldots,\tilde{h}_m,\tilde{g}_1]$ is regular. If $M[\tilde{h}_1,\ldots,\tilde{h}_m,\tilde{g}_1]$ is empty, then we are done. Otherwise we may proceed and consider a function $\tilde{g}_2 \in V^k_{\phi,g_2} \cap E(M[\tilde{h}_1,\ldots,\tilde{h}_m,\tilde{g}_1]; M[\tilde{h}_1,\ldots,\tilde{h}_m,\tilde{g}_1])$ etc.

Step 4. Suppose that $X = M[\tilde{h},\tilde{g}]$ is regular and non-empty. In view of Theorem 7.1.3 there exists a smooth function $\tilde{f}$ such that $\tilde{f}_{|X}$ is nondegenerate and is contained in $V^k_{\phi,f}$. This completes the proof of the theorem. □

7.2. Transversal intersection of manifolds in $\mathbb{R}^n$.

Throughout Section 7.2, the word manifold will always refer to C^∞-manifold (although C^1 would be sufficient).

Let $M \subset \mathbb{R}^n$ be a manifold of dimension n-m. This means that, for every $\bar{x} \in M$ there exist an open ($\mathbb{R}^n$-)neighborhood $U_{\bar{x}}$ and functions $h_i \in C^\infty(U_{\bar{x}}, \mathbb{R})$, $i = 1,\ldots,m$, such that the following conditions hold: (cf. also Remark 3.1.1)

(M1) $M \cap U_{\bar{x}} = \{x \in U_{\bar{x}} \mid h_i(x) = 0, \quad i = 1,\ldots,m\}$,

(M2) The set $\{Dh_i(x),\ i=1,\ldots,m\}$ is linearly independent for all $x \in M \cap U_{\bar{x}}$.

Moreover, the tangentspace $T_{\bar{x}}M$ just equals $\bigcap_{i=1}^{m} \mathrm{Ker}\, Dh_i(\bar{x})$. See also Example 3.2.2.

We will refer to a set of functions $(h_1,\ldots,h_m)$ satisfying (M1), (M2) as to a <u>defining system of functions</u> for $M \cap U_{\bar{x}}$. The number m denotes the number of "independent" nonlinear restrictions for the local definition of M in $\mathbb{R}^n$. We call this number the <u>codimension</u> of M and we emphasize that the codimension is always related to the dimension of the embedding space ($\mathbb{R}^n$ in this case). We call a set N a <u>submanifold</u> of M of codimension k, if $N \subset M$ and N is a manifold in $\mathbb{R}^n$ of codimension: $k + \mathrm{codim}(M)$.

<u>Definition 7.2.1</u>. Let M_1, M_2 be manifolds in $\mathbb{R}^n$ of codimension m_1, m_2. We say that M_1, M_2 intersect <u>transversally</u> (notation $M_1 \bar{\pitchfork} M_2$), if for every $\bar{x} \in M_1 \cap M_2$ the following holds:
There exist an open ($\mathbb{R}^n$-)neighborhood $U_{\bar{x}}$ of $\bar{x}$ and defining systems of functions $(h_1,\ldots,h_{m_1})$, $(\rho_1,\ldots,\rho_{m_2})$ for $M_1 \cap U_{\bar{x}}$, $M_2 \cap U_{\bar{x}}$ such that the set $\{Dh_i(x), D\rho_j(x), i = 1,\ldots,m_1,\ j = 1,\ldots,m_2\}$ is linearly independent for all $x \in M_1 \cap M_2 \cap U_{\bar{x}}$.

<u>Remark 7.2.1</u>. Note that $M_1 \cap M_2 = \emptyset$ logically implies $M_1 \bar{\pitchfork} M_2$. If $M_1 \bar{\pitchfork} M_2$ and $M_1 \cap M_2 \neq \emptyset$, then $M_1 \cap M_2$ is a submanifold of M_1, M_2, $\mathbb{R}^n$ of codimension m_2, m_1, m_1+m_2 respectively; note that the tangent space $T_{\bar{x}}(M_1 \cap M_2)$ at an $\bar{x} \in M_1 \cap M_2$ equals $T_{\bar{x}}M_1 \cap T_{\bar{x}}M_2$.

The concept of transversal intersection of M_1, M_2 merely refers to a condition on the tangentspace at the intersection points:

<u>Theorem 7.2.1</u>. Let M_1, M_2 be manifolds in $\mathbb{R}^n$. Then, $M_1 \pitchfork M_2$ iff at every point $\bar{x} \in M_1 \cap M_2$ the following condition on the tangentspace holds:

$$T_{\bar{x}}M_1 + T_{\bar{x}}M_2 = \mathbb{R}^n, \tag{7.2.1}$$

(i.e. the tangentspaces $T_{\bar{x}}M_1$, $T_{\bar{x}}M_2$ together span the whole embedding space

<u>Proof</u>. Let $M_1 \pitchfork M_2$, $\bar{x} \in M_1 \cap M_2$ and $m_i = \operatorname{codim}(M_i)$, $i = 1,2$. Since $T_{\bar{x}}M_1$, $T_{\bar{x}}M_2$ are linear subspaces of $\mathbb{R}^n$ we have:

$$\dim(T_{\bar{x}}M_1 + T_{\bar{x}}M_2) = \dim T_{\bar{x}}M_1 + \dim T_{\bar{x}}M_2 - \dim(T_{\bar{x}}M_1 \cap T_{\bar{x}}M_2). \tag{7.2.2}$$

From Remark 7.2.1 we obtain:

$$\dim(T_{\bar{x}}M_1 \cap T_{\bar{x}}M_2) = \dim(T_{\bar{x}}(M_1 \cap M_2)) = n - (m_1+m_2).$$

Consequently, the right side of (7.2.2) equals n and thus (7.2.1) holds.

Conversely, let $\bar{x} \in M_1 \cap M_2$ and suppose that (7.2.1) holds. Formula (7.2.2) yields now:

$$\dim(T_{\bar{x}}M_1 \cap T_{\bar{x}}M_2) = n - (m_1+m_2). \tag{7.2.3}$$

We may choose an open ($\mathbb{R}^n$-)neighborhood $U_{\bar{x}}$ of $\bar{x}$ and $h_i, \rho_j \in C^\infty(U_{\bar{x}}, \mathbb{R})$ such that $(h_1,\ldots,h_{m_1})$, $(\rho_1,\ldots,\rho_{m_2})$ is a defining system of functions for $M_1 \cap U_{\bar{x}}$, $M_2 \cap U_{\bar{x}}$ respectively. For $A \subset \mathbb{R}^n$ let $A^\perp$ denote the orthogonal complement of A in $\mathbb{R}^n$. Then we have:

$$\begin{aligned} T_{\bar{x}}M_1 \cap T_{\bar{x}}M_2 &= \{Dh_i^\top(\bar{x}),\ i = 1,\ldots,m_1\}^\perp \cap \{D\rho_j^\top(\bar{x}), j = 1,\ldots,m_2\}^\perp = \\ &= \{Dh_i^\top(\bar{x}), D\rho_j^\top(\bar{x}), i = 1,\ldots,m_1, j = 1,\ldots,m_2\}^\perp := \{\ldots\}^\perp. \end{aligned} \tag{7.2.4}$$

We contend that the set $\{\ldots\}$ is linearly independent. In fact, suppose that $\{\ldots\}$ is a linearly dependent system. Then $\dim\{\ldots\}^\perp > n - (m_1+m_2)$, which contradicts the validity of (7.2.3) and (7.2.4). By continuity there exists an open ($\mathbb{R}^n$-)neighborhood $\tilde{U}_{\bar{x}}$ of $\bar{x}$, $\tilde{U}_{\bar{x}} \subset U_{\bar{x}}$ such that the set $\{Dh_i(x), D\rho_j(x), i = 1,\ldots,m_1, j = 1,\ldots,m_2\}$ is linearly independent for all $x \in \tilde{U}_{\bar{x}}$. Consequently, the restriction of the functions h_i, ρ_j to $\tilde{U}_{\bar{x}}$ generate defining systems of functions for $M_1 \cap \tilde{U}_{\bar{x}}$, $M_2 \cap \tilde{U}_{\bar{x}}$ respectively and satisfy the transversality condition as stated in Definition 7.2.1. □

<u>Remark 7.2.2</u>. If M_1, M_2 are manifolds in $\mathbb{R}^n$ and $M_1 \cap M_2 \neq \emptyset$, then $M_1 \,\bar{\pitchfork}\, M_2$ implies that $\dim M_1 + \dim M_2 \geq n$. In other words, $\operatorname{codim} M_2 \leq \dim M_1$, i.e. the number of admissible "transversal restrictions" on M_1 is at most equal to the dimension of M_1 ($\dim M_1$ = number of "degrees of freedom" on M_1).

<u>Remark 7.2.3</u>. If M_1, M_2 are manifolds in $\mathbb{R}^n$ and $M_1 \,\bar{\pitchfork}\, M_2$, then intersection points will move smoothly under smooth (local) perturbations of M_1, M_2. This is basically an implication of the Implicit Function Theorem. However, if M_1 is not closed as a subset of $\mathbb{R}^n$, then under arbitrary "small" local perturbations of M_2 non-transversal intersection points may appear if $\bar{M}_1 \cap M_2 \neq \emptyset$. See Fig. 7.2.1 for a situation sketch in $\mathbb{R}^2$ and consider an arbitrarily small shift of M_2 to the right.

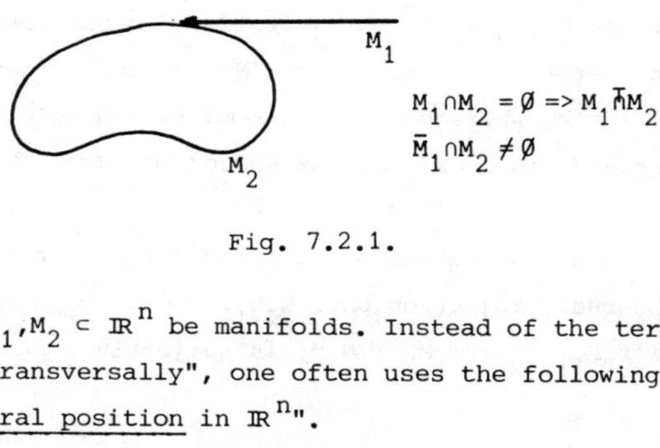

Fig. 7.2.1.

<u>Remark 7.2.4</u>. Let $M_1, M_2 \subset \mathbb{R}^n$ be manifolds. Instead of the terminology "M_1, M_2 intersect transversally", one often uses the following: "M_1, M_2 are <u>in general position</u> in $\mathbb{R}^n$".
In order to define the concept "in general position" for a system of (more than two) manifolds, the formulation via (7.2.1) is rather clumsy and an analogous approach according to Definition 7.2.1 is more convenient.

Let $M \subset \mathbb{R}^n$ be a manifold. The <u>normal space</u> $N_{\bar{x}}M$ at $\bar{x} \in M$ is defined as follows:

$$N_{\bar{x}}M = \{\xi \in \mathbb{R}^n | \xi \perp T_{\bar{x}}M\} \tag{7.2.5}$$

In the terminology of the conditions (M1), (M2) in the beginning of this section we have: $N_{\bar{x}}M = \text{span}\{D^T h_i(\bar{x}),\ i = 1,\dots,m\}$. Now, let $M_1,\dots,M_s \subset \mathbb{R}^n$ be manifolds. Then, $M_1,\dots,M_s$ are said to be <u>in general position at $\bar{x}$</u> if the following formula holds in case that $\bar{x} \in M_1 \cap \dots \cap M_s$:

$$\dim\left(\sum_{i=1}^{s} N_{\bar{x}}M_i\right) = \sum_{i=1}^{s} \dim N_{\bar{x}}M_i. \tag{7.2.6}$$

The system $\{M_1,\dots,M_s\}$ is said to be <u>in general position</u> if we have for every $\bar{x} \in \mathbb{R}^n$:

- either: $\bar{x} \notin \bigcup_{i=1}^{s} M_i$
- or : $M_{i_1},\dots,M_{i_r}$ are in general position at $\bar{x}$, where $\bar{x} \in M_i$ iff $i \in \{i_1,\dots,i_r\}$.

<u>Exercise 7.2.1</u>. Let $M_1, M_2 \subset \mathbb{R}^n$ be manifolds. Prove: M_1, M_2 intersect transversally iff the system $\{M_1,M_2\}$ is in general position.

<u>Exercise 7.2.2</u>. Let $M_1, M_2 \subset \mathbb{R}^n$ be manifolds. Suppose that codim $(M_1) = k$, dim $(M_2) = r$ and let $\bar{x} \in M_1 \cap M_2$. Prove: M_1, M_2 are in general position at $\bar{x}$ iff $\text{rank}(V_1^T.V_2) = k$, where V_1 is an $n\times k$ matrix whose columns form a basis for $N_{\bar{x}}M_1$ and where V_2 is an $n\times r$ matrix whose columns form a basis for $T_{\bar{x}}M_2$.

<u>Exercise 7.2.3</u>. In the situation of Exercise 7.2.2, let $P_1\colon \mathbb{R}^n \to N_{\bar{x}}M_1$ denote the orthogonal projection onto $N_{\bar{x}}M_1$. Prove: M_1, M_2 are in general position at $\bar{x}$ iff $P_{1|T_{\bar{x}}M_2}\colon T_{\bar{x}}M_2 \to N_{\bar{x}}M_1$ is surjective.

7.3. Transversality of mappings

Throughout Section 7.3, the word manifold will again refer to C^∞-manifold.

Let $f \in C^\infty(\mathbb{R}^n,\mathbb{R}^m)$, $f(x_1,\dots,x_n) = (f_1(x),\dots,f_m(x))^T$. For convenience we will use the notations $\mathbb{R}^{n+m}$ and $\mathbb{R}^n \times \mathbb{R}^m$ for $\mathbb{R}^k$, $k = n+m$. In the space $\mathbb{R}^n \times \mathbb{R}^m$ we consider the set Graph(f) (see Fig. 7.3.1):

$$\text{Graph}(f) = \{(x,f(x)) \in \mathbb{R}^n \times \mathbb{R}^m | x \in \mathbb{R}^n\}. \tag{7.3.1}$$

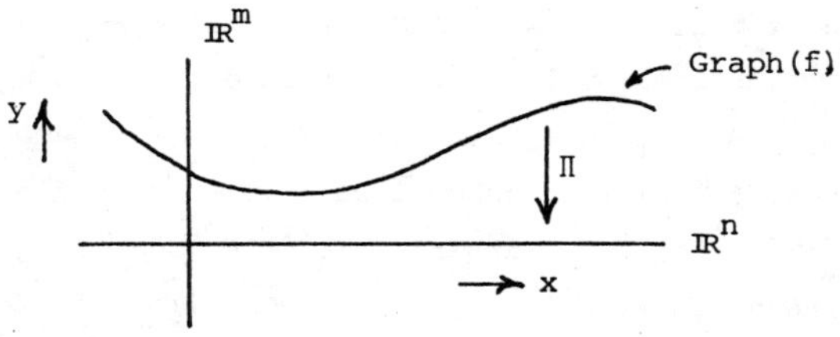

Fig. 7.3.1

Let us consider the C^∞-mapping G_f (cf. also Definition 3.1.4):

$$G_f\colon \mathbb{R}^n \to \text{Graph}(f), \quad x \mapsto (x,f(x)). \tag{7.3.2}$$

Then, G_f is a C^∞-diffeomorphism and its inverse G_f^{-1} is simply:

$$G_f^{-1} = \Pi_{|\text{Graph}(f)}, \tag{7.3.3}$$

where Π is the projection $\Pi\colon \mathbb{R}^n \times \mathbb{R}^m \to \mathbb{R}^m$, $\Pi(x,y) = x$. (7.3.4)

Note, on the other hand, that Graph(f) is a Regular Constraint Set. In fact, Graph(f) is determined by means of the equations $h_i(x,y) = 0$, $i = 1,\dots,m$, where each h_i is the (special) function $h_i(x,y) = y_i - f_i(x_1,\dots,x_n)$.

Let N be a submanifold of Graph(f) of codimension k. Then, by means of the diffeomorphism G_f (in particular (7.3.3)), we see that $\Pi(N)$ is a manifold in $\mathbb{R}^n$ of the same codimension k. If $M \subset \mathbb{R}^m$ is a manifold of codimension m_1, then the product $\mathbb{R}^n \times M$ is a manifold in $\mathbb{R}^n \times \mathbb{R}^m$ of the same codimension m_1 (exercise).

Definition 7.3.1. Let $M \subset \mathbb{R}^m$ be a manifold and $f \in C^\infty(\mathbb{R}^n,\mathbb{R}^m)$. We say that f meets M transversally (notation: $f \pitchfork M$), if the following two manifolds M_1, M_2 in $\mathbb{R}^n \times \mathbb{R}^m$ intersect transversally:

$$M_1 = \text{Graph}(f) \quad , \quad M_2 = \mathbb{R}^n \times M .$$

Example 7.3.1. Let $f \in C^\infty(\mathbb{R}^2,\mathbb{R})$, $f(x) = x_1^2 + x_2^2 + 1$ and $c \in \mathbb{R}$. Then $M_c := \{c\}$ is a manifold in $\mathbb{R}$ of codimension 1 and $\mathbb{R}^2 \times M_c$ is a plane parallel to the (x_1,x_2)-plane. Now, $f \pitchfork M_c$ iff $c \neq 1$ (see Fig. 7.3.2.a). Note that $c = 1$ is a critical value for f. Moreover, if $f \pitchfork M_c$ and $\text{Graph}(f) \cap (\mathbb{R}^2 \times M_c) \neq \emptyset$, then $\text{Graph}(f) \cap (\mathbb{R}^2 \times M_c)$ is a nonempty submanifold of Graph(f) and thus, $f^{-1}(c)$ is a submanifold in $\mathbb{R}^2$ of codimension 1 (= codim. M_c); note: $f^{-1}(c) = \Pi(\text{Graph}(f) \cap (\mathbb{R}^2 \times M_c))$. Although, in this case, $f^{-1}(1)$ - being a single point - is also a manifold in $\mathbb{R}^2$, this situation is very exceptional. See Fig. 7.3.2.b for another function $g \in C^\infty(\mathbb{R}^2,\mathbb{R})$ which has the property that g does not meet M_c transversally and that $g^{-1}(c)$ is not a manifold in $\mathbb{R}^2$.

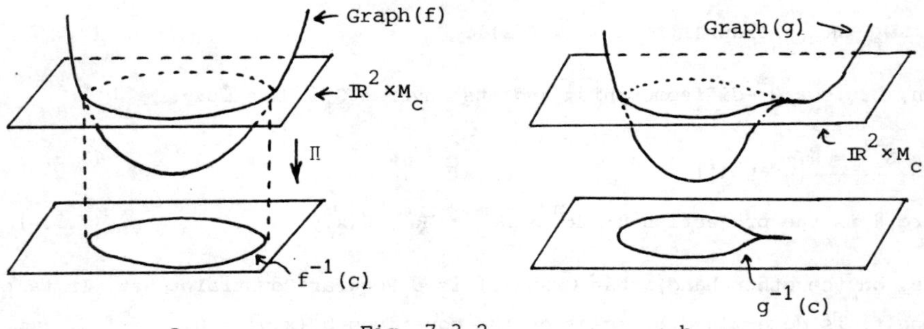

Fig. 7.3.2

Theorem 7.3.1. Let $M \subset \mathbb{R}^m$ be a manifold of codimension m_1. Let $f \in C^\infty(\mathbb{R}^n,\mathbb{R}^m)$ and suppose that $f \pitchfork M$. Then, either $f^{-1}(M) = \emptyset$, or, otherwise, $f^{-1}(M)$ is a manifold in $\mathbb{R}^n$ of codimension m_1. Moreover, for $\bar{x} \in f^{-1}(M)$ we have: $T_{\bar{x}}f^{-1}(M) = Df(\bar{x})^{-1}T_{f(\bar{x})}M$.

Proof. Suppose $f \pitchfork M$ and $f^{-1}(M) \neq \emptyset$. Then $\text{Graph}(f) \cap (\mathbb{R}^n \times M)$ is a submanifold of Graph(f) of codimension m_1. Consequently, $f^{-1}(M)$ - being equal to $\Pi(\text{Graph}(f) \cap (\mathbb{R}^n \times M))$ - is a submanifold in $\mathbb{R}^n$ of the same codimension m_1. For the proof of the last statement, use Remark 7.2.1. $\square$

Theorem 7.3.2. Let $M \subset \mathbb{R}^m$ be a manifold and $f \in C^\infty(\mathbb{R}^n,\mathbb{R}^m)$. Then, $f \pitchfork M$ iff at every $x \in \mathbb{R}^n$ for which $f(x) \in M$ the following holds:

$$Df(x)[\mathbb{R}^n] + T_{f(x)}M = \mathbb{R}^m. \tag{7.3.5}$$

Proof. From Definition 7.3.1 and Theorem 7.2.1 we have:

$$\begin{cases} f \pitchfork M \text{ iff for all } (\bar{x},\bar{y}) \in \text{Graph}(f) \cap (\mathbb{R}^n \times M): \\ T_{(\bar{x},\bar{y})}\text{Graph}(f) + T_{(\bar{x},\bar{y})}(\mathbb{R}^n \times M) = \mathbb{R}^n \times \mathbb{R}^m. \end{cases} \tag{7.3.6}$$

By means of (7.3.2) we see that $T_{(\bar{x},\bar{y})}\text{Graph}(f)$ equals $DG_f(\bar{x})[\mathbb{R}^n]$ and that the latter space is spanned by the columns of the matrix $\begin{pmatrix} I \\ Df(\bar{x}) \end{pmatrix}$, where I stands for the n×n-identity matrix.
The tangent space $T_{(\bar{x},\bar{y})}(\mathbb{R}^n \times M)$ is spanned by both $\{o\} \times T_{\bar{y}}M$ and the columns of the matrix $\begin{pmatrix} I \\ 0 \end{pmatrix}$.
Consequently, (7.3.6) is valid iff the columns of $\begin{pmatrix} 0 \\ Df(\bar{x}) \end{pmatrix}$ together with $\{o\} \times T_{\bar{y}}M$ span $\{o\} \times \mathbb{R}^m$; and since the columns of $\begin{pmatrix} 0 \\ Df(\bar{x}) \end{pmatrix}$ span $\{o\} \times (Df(\bar{x})[\mathbb{R}^n])$, the assertion of the theorem follows. □

Remark 7.3.1. Let $M \subset \mathbb{R}^m$ be a manifold of codimension k, and $f \in C^\infty(\mathbb{R}^n, \mathbb{R}^m)$. Another useful formulation for "$f \pitchfork M$" is the following (exercise):

$f \pitchfork M$ iff for every $\bar{x} \in \mathbb{R}^n$ with $f(\bar{x}) \in M$ we have: if $\{h_1,\ldots,h_k\}$ is a defining system of functions for $M \cap U_{f(\bar{x})}$, $U_{f(\bar{x})}$ an open neighborhood of $f(\bar{x})$ (cf. Section 7.2), then $h_1 \circ f,\ldots,h_k \circ f$ are independent at $\bar{x}$ (cf. Remark 3.1.1).

Remark 7.3.2. In connection with Example 7.3.1 we have the following simple application of Theorem 7.3.2. Let $f \in C^\infty(\mathbb{R}^n, \mathbb{R}^m)$. Then, $q \in \mathbb{R}^m$ is a regular value of f iff $f \pitchfork \{q\}$. In fact, note that the tangent space at $\{q\}$ - as a zero dimensional manifold - is just the zero vector; then, apply (7.3.5). On the other hand, we remark that $f \pitchfork \{q\}$ iff $\text{Graph}(q) \pitchfork \text{Graph}(f)$, where Graph(q) is the graph of the constant map $x \mapsto q$.

Remark 7.3.3. Let $q \in \mathbb{R}^m$ denote again the constant map $\mathbb{R}^n \to \mathbb{R}^m$, $x \mapsto q$, and let M be a manifold in $\mathbb{R}^n \times \mathbb{R}^m$. Then, for almost all $q \in \mathbb{R}^m$ we have: $\text{Graph}(q) \pitchfork M$. This follows from Sard's Theorem and the following observation (cf. also Remark 7.1.5): $\text{Graph}(q) \pitchfork M$ iff q is regular value for $\Pi_{2\,|M}: M \to \mathbb{R}^m$, where Π_2 is the projection $\Pi_2(x,y) = y$.

In the special case that M = Graph(f), $f \in C^\infty(\mathbb{R}^n \times \mathbb{R}^m)$, we are back in the situation of Remark 7.3.2. □

Now we proceed with transversality theorems for mappings. Originally, these ideas are due to R. Thom.

The space $C^\infty(\mathbb{R}^n, \mathbb{R}^m)$ will be topologized by means of the C^k-topology for the product $\displaystyle\mathop{\mathrm{X}}_{i=1}^{m} C^\infty(\mathbb{R}^n, \mathbb{R})$; cf. Definition 6.1.1.

Theorem 7.3.3. (Transversality theorem for mappings).

Let $M \subset \mathbb{R}^m$ be a manifold and denote by $\pitchfork M$ the set of all $f \in C^\infty(\mathbb{R}^n, \mathbb{R}^m)$ with $f \pitchfork M$.
Then, $\pitchfork M$ is C^k-dense for all k; moreover, if M is closed as a subset of $\mathbb{R}^m$, then $\pitchfork M$ is C^k-open for all $k \geq 1$. □

Remark 7.3.4. Let $m \geq 1$ be given and let H be the subset of $C^\infty(\mathbb{R}^n, \mathbb{R})^m$ defined by: $h = (h_1, \ldots, h_m) \in H$ iff M[h] is regular. From Theorem 7.3.3 it follows at once, that H is C^k-dense for all k and C^k-open for $k \geq 1$. To see this, put $M = \{o\}$, $o \in \mathbb{R}^m$ and note that $H = \pitchfork M$. (Observation: if $m > n$, then $h \in H$ necessarily implies that $M[h] = \emptyset$).

Instead of proving Theorem 7.3.3, we prove a bit more general result. We do so, because of the fact that the geometrical interpretation of the next result is even easier and, on the other hand, it is closer to the generalizations we have in mind (cf. also Remark 7.3.6).

Theorem 7.3.4. Let M be a manifold in $\mathbb{R}^{n+m}$ and denote by F the set of all $f \in C^\infty(\mathbb{R}^n, \mathbb{R}^m)$ with Graph(f) $\pitchfork M$.
Then, F is C^k-dense for all k; moreover, if M is closed as a subset of $\mathbb{R}^{n+m}$, then F is C^k-open for all $k \geq 1$. □

Remark 7.3.5. Theorem 7.3.3 follows immediately from Theorem 7.3.4. In fact, put $M = \mathbb{R}^n \times M$ and note that M is a closed subset of $\mathbb{R}^{n+m}$ iff M is a closed subset of $\mathbb{R}^m$.

<u>Remark 7.3.6.</u> For an $f \in C^\infty(\mathbb{R}^n, \mathbb{R}^m)$ let us define $j^0 f \in C^\infty(\mathbb{R}^n, \mathbb{R}^n \times \mathbb{R}^m)$ as follows: $j^0 f(x) = (x, f(x))$. Note that $j^0 f(x) = G_f(x)$ (cf. (7.3.2)). Getting ahead of Section 7.4, we might call $j^0 f$ the o-jet extension of f. Let $M \subset \mathbb{R}^{n+m}$ be a manifold. As a consequence of the fact that G_f is a diffeomorphism, we have:

$$\text{Graph}(f) \pitchfork M \leftrightarrow j^0 f \pitchfork M.$$

<u>Proof of Theorem 7.3.4.</u> It will be convenient to define a special <u>continuous</u> function $\phi_f: M \to \mathbb{R}$, depending on f and its first order partial derivatives, with the following property:

$$\text{Graph}(f) \pitchfork M \quad \text{iff} \quad \phi_f(z) > 0 \text{ for all } z \in M. \tag{7.3.7}$$

In fact, let P(z) be the orthogonal-projection matrix onto the normal space $N_z M$, $z \in M$. Then, P(z) is a C^∞-mapping from M to the space of $(n+m)\times(n+m)$-matrices. To see this, choose (locally) a defining system of C^∞-functions for M, say $h_1, \ldots, h_k$, k being the codimension of M in $\mathbb{R}^{n+m}$; then, put $P(z) = [H(H^T H)^{-1} H^T](z)$, where $H(z) = (D^T h_1(z) \vdots \ldots \vdots D^T h_k(z))$.
Note that the columns of $\begin{pmatrix} I \\ Df(x) \end{pmatrix}$ form a basis for the tangent space of Graph(f) at (x,f(x)). Now, recalling Exercise 7.2.3, it is easy to see that the following function ϕ_f satisfies (7.3.7):

$$\phi_f(z) = \|y - f(x)\| + \sum_{\sigma(z)} |\det \sigma(z)| \,, \quad z = (x,y) \in M \tag{7.3.8}$$

where $\sigma(z)$ in (7.3.8) ranges over all k×k submatrices of $\Sigma(z) := P(z) \cdot \begin{pmatrix} I \\ Df(x) \end{pmatrix}$.

<u>"Open part".</u> Now, suppose that M is closed as a subset of $\mathbb{R}^n \times \mathbb{R}^m$ and that Graph(f) $\pitchfork M$. Then, for every compact subset $K \subset \mathbb{R}^n \times \mathbb{R}^m$ with $K \cap M \neq \emptyset$ we have: $\inf\limits_{z \in K \cap M} \phi_f(z) > 0$.

Let $B(\bar{x}, 1)$ be the (Euclidean) ball in $\mathbb{R}^n$ with center $\bar{x}$ and radius 1. Then, for every $\alpha > 0$ there exists a $\delta > 0$ such that, for every $g \in C^\infty(\mathbb{R}^n, \mathbb{R}^m)$ with $\max\limits_{x \in B(\bar{x},1)} \|g(x) - f(x)\| \leq \alpha$ we have:
$\text{Graph}(g) \cap (B(\bar{x},1) \times \mathbb{R}^m) \subset B(\bar{x},1) \times B(0,\delta)$, $B(0,\delta)$ being the ball in $\mathbb{R}^m$ with center $o \in \mathbb{R}^m$ and radius δ. Now we fix $\alpha > 0$ and suppose that $(B(\bar{x},1) \times B(0,\delta)) \cap M \neq \emptyset$.

Define $\eta = \min_{z\in(B(\bar{x},1)\times B(0,\delta))\cap M} \phi_f(z)$ (Thus, $\eta > 0$).

Then there exists an $\varepsilon > 0$ such that for every $g \in C^\infty(\mathbb{R}^n,\mathbb{R}^m)$ with

$\max_{x\in B(\bar{x},1)} \{\|g(x)-f(x)\| + |||Dg(x)-Df(x)|||\} \leq \varepsilon$ we have:

$\min_{z\in(B(\bar{x},1)\times B(0,\delta))\cap M} \phi_g(z) \geq \frac{1}{2}\eta > 0$, and hence, $\min_{z\in(B(\bar{x},1)\times\mathbb{R}^m)\cap M} \phi_g(z) \geq \frac{1}{2}\eta$.

Here, $|||Dg(x)-Df(x)||| = \sum_{i=1}^{n}\sum_{j=1}^{m} |\frac{\partial}{\partial x_i} g_j(x) - \frac{\partial}{\partial x_i} f_j(x)|$.

Consequently, to every $\bar{x} \in \mathbb{R}^n$ we may assign an $\varepsilon_{\bar{x}} > 0$ such that $g \in C^\infty(\mathbb{R}^n,\mathbb{R}^m)$ and $\max_{x\in B(\bar{x},1)} \{\|g(x)-f(x)\| + |||Dg(x)-Df(x)|||\} \leq \varepsilon_{\bar{x}}$ implies that, in case $(B(\bar{x},1) \times \mathbb{R}^m) \cap M \neq \emptyset$, $\phi_g(z) > 0$ for all $z \in (B(\bar{x},1) \times \mathbb{R}^m) \cap M$. In case $(B(\bar{x},1) \times \mathbb{R}^m) \cap M = \emptyset$, we "formally" put $\varepsilon_{\bar{x}} = 1$.
Now, take a sequence $\bar{x}_i$, $i = 1,2,...$ in $\mathbb{R}^n$ such that the set of open balls $\{\mathring{B}(\bar{x}_i,1), i = 1,2,...\}$ covers $\mathbb{R}^n$ and is locally finite. Finally, a continuous positive function $\varepsilon(\cdot)$: $\mathbb{R}^n \to \mathbb{R}$ is easily constructed such that for every i we have: $\varepsilon(x) < \varepsilon_{\bar{x}_i}$ if $x \in B(\bar{x}_i,1)$. (See also "Open part" in the proof of Theorem 7.1.3). So, the function $\varepsilon(\cdot)$ indicates how much f may be varied in the C^1-sense in order that Graph(f) remains intersecting M transversally. This completes the "open-part".

<u>"Dense-part"</u>. The dense-part of the proof consists of a local argument and a globalization of this.
<u>Local argument</u>. Roughly speaking, the key is to "move" Graph(f) in such a way that, after a suitable move, Graph(f) will intersect M transversally. For this movement, we parametrize the whole $\mathbb{R}^n \times \mathbb{R}^m$ by means of an m-parameter family of graphs of functions by putting

$$F_c(x) = f(x) + c \quad , \quad c \in \mathbb{R}^m. \tag{7.3.9}$$

To each $z \in \mathbb{R}^n \times \mathbb{R}^m$, $z = (x,y)$, we assign that parameter value $\psi(z)$ of c having the property that $z \in \text{Graph}(F_c)$, thus $\psi(z) = y-f(x) \in \mathbb{R}^m$.
Suppose that dim M = p. We take a $\tilde{z} \in M$ and choose an open ($\mathbb{R}^{n+m}$-)neighborhood $\mathcal{O}$ of $\tilde{z}$ such that there exists a C^∞-diffeomorphism Φ: $\mathcal{O} \to \mathbb{R}^{n+m}$, with $\Phi(\tilde{z}) = 0$ and $\Phi(\mathcal{O} \cap M) = \mathbb{R}^p \times \{o\}$, $o \in \mathbb{R}^{n+m-p}$. Let us denote a point

in $\mathbb{R}^p \times \{o\}$ by $(\xi, 0)$, $\xi \in \mathbb{R}^p$. Next, we define the mapping $\tilde{\psi}: \mathbb{R}^p \to \mathbb{R}^m$ as follows:

$$\tilde{\psi}(\xi) = \psi \circ \Phi^{-1}(\xi, 0). \tag{7.3.10}$$

Let $\bar{c} \in \mathbb{R}^m$ be a regular value of $\tilde{\psi}$. (So, at this point, Sard's theorem comes into play). We contend, that - restricting ourselves to the neighborhood $\mathcal{O}$ of $\tilde{z}$ - $\mathrm{Graph}(F_{\bar{c}}) \pitchfork M$. To see this, suppose that $\bar{z} \in \mathrm{Graph}(F_{\bar{c}}) \cap M \cap \mathcal{O}$; thus $\bar{y} = f(\bar{x}) + \bar{c}$. Then, we have to show that

$$\begin{pmatrix} I_n \\ DF_{\bar{c}}(\bar{x}) \end{pmatrix}[\mathbb{R}^n] + T_{\bar{z}}M = \mathbb{R}^n \times \mathbb{R}^m, \tag{7.3.11}$$

where I_n stands for the $n \times n$ identity matrix.
Since $\bar{c}$ is a regular value for $\tilde{\psi}$ we obtain from (7.3.10) at $(\bar{\xi}, 0) = \Phi(\bar{z})$:

$$D\psi \cdot D\Phi^{-1}[\mathbb{R}^p \times \{o\}] = \mathbb{R}^m. \tag{7.3.12}$$

Note, that $D\Phi^{-1}[\mathbb{R}^p \times \{o\}] = T_{\bar{z}}M$. Consequently, (7.3.12) implies that $D\psi(\bar{z})[T_{\bar{z}}M] = \mathbb{R}^m$ (and thus $p \geq m$!).
From the p-dim. linear subspace $T_{\bar{z}}M \subset \mathbb{R}^n \times \mathbb{R}^m$ we choose m vectors $v_1, \ldots, v_m$ such that

$$D\psi(\bar{z})[\mathrm{Span}\{v_1, \ldots, v_m\}] = \mathbb{R}^m. \tag{7.3.13}$$

By V we denote the $(m+n) \times m$ matrix whose columns are $v_1, \ldots, v_m$ (in that order). Since $\psi(x,y) = y - f(x)$ we have

$$D\psi(\bar{z}) = (-Df(\bar{x}) \,|\, I_m). \tag{7.3.14}$$

Then, (7.3.13), (7.3.14) imply that $W := (-Df(\bar{x}) \,|\, I_m)V$ is a nonsingular $m \times m$ matrix.
Finally we show that the $(m+n) \times (m+n)$ matrix $\left(V \left| \begin{array}{c} I_n \\ \hline Df(\bar{x}) \end{array} \right. \right)$ is nonsingular..
This implies (7.3.11). To this aim, consider the following matrix-equation:

$$\left(\begin{array}{c|c} I_n & 0 \\ \hline Df(\bar{x}) & -I_m \end{array} \right) \left(V \left| \begin{array}{c} I_n \\ \hline Df(\bar{x}) \end{array} \right. \right) = \left(\begin{array}{c|c} * & I_n \\ \hline -W & 0 \end{array} \right). \tag{7.3.15}$$

Since W is nonsingular, the matrix on the right in (7.3.15) is nonsingular. Therefore, both (quadratic) matrices determining the product on the left-handside of (7.3.15) are nonsingular.

Globalization. The globalization part could be done directly, by using an analogous globalization argument as we used in proving the "dense-part" of Theorem 7.1.3. In a certain sense such a reasoning already "reflects" a (Baire-)category-argument. Therefore we will proceed by applying such a category-argument. In fact, in the subsequent Lemma 7.3.1 we will prove the following:

Let $A_i \subset C^\infty(\mathbb{R}^n,\mathbb{R}^m)$ be a C^∞-open, dense subset, $i = 1,2,\ldots$.
Then $\bigcap_{i=1}^{\infty} A_i$ is C^∞-dense in $C^\infty(\mathbb{R}^n,\mathbb{R}^m)$.

Taking this lemma for granted at the moment, we proceed with the globalization part. For each $\tilde{z} \in M$, $\tilde{z} = (\tilde{x},\tilde{y})$, we choose such an open ($\mathbb{R}^n \times \mathbb{R}^m$-) neighborhood $\mathcal{O}$ of $\tilde{z}$ as we did in the "local argument" before. Then we assign to $\tilde{z}$ numbers $\tilde{\varepsilon} > 0$, $\tilde{\delta} > 0$ such that $B(\tilde{x},\tilde{\varepsilon}) \times B(\tilde{y},\tilde{\delta}) \subset \mathcal{O}$; it follows that $(B(\tilde{x},\tilde{\varepsilon})) \times B(\tilde{y},\tilde{\delta})) \cap M$ is compact. The next step is the choice of a sequence $z_i \in M$, $i = 1,2,\ldots$ such that the family of open ball-products $\{\mathring{B}(x_i,\frac{1}{3}\varepsilon_i) \times \mathring{B}(y_i,\frac{1}{3}\delta_i),\ i = 1,2,\ldots\}$ forms a covering of M. For $i = 1,2,\ldots$ let $\xi_i \in C^\infty(\mathbb{R}^n,\mathbb{R})$ be a function having the following properties:

(1) $\xi_i = 1$ on $B(x_i,\frac{2}{3}\varepsilon_i)$

(2) $\xi_i = 0$ outside $B(x_i,\varepsilon_i)$.

Write $f \in C^\infty(\mathbb{R}^n,\mathbb{R}^m)$ as $f = (f_1,\ldots,f_m)$ and select a base neighborhood $\overset{m}{\underset{j=1}{\mathsf{X}}} V^k_{\phi,f_j}$ of f. Then, we may choose a $v = (v_1,\ldots,v_m)$ (cf. "local argument") such that we have with $g(x) = (f_1(x) + \xi_i(x)v_1,\ldots,f_m(x) + \xi_i(x)v_m)$:

(a) $g \in \overset{m}{\underset{j=1}{\mathsf{X}}} V^k_{\phi,f_j}$

(b) $\phi_g(z) > 0$ for all $z \in (B(x_i,\frac{1}{3}\varepsilon_i) \times B(y_i,\frac{1}{3}\delta_i)) \cap M$

(recall the meaning of ϕ_g; cf. (7.3.7), (7.3.8)).

Thus, recalling that $(B(x_i,\frac{1}{3}\varepsilon_i) \times B(y_i,\frac{1}{3}\delta_i)) \cap M$ is compact, we have proved that A_i is <u>C^∞-open, dense</u> in $C^\infty(\mathbb{R}^n,\mathbb{R}^m)$, where

$$A_i = \{h \in C^\infty(\mathbb{R}^n,\mathbb{R}^m) \mid \phi_h(z) > 0 \text{ for all } z \in (B(x_i,\tfrac{1}{3}\varepsilon_i) \times B(y_i,\tfrac{1}{3}\delta_i)) \cap M\}.$$

Then, by the subsequent Lemma 7.3.1 we have: $A := \bigcap_{i=1}^{\infty} A_i$ is C^∞-dense (and thus C^k-dense for every k). Note that $f \in A$ implies Graph(f) $\pitchfork M$. This completes the proof of Theorem 7.3.4. □

<u>Remark 7.3.7.</u> If $M \subset \mathbb{R}^n \times \mathbb{R}^m$ is a manifold which is not closed as a subset of $\mathbb{R}^n \times \mathbb{R}^m$, then there might be a compact subset $K \subset \mathbb{R}^n \times \mathbb{R}^m$, $K \cap M \neq \emptyset$, and an $f \in C^\infty(\mathbb{R}^n,\mathbb{R}^m)$ such that we have $\phi_f(z) > 0$ for all $z \in K \cap M$ but $\inf_{z \in K\cap M} \phi_f(z) = 0$. Cf. Formula (7.3.8) for the definition of ϕ_f; see Fig. 7.3.3. For an analogous phenomenon see also Remark 7.2.3.

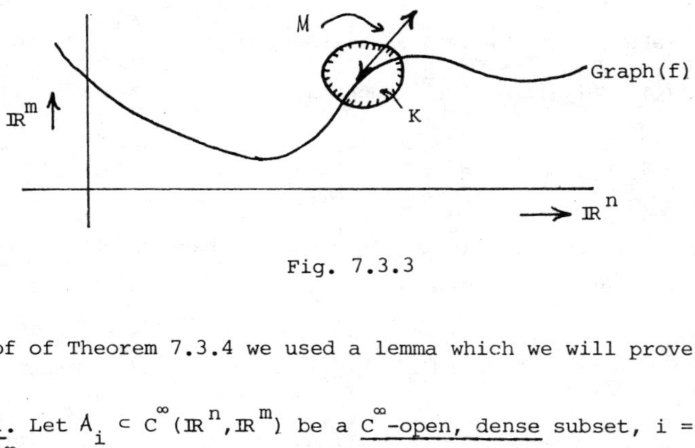

Fig. 7.3.3

In the proof of Theorem 7.3.4 we used a lemma which we will prove now.

<u>Lemma 7.3.1.</u> Let $A_i \subset C^\infty(\mathbb{R}^n,\mathbb{R}^m)$ be a <u>C^∞-open, dense</u> subset, $i = 1,2,\ldots$. Then $A := \bigcap_{i=1}^{\infty} A_i$ is C^∞-dense.

<u>Proof.</u> In order to simplify the notations we carry out the proof for the case m = 1, the proof in case m > 1 running essentially along the same lines.
It suffices to prove that $A \cap U \neq \emptyset$ for any nonempty C^∞-open subset U of $C^\infty(\mathbb{R}^n,\mathbb{R})$. Note that U contains a base-neighborhood $V^{\ell_0}_{\phi_0,f_0}$ for some $f_0 \in U$

and without loss of generality we may assume that $\phi_0(x) \le 1$ for all $x \in \mathbb{R}^n$. By $\tilde{V}^{\ell}_{\phi,f}$ we denote the following set:

$$\tilde{V}^{\ell}_{\phi,f} = \{g \in C^{\infty}(\mathbb{R}^n,\mathbb{R}) \mid |\partial^{\alpha}f(x)-\partial^{\alpha}g(x)| \underset{\uparrow}{\le} \phi(x) \text{ for all } x \in \mathbb{R}^n, \text{ for all } \alpha \text{ with } |\alpha| \le \ell\}.$$

Note: $\tilde{V}^{\ell}_{\phi,f} \subset V^{k}_{\psi,f}$ if $\ell \ge k$ and $\phi(x) < \psi(x)$ for all $x \in \mathbb{R}^n$ (see also Remark 6.1.4).

The set A_1 is C^{∞}-open, dense. Consequently, $A_1 \cap V^{\ell_0}_{\phi_0,f_0}$ is C^{∞}-open and nonempty. It follows that there exists a base-neighborhood $V^{\ell_1}_{\phi_1,f_1}$ with $\ell_1 > \ell_0$ and $\phi_1(x) \le \frac{1}{2}$ for all $x \in \mathbb{R}^n$, such that

$$\tilde{V}^{\ell_1}_{\phi_1,f_1} \subset A_1 \cap V^{\ell_0}_{\phi_0,f_0} .$$

We proceed with $V^{\ell_1}_{\phi_1,f_1}$ and obtain that $A_2 \cap V^{\ell_1}_{\phi_1,f_1}$ is C^{∞}-open and nonempty. Reasoning as above, there exists a base-neighborhood $V^{\ell_2}_{\phi_2,f_2}$ with $\ell_2 > \ell_1$ and $\phi_2(x) \le (\frac{1}{2})^2$ for all $x \in \mathbb{R}^n$, such that

$$\tilde{V}^{\ell_2}_{\phi_2,f_2} \subset A_2 \cap V^{\ell_1}_{\phi_1,f_1} \text{, etc.}$$

In this way we obtain recursively a sequence of base-neighborhoods $V^{\ell_i}_{\phi_i,f_i}$ with the properties:

$$\tilde{V}^{\ell_i}_{\phi_i,f_i} \subset A_i \cap V^{\ell_{i-1}}_{\phi_{i-1},f_{i-1}}, \; \ell_i > \ell_{i-1}, \; \phi_i(x) \le (\tfrac{1}{2})^i .$$

Now, we consider the sequence of functions $\{f_i, i = 0,1,\dots\}$. For every partial derivative (including the zero-th) up to order $\underline{|\alpha| \le \ell_i}$ we have:

$$|\partial^{\alpha}f_k(x) - \partial^{\alpha}f_i(x)| < \phi_i(x) \le (\tfrac{1}{2})^i, \; k \ge i, \text{ all } x. \tag{7.3.16}$$

Let ∂^{α} be a fixed chosen partial derivation of order $|\alpha|$. Choose an i^* with $|\alpha| \le \ell_{i^*}$ (Note: $\lim_{i\to\infty} \ell_i = \infty$). From (7.3.16) we obtain for $m_1 \ge m_2 \ge i^*$:

$$|\partial^{\alpha}f_{m_1}(x) - \partial^{\alpha}f_{m_2}(x)| \le (\tfrac{1}{2})^{m_2} \text{ for all } x \in \mathbb{R}^n . \tag{7.3.17}$$

From (7.3.17) it follows that $\partial^\alpha f_i$ converges <u>uniformly</u> on $\mathbb{R}^n$ to a continuous limit-function. Consequently, we have with $f := \lim f_i$ that $f \in C^\infty(\mathbb{R}^n, \mathbb{R})$ and $\partial^\alpha f = \lim \partial^\alpha f_i$. (Note, that we proved a certain completeness-aspect of $C^\infty(\mathbb{R}^n, \mathbb{R})$ endowed with the C^∞-topology!). If we are able to show that f lies in all sets $\tilde{V}^{\ell_i}_{\phi_i, f_i}$, $i \geq 1$, then - because of the fact that $\tilde{V}^{\ell_i}_{\phi_i, f_i} \subset A_i \cap V^{\ell_{i-1}}_{\phi_{i-1}, f_{i-1}}$, $i \geq 1$ - we have $f \in U$ and $f \in \bigcap_{i=1}^\infty A_i$ and thus $A \cap U \neq \emptyset$. To this end, let us consider a set $\tilde{V}^{\ell_i}_{\phi_i, f_i}$, $i \geq 1$. Let ∂^α be a partial derivation of order $|\alpha| \leq \ell_i$. For every fixed $x \in \mathbb{R}^n$ we have:

$$|\partial^\alpha f_k(x) - \partial^\alpha f_i(x)| < \phi_i(x),\ k \geq i \text{ and } \lim_{k \to \infty} \partial^\alpha f_k(x) = \partial^\alpha f(x).$$

Consequently, $|\partial^\alpha f(x) - \partial^\alpha f_i(x)| \leq \phi_i(x)$ and thus $f \in \tilde{V}^{\ell_i}_{\phi_i, f_i}$. □

The next theorem is another type of generalization of Theorem 7.3.3.

<u>Theorem 7.3.5</u>. Let $N \subset \mathbb{R}^n$, $M \subset \mathbb{R}^m$ be manifolds and define $\pitchfork M = \{f \in C^\infty(\mathbb{R}^n, \mathbb{R}^m) \mid f \pitchfork M\}$.
Let $F \subset C^\infty(\mathbb{R}^n, \mathbb{R}^m)$ be the following subset of $\pitchfork M$:

$f \in \pitchfork M$ is an element of F iff $f^{-1}(M) \pitchfork N$.

Then, F is C^k-dense in $C^\infty(\mathbb{R}^n, \mathbb{R}^m)$ for all k. Furthermore, if both N is closed as a subset of $\mathbb{R}^n$ and M is closed as a subset of $\mathbb{R}^m$, then F is C^k-open for all $k \geq 1$.

<u>Proof</u>. We will restrict ourselves to the key-part which consists of a framework in which Sard's theorem is applicable. (The remainder of the proof runs basically along analogous lines as in the proof of Theorem 7.3.4 and we will omit these details).

Suppose that $f \pitchfork M$ and $f^{-1}(M) \neq \emptyset$; let k be the codimension of M (in $\mathbb{R}^m$). Consider a point $\bar{x} \in f^{-1}(M)$; then $\bar{y} = f(\bar{x}) \in M$. Choose an open ($\mathbb{R}^m$-) neighborhood $U_{\bar{y}}$ of $\bar{y}$ and suppose that $(h_1, \ldots, h_k)$, where $h_i \in C^\infty(U_{\bar{y}}, \mathbb{R})$, $i = 1, \ldots, k$, is a defining system of functions for $M \cap U_{\bar{y}}$ (cf. beginning of Section 7.2). Put $h = (h_1, \ldots, h_k)^T$, thus $h: U_{\bar{y}} \to \mathbb{R}^k$.

Note that rank $Dh(\bar{y}) = k$ and $Dh(\bar{y})v = 0$ for all $v \in T_{\bar{y}}M$. Since $f \pitchfork M$, we have $Df(\bar{x})[\mathbb{R}^n] + T_{\bar{y}}M = \mathbb{R}^m$ (cf. Formula (7.3.5))) and thus $Dh(\bar{y})Df(\bar{x})[\mathbb{R}^n] = Dh(\bar{y})[\mathbb{R}^m]$ and it follows:

$$\text{rank } Dh(\bar{y})Df(\bar{x}) = k. \tag{7.3.18}$$

Note that (7.3.18) is an equivalent way of saying that Graph(f) intersects $\mathbb{R}^n \times M$ transversally at $(\bar{x},\bar{y}) \in \mathbb{R}^n \times M$, $\bar{y} = f(\bar{x})$ (see Remark 7.3.1). In the "local argument" of the dense-part of the proof of Theorem 7.3.4 we used for the "movement" of Graph(f) the following m-parameter family:

$$F_c(x) = f(x) + c, \quad c \in \mathbb{R}^m.$$

Now, the idea is to select k vectors $c_1,\ldots,c_k \in \mathbb{R}^m$ such that $F_c^{-1}(M)$ suitably parametrizes a whole $\mathbb{R}^n$-neighborhood of $\bar{x}$ as c varies in a neighborhood of the origin in Span $\{c_1,\ldots,c_k\}$. (Note that codim $F_o^{-1}(M) = k$, so the "missing number of dimensions" in order to fill up an $\mathbb{R}^n$-neighborhood of $\bar{x}$ just equals k).
We select $c_i = D^T h_i(\bar{y})$, $i = 1,\ldots,k$ (i.e., directions, normal to M at $\bar{y}$!). Consider the function g which is defined on an open neighborhood of $(\bar{x},0)$ in $\mathbb{R}^n \times \mathbb{R}^k$ and whose range space is $\mathbb{R}^k$:

$$g(x,w) = h[f(x) + D^T h(\bar{y})w]. \tag{7.3.19}$$

Note that $g(x,w) = 0$ means that $f(x) + D^T h(\bar{y})w \in M$, thus $g(\bar{x},0) = 0$ in particular. Furthermore, $D_w g(\bar{x},0)$ is nonsingular since $D_w g(\bar{x},0)$ is equal to the Gram-matrix $Dh(\bar{y})D^T h(\bar{y})$ and this Gram-matrix is nonsingular because of the fact that rank $Dh(\bar{y}) = k$. Then, by the Implicit Function Theorem, there exist a unique C^∞-mapping $w(x)$, defined on an open neighborhood $U_{\bar{x}}$ of $\bar{x}$ with range space $\mathbb{R}^k$ and satisfying

$$g(x,w(x)) \equiv 0. \tag{7.3.20}$$

In this way we have obtained a smooth parameter distribution $w(x)$ in a neighborhood of $\bar{x}$. On the other hand, $w(U_{\bar{x}})$ covers a whole neighborhood of $o \in \mathbb{R}^k$. In order to see this, noting that $w(\bar{x}) = o \in \mathbb{R}^k$, it is sufficient to show that $Dw(\bar{x})$ has rank k. From (7.3.20) we have

$$D_x g + D_w g . Dw \equiv 0. \qquad (7.3.21)$$

At $(\bar{x},0)$ we obtain in particular, using (7.3.19):

$$Dw(\bar{x}) = -(Dh(\bar{y})D^T h(\bar{y}))^{-1} Dh(\bar{y})Df(\bar{x}). \qquad (7.3.22)$$

A combination of (7.3.18) and (7.3.22) yields rank $Dw(\bar{x}) = k$.

From a continuity argument it follows that rank $Dh(\tilde{y})Df(\tilde{x}) = k$ and rank $Dh(\tilde{y})D^T h(\tilde{y}) = k$ for $\tilde{y}$ in a neighborhood of $\bar{y}$ and $\tilde{x}$ in a neighborhood of $\bar{x}$.

<u>As a resumé</u> we have: there is an open neighborhood $V_{\bar{x}}$ of $\bar{x}$ such that - if we restrict our attention to $V_{\bar{x}}$ - $F_c \pitchfork M$ and $F_c^{-1}(M) \cap V_{\bar{x}}$ are (n-k)dimensional manifolds which form a "parametrization" of $V_{\bar{x}}$ as c varies in a neighborhood of the origin in $\mathrm{Span}\{D^T h_1(\bar{y}),\dots,D^T h_k(\bar{y})\}$.

Finally we check that this framework is the right one for applying Sard's theorem for our purposes.

Let $N \subset \mathbb{R}^n$ be a manifold of dimension p and suppose that $\bar{x} \in N \cap f^{-1}(M)$. Take an open $\mathbb{R}^n$-neighborhood $\mathcal{O}$ of $\bar{x}$, $\mathcal{O} \subset V$ (V as in the above resumé) such that there exists a C^∞-diffeomorphism Φ: $\mathcal{O} \to \mathbb{R}^n$, with $\Phi(\bar{x}) = 0$ and $\Phi(\mathcal{O} \cap N) = \mathbb{R}^p \times \{o\}$, $o \in \mathbb{R}^{n-p}$. Let us denote a point in $\mathbb{R}^p \times \{o\}$ by $(\xi,0)$, $\xi \in \mathbb{R}^p$. Define the mapping ψ: $\mathbb{R}^p \to \mathbb{R}^k$ as follows:

$$\psi(\xi) = w \circ \Phi^{-1}(\xi,0). \qquad (7.3.23)$$

Let $\hat{w} \in \mathbb{R}^k$, $\|\hat{w}\|$ sufficiently small, be a regular value of ψ (<u>Sard's theorem!</u>). We contend that - restricting ourselves to the neighborhood $\mathcal{O}$ of $\bar{x}$ - $F_{\hat{c}}^{-1}(M) \pitchfork N$, where $\hat{c} = D^T h(\bar{y})\hat{w}$, $\bar{y} = f(\bar{x})$.
To see this, suppose that $\tilde{x} \in F_{\hat{c}}^{-1}(M) \cap N \cap \mathcal{O}$. Then, we have to show that

$$T_{\tilde{x}}(F_{\hat{c}}^{-1}(M) \cap \mathcal{O}) + T_{\tilde{x}}N = \mathbb{R}^n. \qquad (7.3.24)$$

Put $\Phi(\tilde{x}) = (\tilde{\xi},0)$ and $\tilde{y} = f(\tilde{x}) + \hat{c}$. Firstly, we calculate the tangentspace $T_{\tilde{x}}(F_{\hat{c}}^{-1}(M) \cap \mathcal{O})$. For this, note that $F_{\hat{c}}^{-1}(M) \cap \mathcal{O}$ is equal to $\{x \mid g(x,\hat{w}) = 0\} \cap \mathcal{O}$. Consequently,

$$T_{\tilde{x}}(F_{\hat{c}}^{-1}(M) \cap \mathcal{O}) = \{\eta \in \mathbb{R}^n \mid D_x g(\tilde{x},\hat{w})\eta = 0\} =$$

$$= \{\eta \in \mathbb{R}^n \mid Dh(\tilde{y}) \cdot Df(\tilde{x})\xi = 0\}$$

$$:= \{\eta \in \mathbb{R}^n \mid B\eta = 0\}. \tag{7.3.25}$$

Since $\hat{w}$ is a regular value for ψ we obtain

$$Dw(\tilde{x})\underbrace{D\Phi^{-1}(\tilde{\xi},0)[\mathbb{R}^p \times \{o\}]}_{=T_{\tilde{x}}N} = \mathbb{R}^k. \tag{7.3.26}$$

Now, from (7.3.21) we obtain

$$\begin{aligned} Dw(\tilde{x}) &= -[Dh(\tilde{y})D^T h(\bar{y})]^{-1} Dh(\tilde{y})Df(\tilde{x}). \\ &:= AB. \end{aligned} \tag{7.3.27}$$

From $T_{\tilde{x}}N$ we choose k vectors $v_1,\ldots,v_k$ such that

$$Dw(\tilde{x})[\mathrm{Span}\{v_1,\ldots,v_k\}] = \mathbb{R}^k.$$

Choose n-k vectors $v_{k+1},\ldots,v_n$ which span $\{\eta \in \mathbb{R}^n \mid B\eta = 0\}$ (cf. (7.3.25)). Let V be the n×n matrix whose columns are $v_1,\ldots,v_n$ (in that order). If we can show that V is nonsingular, then (7.3.24) is valid and we are done. To this aim, consider the following matrix equation:

$$\begin{pmatrix} v_{k+1}^T \\ \vdots \\ v_n^T \\ \hline AB \end{pmatrix} \begin{pmatrix} \\ V \\ \\ \end{pmatrix} = \left(\begin{array}{c|c} * & D \\ \hline C & 0 \end{array}\right). \tag{7.3.28}$$

In (7.3.28) the zero-submatrix on the righthand side appears since $v_{k+1},\ldots,v_n$ are annihilated by B (and thus by AB). Furthermore, C and D are nonsingular by the very construction. Consequently, the matrix on the righthand side of (7.3.28) is nonsingular and this implies that V is nonsingular. □

Remark 7.3.8. Let $f \in C^{\infty}(\mathbb{R}^n,\mathbb{R}^m)$, $N \subset \mathbb{R}^n$ a manifold, $M \subset \mathbb{R}^{n+m}$ a manifold and $F(x) = (x,f(x))$. Suppose that $F \pitchfork M$ and $F^{-1}(M) \neq \emptyset$. Then $F^{-1}(M)$ is a nonempty manifold in $\mathbb{R}^n$. However, we cannot expect in general that $F^{-1}(M) \pitchfork N$ since M - as a manifold in $\mathbb{R}^{n+m}$ - might impose certain constraints on $\mathbb{R}^n$. See Fig. 7.3.4.
On the other hand, from the proof of Theorem 7.3.5 it is not difficult to see which possibilities there are to move $F^{-1}(M)$ in $\mathbb{R}^n$ by considering locally the function-family $F_c(x) = (x,f(x)+c)$, $c \in \mathbb{R}^m$.

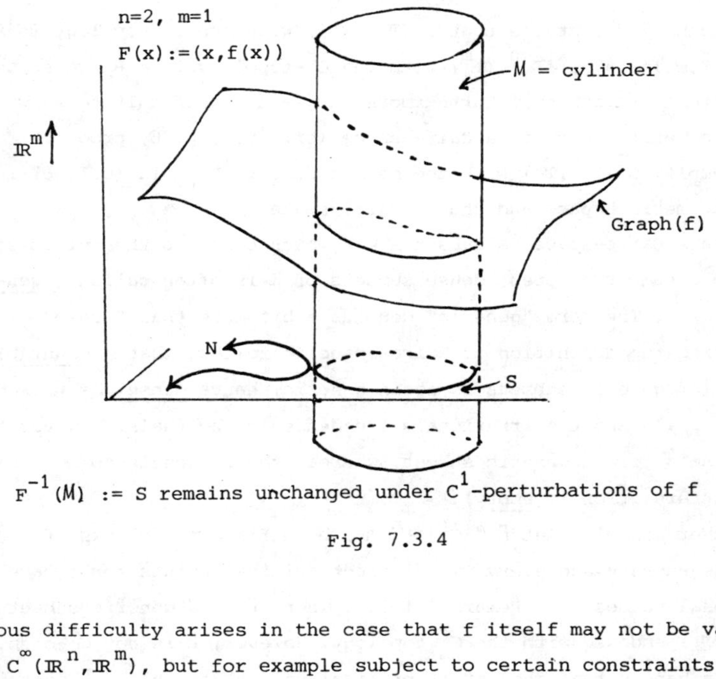

$F^{-1}(M) := S$ remains unchanged under C^1-perturbations of f

Fig. 7.3.4

An analogous difficulty arises in the case that f itself may not be varied freely in $C^{\infty}(\mathbb{R}^n,\mathbb{R}^m)$, but for example subject to certain constraints on its components $(f_1,\ldots,f_m)$.
However, the basic idea will be always to look for the "degree of variational freedom" on which one builts a framework in which Sard's theorem can be applied in an exhausting way. So, roughly speaking, transversality theory is "self-refining". In a concrete form this will become appearant in Section 7.4, in which we discuss jet - transversality.

We proceed with a remark on Lemma 7.3.1.

Remark 7.3.9. A topological space X is called a Baire-space if for every sequence $A_1, A_2, \ldots$ of subsets of X with A_i open and dense in X, $i = 1,2,\ldots$, we have $\bigcap_{i=1}^{\infty} A_i$ is dense in X. Or, equivalently:

X is called a Baire-space if for every sequence $C_1, C_2, \ldots$ of subsets of X with the property: C_i is closed and contains no interior points, $i = 1,2,\ldots$, we have: $\bigcup_{i=1}^{\infty} C_i$ has no interior points.

Then, Lemma 7.3.1 states that $C^{\infty}(\mathbb{R}^n, \mathbb{R}^m)$ with the C^{∞}-topology is a Baire-space. Similarly, $C^k(\mathbb{R}^n, \mathbb{R}^m)$ with the C^k-topology is a Baire-space (completeness-argument). Furthermore, it is not difficult to show that any complete metric space is a Baire-space (cf. [56 , p.26, proof of "Satz 3"]). For example, $C^2(\mathbb{R}^n, \mathbb{R})$ with the metric d as in Formula (6.1.36) is a complete metric space and thus a Baire-space.

Let X be a Baire-space. A subset $A \subset X$ which contains the intersection of a countable number of open, dense subsets of X is often called a generic subset of X. The term "generic" denotes a bit more than "dense".

Note: from the definition of Baire-space it follows that a countable intersection of generic subsets is again generic, hence dense. As an example both the rationals and the irrationals are dense in the reals. However the irrationals form a generic subset, whereas the rationals do not (note that their intersection is empty).

As another example, let $F \subset C^{\infty}(\mathbb{R}^n, \mathbb{R})$ be defined as follows: $f \in F$ iff f is nondegenerate and every two distinct critical points for f have distinct functional values (cf. Remark 7.1.12). Then, F is a generic subset of $C^{\infty}(\mathbb{R}^n, \mathbb{R})$ endowed with the C^{∞}-topology. However, F is not C^{∞}-open, since it might happen that the set of critical values for an $\tilde{f} \in F$ contains all rational numbers. For the construction of such an $\tilde{f}$, we consider the periodic function $g \in C^{\infty}(\mathbb{R}, \mathbb{R})$ as sketched in Fig. 7.3.5. The functional values at the local minima/maxima of g can be decreased/increased freely and independently from each other without disturbing the nondegeneracy. In this way we can obtain a function $\tilde{g} \in C^{\infty}(\mathbb{R}, \mathbb{R})$ whose set of critical values contains all rational numbers. For an example in $\mathbb{R}^n$, just put $\tilde{f}(x_1, x_2, \ldots, x_n) = \tilde{g}(x_1) + \sum_{i=2}^{n} x_i^2$.

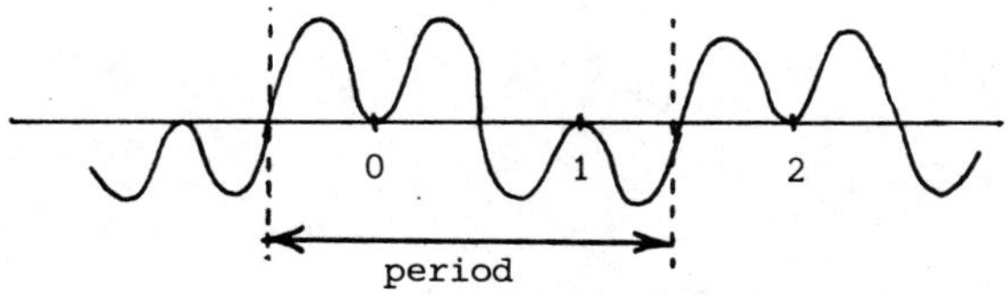

Fig. 7.3.5

Finally, we emphasize that - in terms of genericity - Theorems 7.3.3, 7.3.4 and 7.3.5 can slightly be sharpened as follows. Let $C^\infty(\mathbb{R}^n,\mathbb{R}^m)$ be endowed with the C^∞-topology; then the entry "C^k-dense for all k" can be replaced by "generic". □

Let us return to the statement of Theorem 7.3.5. We will give another interpretation of the set F. To this aim, let $N \subset \mathbb{R}^n$ and $M \subset \mathbb{R}^m$ be manifolds. In a natural way the concept "$f_{|N}$ meets M transversally" ($f_{|N} \pitchfork M$) is defined as follows:

$$f_{|N} \pitchfork M \quad \text{if} \quad \begin{cases} Df(x)[T_xN] + T_{f(x)}M = \mathbb{R}^m \\ \\ \text{for all } x \in N \text{ with } f(x) \in M. \end{cases} \tag{7.3.29}$$

<u>Theorem 7.3.6</u>. Let $N \subset \mathbb{R}^n$, $M \subset \mathbb{R}^m$ be manifolds, $f \in C^\infty(\mathbb{R}^n,\mathbb{R}^m)$, and suppose that $f \pitchfork M$. Then we have:

$$f^{-1}(M) \pitchfork N \quad \text{iff} \quad f_{|N} \pitchfork M.$$

<u>Proof</u>. Firstly, we assume $f^{-1}(M) \pitchfork N$ and we choose $x \in f^{-1}(M) \cap N$. So, we have:

$$T_xf^{-1}(M) + T_xN = \mathbb{R}^n, \tag{7.3.30}$$

and we must show (cf. (7.3.29)):

$$Df(x)[T_xN] + T_{f(x)}M = \mathbb{R}^m. \tag{7.3.31}$$

Since $f \pitchfork M$ we have (cf. (7.3.5)):

$$Df(x)[\mathbb{R}^n] + T_{f(x)}M = \mathbb{R}^m, \tag{7.3.32}$$

and, from Theorem 7.3.1:

$$T_x f^{-1}(M) = Df(x)^{-1} T_{f(x)} M. \tag{7.3.33}$$

Now, substitute (7.3.33) in (7.3.30) and take the image under Df(x) on both sides. This yields:

$$T_{f(x)} M + Df(x)[T_x N] \supset Df(x)[\mathbb{R}^n]. \tag{7.3.34}$$

Adding $T_{f(x)}M$ on both sides of (7.3.34) and then using (7.3.32) gives (7.3.31):

$$\mathbb{R}^m = Df(x)[\mathbb{R}^n] + T_{f(x)}M \subset Df(x)[T_x N] + T_{f(x)}M \subset \mathbb{R}^m.$$

Conversely, suppose that $f|_N \pitchfork M$ and choose $x \in f^{-1}(M) \cap N$. We have to show that (7.3.30) holds; note that we may use (7.3.33) again. So, let $\xi \in \mathbb{R}^n$ be arbitrarily chosen. Then, $Df(x)\xi \in \mathbb{R}^m$ and hence, from (7.3.31) we read:

$$Df(x)\xi = Df(x)\xi_1 + \eta, \quad \text{where } \xi_1 \in T_x N,\ \eta \in T_{f(x)}M.$$

Consequently, $Df(x)(\xi-\xi_1) \in T_{f(x)}M$ and thus, $\xi-\xi_1 \in Df(x)^{-1}T_{f(x)}M$. Finally, we use (7.3.33) and obtain $\xi-\xi_1 \in T_x f^{-1}(M)$ and hence, $\xi \in T_x N + T_x f^{-1}(M)$. This shows (7.3.30). □

Remark 7.3.10. As an application of Theorems 7.3.5 and 7.3.6 we consider a statement which is "dual" to Remark 7.1.6. In fact, let X be a k-dim. MGB in $\mathbb{R}^n$ of the class C^∞ and $\alpha \in \mathbb{R}$ be fixed. Let F be the following subset of $C^\infty(\mathbb{R}^n,\mathbb{R})$:

$f \in F$ iff the sets $f^{-1}((-\infty,\alpha]) \cap X$ and $f^{-1}(\alpha) \cap X$ are MGB's of class C^∞ of dimension k resp. k-1 (if not empty).

Statement: F is a generic subset of $C^\infty(\mathbb{R}^n,\mathbb{R})$ in the C^∞-topology.

To see this, take a stratum Σ of X and define:

$$F(\Sigma) = \{f \in C^\infty(\mathbb{R}^n,\mathbb{R}) \mid f \pitchfork \{\alpha\} \text{ and, moreover, } f^{-1}(\alpha) \pitchfork \Sigma\}.$$

From Theorem 7.3.5 and Remark 7.3.9 we see that $F(\Sigma)$ is generic. Note that F contains the intersection of the sets $F(\Sigma)$, Σ ranging over the (countable) strata of X. But the intersection of a countable number of generic subsets is obviously generic.

<u>Remark 7.3.11</u>. Let $M_1, M_2, \ldots$ be a <u>countable</u> number of manifolds in $\mathbb{R}^{n+m}$. Let F denote the set of all $f \in C^\infty(\mathbb{R}^n, \mathbb{R}^m)$ with Graph(f) $\pitchfork M_i$, $i = 1,2,\ldots$. As a slight generalization of Theorem 7.3.4 we have: F is a generic subset of $C^\infty(\mathbb{R}^n, \mathbb{R}^m)$ endowed with the C^∞-topology.
To see this, recall that the set F_i consisting of all $f \in C^\infty(\mathbb{R}^n, \mathbb{R}^m)$ with Graph(f) $\pitchfork M_i$ is <u>generic</u> (cf. Remark 7.3.9). But $F = \bigcap_i F_i$, and, as a countable intersection of generic sets, F is generic.

The rest of this section will be devoted to a more careful analysis and generalization of the "open-part" in Theorem 7.3.4. The next corollary is a first generalization and it is easily verified by copying the proof of Theorem 7.3.4 and using the observation: if $\{V_j, j \in J\}$ is a locally finite family of subsets of $\mathbb{R}^n$ and $K \subset \mathbb{R}^n$ a compact subset, then $V_j \cap K$ is non-empty for at most a finite number of indices j.

<u>Corollary 7.3.1</u>. Let $\{M_j, j \in J\}$ be a locally finite family of manifolds in $\mathbb{R}^{n+m}$ and let F denote the set of all $f \in C^\infty(\mathbb{R}^n, \mathbb{R}^m)$ with Graph(f) $\pitchfork M_j, j \in J$. If, in addition, every manifold M_j is a closed subset of $\mathbb{R}^{n+m}$, then F is C^k-open for all $k \geq 1$. □

In the above corollary we assumed that every manifold M_j is closed. We emphasize that the assertion of the corollary need not be true if we only assume the weaker condition that the union of all sets M_j is closed. This will be clarified in the following example.

<u>Example 7.3.2</u>. Let S be the zero-set in $\mathbb{R}^3$ of the polynomial $p(x_1,x_2,x_3) := x_2^2 - x_1^2 x_3$ (cf. also [16]). Note that $Dp = (-2x_1x_3, 2x_2, -x_1^2)$. Hence, Dp vanishes exactly on the x_3-axis M_1 ($\subset$ S). It follows that the set $M_2 := S \backslash M_1$ is a 2-dim. manifold. Altogether we conclude that S is a <u>closed</u> set consisting of the union of the disjoint manifolds M_1, M_2. See Fig. 7.3.6.a for a picture of S. (For a verification of the picture, consider the intersection of S with planes x_3 = constant, resp. x_1 = constant; note, in particular, that the manifold M_2 is the image of the map $\phi(u,v) = (u, uv, v^2)$, $u \neq 0$).

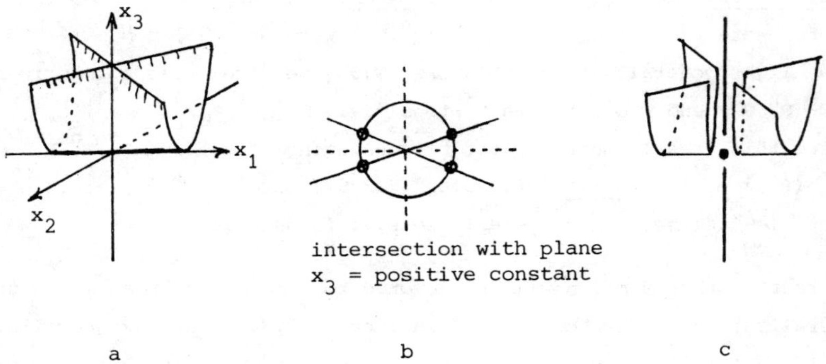

Fig. 7.3.6

Next, consider the map f, $f(x_1,x_2) = x_1^2 + x_2^2$. Note that Graph(f) $\pitchfork M_i$, $i = 1,2$. In fact, the intersection Graph(f) $\cap M_1$ is exactly the origin, the transversality being obvious. In order to see that Graph(f) $\pitchfork M_2$, note that the intersection of Graph(f) with any plane x_3 = positive constant is a circle, whereas the (closure of the) intersection of M_2 with the latter plane are two straight lines; cf. Fig. 7.3.6.b. Although Graph(f) $\pitchfork M_i$, $i = 1,2$, this is false for every C^k-neighborhood $\mathcal{O}$ of f. To see this, just "shift" Graph(f) along the x_1-axis, thereby noting that the tangent space of M_2 at every point $(x_1,0,0)$, $x_1 \neq 0$, is the plane "$x_3 = 0$". However, such a global shift might exceed a given C^k-neighborhood $\mathcal{O}$ of f. So, in order to stay within $\mathcal{O}$, we have to "localize" this shift. To this aim, let $\psi(x_1,x_2)$ be a C^∞-function with compact support and $\psi(x_1,x_2) \equiv 1$ in a neighborhood of the origin (cf. Section 2.6). Put $f_\varepsilon(x_1,x_2) = f(x_1-\varepsilon\psi(x_1,x_2),x_2)$. Then, for $\varepsilon > 0$ and sufficiently small, we have: $f_\varepsilon \in \mathcal{O}$ (recall that ψ has compact support) and Graph(f_ε) does <u>not</u> intersect M_2 transversally. (In fact, consider the intersection-point $(\varepsilon,0,0)$).

It turns out that the origin in the set S is the only point which causes difficulties. We shall come back to this point later on; here we remark that the refinement of the partitioning of S into $M_1^{(1)}, M_1^{(2)}, M_2$ with $M_1^{(1)}$ = origin, and $M_1^{(2)} = M_1 \setminus M_1^{(1)}$ (cf. Fig. 7.3.6.c) has the property: if Graph(f) intersects $M_1^{(1)}, M_1^{(2)}, M_2$ transversally, then this holds for a whole C^1-neighborhood of f.

Apart from the illustrative set S in Example 7.3.2, sets which can be partitioned into manifolds of several dimensions often occur in a natural way. A first important example is a Manifold with Generalized Boundary (MGB). We give two more examples, dealing with familiar objects, namely (real) matrices.

Example 7.3.3. Let m,n be given integers with $m \geq n \geq 1$ and denote by $\mathcal{R}$ the set of all real n×m matrices. We identify the set $\mathcal{R}$ with the space $\mathbb{R}^{nm}$. Let $\mathcal{R}_i$ be the set of all n×m matrices with rank equal to i, $i = 0,1,\dots,n$. Consequently, the set $\mathcal{R}$ is the disjoint union of the sets $\mathcal{R}_i$.
We contend: $\mathcal{R}_i$ is a submanifold of $\mathbb{R}^{nm}$ with codimension $(n-i)(m-i)$.
To this aim, we firstly note: given a nonsingular n×n matrix E and a nonsingular m×m matrix F, the mapping $\Phi: \mathcal{R} \to \mathcal{R}$, $M \mapsto EMF$, induces a C^∞-diffeomorphism from $\mathbb{R}^{nm}$ onto $\mathbb{R}^{nm}$; moreover, Φ leaves $\mathcal{R}_i$ invariant. Now, let i be fixed. In order to study the local structure of $\mathcal{R}_i$, we take an $M \in \mathcal{R}_i$ and may assume that the upper left i×i submatrix of M is nonsingular (otherwise, choose for E,F permutation matrices such that EMF has this property). Now, let $i < n$. Decompose M into the matrices A, B, C, D:

$$M = \left(\begin{array}{c|c} A & C \\ \hline B & D \end{array}\right) \qquad \text{A: i×i matrix} \tag{7.3.35}$$

Let $\mathcal{O}$ be an open $\mathbb{R}^{nm}$-neighborhood of M such that for all $\tilde{M} \in \mathcal{O}$ the corresponding matrix $\tilde{A}$ is nonsingular. Then, $\tilde{M} \in \mathcal{O}$ is an element of $\mathcal{R}_i$ iff the following relation holds:

$$\tilde{D} = \tilde{B}\tilde{A}^{-1}\tilde{C} \quad . \tag{7.3.36}$$

To see this, let $\tilde{c}_{\cdot k}$, resp. $\tilde{d}_{\cdot k}$ denote the k-th column of $\tilde{C}$ resp. $\tilde{D}$. Since $\tilde{A}$ is nonsingular, there exists a unique vector $\tilde{\lambda}_k \in \mathbb{R}^i$ such that $\tilde{c}_{\cdot k} = \tilde{A}\tilde{\lambda}_k$. Now, $\tilde{M} \in \mathcal{R}_i$ iff the columns of the matrix $\begin{pmatrix} \tilde{C} \\ \tilde{D} \end{pmatrix}$ lie in the span of the columns of the matrix $\begin{pmatrix} \tilde{A} \\ \tilde{B} \end{pmatrix}$. Consequently, we have $\tilde{d}_{\cdot k} = \tilde{B}\tilde{\lambda}_k$ and thus, $\tilde{d}_{\cdot k} = \tilde{B}\tilde{A}^{-1}\tilde{c}_{\cdot k}$. This proves (7.3.36). From the same formula it follows, in addition, that each element of $\tilde{D}$ is a rational (and thus C^∞-) function of the elements of $\tilde{A}$, $\tilde{B}$, $\tilde{C}$. So, $\mathcal{R}_i$ can locally be defined by $(n-i)(m-i)$ functions which form a defining system of functions (cf. Section 7.2) for

$R_i \cap O$. Finally, the case $i = n$ is accomplished by noting that R_n is open in R.

<u>Example 7.3.4</u>. Let A denote the set of all <u>symmetric</u> $n\times n$ matrices. A symmetric matrix is completely determined by its diagonal and subdiagonal elements; the number of these elements equals $1 + 2 +\ldots+ n = \frac{1}{2}n(n+1)$. So, we can identify the set A with the space $\mathbb{R}^{\frac{1}{2}n(n+1)}$.
Let A_i be the set of all symmetric $n\times n$ matrices with rank equal to i, $i = 0,1,\ldots,n$. Consequently, the set A is the disjoint union of the sets A_i.
<u>We contend</u>: A_i is a submanifold of $\mathbb{R}^{\frac{1}{2}n(n+1)}$ with codimension $\frac{1}{2}(n-i)(n-i+1)$. To this aim, we firstly note: given a nonsingular $n\times n$ matrix E, the mapping $\Phi\colon A \to A$, $M \mapsto E^TME$, induces a C^∞-diffeomorphism from $\mathbb{R}^{\frac{1}{2}n(n+1)}$ onto itself; moreover, Φ leaves A_i invariant. Now, let i be fixed and choose $M \in A_i$. In view of the diffeomorphism Φ we may assume that the left upper $i\times i$ submatrix of M is nonsingular. So, we can decompose M according to (7.3.35) with $B = C^T$. Then, (7.3.36) now reads:

$$\tilde{D} = \tilde{C}^T\tilde{A}^{-1}\tilde{C}\,. \tag{7.3.37}$$

The matrix $\tilde{D}$ is an $(n-i)\times(n-i)$ matrix. Due to symmetry (7.3.37) then contains $\frac{1}{2}(n-i)(n-i+1)$ independent equations and this number constitutes the codimension of A_i. This shows the contention.□ Finally, note that a <u>symmetric</u> $n\times n$ matrix M has rank i iff exactly $n-i$ eigenvalues of M vanish. So, if we denote by B_k the set of all symmetric $n\times n$ matrices with exactly k vanishing eigenvalues, then $B_k = A_{n-k}$. In particular, B_k is a submanifold of $\mathbb{R}^{\frac{1}{2}n(n+1)}$ with codimension $\frac{1}{2}k(k+1)$.

For the statement of the theorem we have in mind as an appropriate generalization of the "open-part" of Theorem 7.3.4 we return to the start of the proof of Theorem 7.3.4. Associated with a manifold $M \subset \mathbb{R}^{n+m}$ and an $f \in C^\infty(\mathbb{R}^n,\mathbb{R}^m)$ we introduced a continuous function ϕ_f on M having the property that Graph(f) $\pitchfork M$ iff $\phi_f(z) > 0$ for all $z \in M$; see Formula (7.3.8). Since we also have to deal with several manifolds simultaneously, we have to put an additional index to ϕ_f. So, we put:

$$\phi_{f,M} = \phi_f \qquad \text{[cf. (7.3.8) with } M \text{ as a manifold].} \tag{7.3.38}$$

There are two extreme cases in relation with the function $\phi_{f,M}$.

Case 1. $\dim(M) < m$: Then, $\mathrm{Graph}(f) \pitchfork M$ just means that $\mathrm{Graph}(f) \cap M = \emptyset$. In fact, the dimension of the normal space to a point of M is greater than the dimension of Graph(f). Hence, in Formula (7.3.8) the r.h.s. reduces to the distance $\|y-f(x)\|$.

Case 2. $\dim(M) = n+m$: Then, M is an open subset of $\mathbb{R}^{n+m}$ and we always have $\mathrm{Graph}(f) \pitchfork M$. In order to translate this into terms appearing in the r.h.s. of (7.3.8) we formally agree that "the sum of the determinants" always equals 1.

Theorem 7.3.7. (The "Openess-Principle").
Let $\{M_j, j \in J\}$ be a locally finite family of manifolds in $\mathbb{R}^{n+m}$. Suppose that $f \in C^\infty(\mathbb{R}^n,\mathbb{R}^m)$ has the property that $\mathrm{Graph}(f) \pitchfork M_j$, $j \in J$. Then, there exists a C^1-neighborhood V of f having the property:

$$\mathrm{Graph}(g) \pitchfork M_j, j \in J, g \in V$$

iff the following condition is fulfilled:

Condition O: For every $j \in J$ and every compact subset $K \subset \mathbb{R}^{n+m}$ with $K \cap M_j \neq \emptyset$ we have:

$$\inf_{z \in K \cap M_j} \phi_{f,M_j}(z) > 0 . \qquad \square$$

The proof of Theorem 7.3.7 will be given later on. In the formula for the function $\phi_{f,M}$ appears the "sum of determinants" in the r.h.s. which basically represents transversality. We will replace this term by another one which is easier to handle and which will present another geometric insight into the concept of transversality. In fact, in Theorem 7.2.1 we described transversality in terms of linear subspaces of the embedding space. This gives rise to a further study of "spaces of linear subspaces". Let $G(n,k)$ (the socalled Grassmann-manifold) denote the set of all k-dim. linear subspaces of $\mathbb{R}^n$. Note that $G(n,n-k)$ can be identified with $G(n,k)$ by taking orthogonal complements.

As an example, $G(2.1)$ can be identified with the unit circle in $\mathbb{R}^2$. In fact, each straight line in $\mathbb{R}^2$ passing through the origin hits the unit circle in two points, one of them being sufficient to represent the line; by means of identifying "antipodal" points we get $G(2,1)$ (see Fig. 7.3.7.a).

The set $G(3,1)$ (= projective plane) is more complicated and cannot be represented as a submanifold of $\mathbb{R}^3$ because of unavoidable selfintersections (cf. Fig. 7.3.7.b).

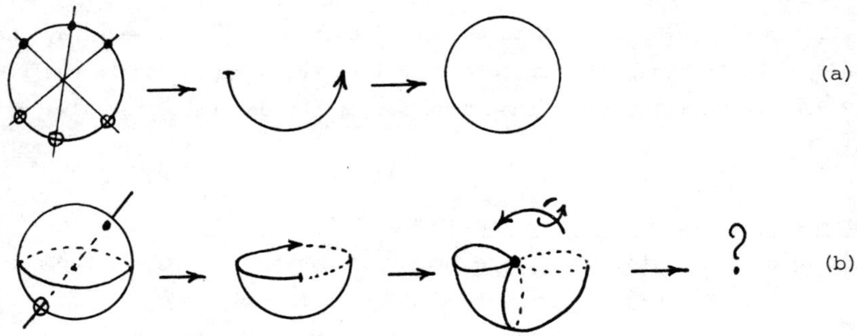

Fig. 7.3.7

We will represent $G(n,k)$ as a submanifold of the space of symmetric matrices by means of orthogonal projections of $\mathbb{R}^n$ to the k-dimensional linear subspaces.

Theorem 7.3.8. Let $A_k \subset \mathbb{R}^{\frac{1}{2}n(n+1)}$ denote the manifold of symmetric $n\times n$ matrices with rank equal to k.
Then, $G(n,k)$ is a compact submanifold of A_k of dimension $k(n-k)$.

Proof. Let P_k denote the set of all $n\times n$ orthogonal-projection matrices of rank k. By means of the obvious bijective correspondence $P_k \longleftrightarrow G(n,k)$ we have a concrete realization of $G(n,k)$. Next, we give three equivalent conditions which are necessary and sufficient for an $n\times n$ matrix P to be an element of P_k:

c1: $P = P^T$, $PP-P = 0$, rank $P = k$.

c2: $P = P^T$, k eigenvalues of P are equal to 1,
n-k eigenvalues of P are equal to 0.

c3: $P = P^T$, rank $P = k$, rank $I_n-P = n-k$ (I_n being the $n\times n$ identity matrix).

From c1 we have $\mathcal{P}_k \subset \mathcal{A}_k$ and c2 implies that $\mathcal{P}_k$ is a closed and bounded subset of $\mathbb{R}^{\frac{1}{2}n(n+1)}$. In fact, note that the (Euclidean) norm $\|P\|$ equals $\sqrt{k}$ for $P \in \mathcal{P}_k$. Consequently, $\mathcal{P}_k$ is compact. Now, pick P_k from $\mathcal{P}_k$ and let us study the local structure of $\mathcal{P}_k$ in an open $\mathbb{R}^{\frac{1}{2}n(n+1)}$-neighborhood $\mathcal{O}$ of P_k. For a fixed orthogonal n×n matrix Q the mapping $M \mapsto Q^T MQ$ induces a C^∞-diffeomorphism of $\mathbb{R}^{\frac{1}{2}n(n+1)}$ onto itself, thereby mapping $\mathcal{A}_k$ onto $\mathcal{A}_k$ and $\mathcal{P}_k$ onto $\mathcal{P}_k$.

So, we may assume that $P_k = \mathrm{diag}(1,1,\ldots,1,0,\ldots,0)$. Decompose a symmetric n×n matrix as follows:

$$M = \left(\begin{array}{c|c} A & C \\ \hline C^T & D \end{array}\right), \text{ A: k×k matrix}$$

If the neighborhood $\mathcal{O}$ of P_k is sufficiently small, then for all $M \in \mathcal{O}$ we have rank $A = k$ and $\mathrm{rank}(I_{n-k}-D) = n-k$. From c3 and Example 7.3.4 (cf. (7.3.37)) we obtain: $M \in \mathcal{O}$ is an element of $\mathcal{P}_k$ iff the following two systems of equations are simultaneously satisfied:

$$\left.\begin{array}{l} D = C^T A^{-1} C \\ I_k - A = C[I_{n-k}-D]^{-1} C^T \end{array}\right\} . \qquad (7.3.39)$$

Because of symmetry, (7.3.39) yields r equations, where $r = \frac{1}{2}(n-k)(n-k+1) + \frac{1}{2}k(k+1)$. These r equations represent a defining system of functions for $\mathcal{P}_k \cap \mathcal{O}$ (the elements of the matrix C can be assumed to be sufficiently small). Thus, $\mathcal{P}_k$ is a manifold in $\mathbb{R}^{\frac{1}{2}n(n+1)}$ of dimension p, where $p = \frac{1}{2}n(n+1) - \frac{1}{2}(n-k)(n-k+1) - \frac{1}{2}k(k+1) = k(n-k)$. □

We proceed by introducing a function, denoted by δ, which represents the transversal intersection of two linear subspaces. To this aim consider the (compact) product-manifold of Grassmannians:

$$\mathcal{G}(n,k,r) = \mathcal{G}(n,k) \times \mathcal{G}(n,r). \qquad (7.3.40)$$

The set $\mathcal{G}(n,k,r)$ is the disjoint union of the following two subsets:

$$\mathcal{E}(n,k,r) = \{(X,Y) \in \mathcal{G}(n,k,r) \,|\, X \not\pitchfork Y\} \qquad (7.3.41)$$

$$B(n,k,r) = G(n,k,r)\setminus E(n,k,r). \tag{7.3.42}$$

Obviously, if $k+r < n$, then $E(n,k,r) = \emptyset$ and, in case $k = n$ or $r = n$, we see that (the "bad set") $B(n,k,r)$ is empty.

Example 7.3.5. In case $n = 2$, $k = r = 1$, the sets $G(n,k,r)$, $E(n,k,r)$ and $B(n,k,r)$ can be visualized in $\mathbb{R}^3$. In fact, the set $G(2,1,1)$ becomes the product $S^1 \times S^1$ of two copies of the circle S^1 (= torus; see Fig. 7.3.8.a). The bad set $B(2,1,1)$ is exactly the diagonal set $\{(x,x) \mid x \in S^1\}$. See Fig. 7.3.8.b.

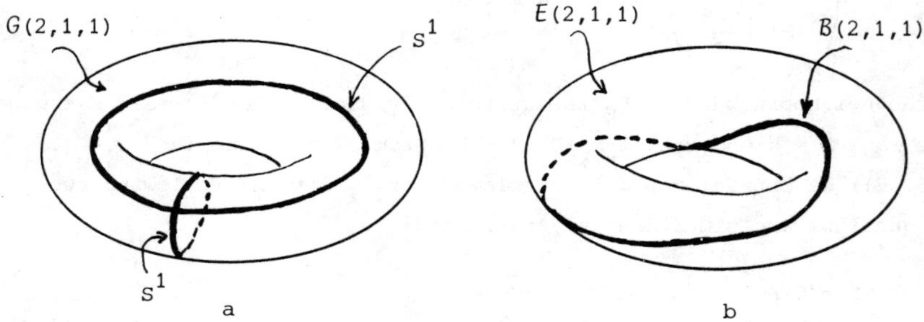

Fig. 7.3.8

We define the functions δ as follows:

$$\left.\begin{aligned} \delta\colon\ & G(n,k,r) \to \mathbb{R} \\ & (X,Y) \mapsto \det(P_X+P_Y) \end{aligned}\right\} , \tag{7.3.43}$$

where P_X stands for the n×n orthogonal-projection matrix w.r.t. the linear subspace X of $\mathbb{R}^n$ (P_Y similar).

Lemma 7.3.2. The function $\delta\colon G(n,k,r) \to \mathbb{R}$ has the properties:

(i) $\delta(X,Y) \geq 0$ for all $(X,Y) \in G(n,k,r)$,

(ii) $\delta(X,Y) > 0$ iff $(X,Y) \in E(n,k,r)$.

Proof.

Statement (i). An orthogonal-projection matrix is symmetric and positive semi-definite. Then, for $(X,Y) \in G(n,k,r)$, the matrix $P_X + P_Y$ is symmetric and positive semi-definite as well. But then, all eigenvalues of $P_X + P_Y$ are real and nonnegative. Now, Statement (i) follows, since the determinant of a square matrix equals the product of its eigenvalues.

Statement (ii). Let $(X,Y) \in G(n,k,r)$. The normal space of X at the origin is nothing else but the null-space Ker P_X of P_X (similar for Y). Then, it follows from Exercise 7.2.1 that: $X \pitchfork Y$ iff $\mathrm{Ker}\, P_X \cap \mathrm{Ker}\, P_Y = \{0\}$. Consequently, in order to show Statement (ii) it is sufficient to establish:

$$\mathrm{Ker}(P_X+P_Y) = \mathrm{Ker}\, P_X \cap \mathrm{Ker}\, P_Y. \qquad (*)$$

Now, let $u \in \mathrm{Ker}(P_X+P_Y)$. Noting that $P_X = P_X^T$ and $P_X^2 = P_X$, we obtain the following chain of implications:

$$(P_X+P_Y)u = 0 \rightarrow P_Xu + P_Yu = 0 \rightarrow P_X^2u + P_Y^2u = 0 \rightarrow$$

$$\rightarrow u^TP_X^2u + u^TP_Y^2u = 0 \rightarrow \|P_Xu\|^2 + \|P_Yu\|^2 = 0 \rightarrow$$

$$\rightarrow P_Xu = 0 \text{ and } P_Yu = 0.$$

Consequently, we have established: $\mathrm{Ker}(P_X+P_Y) \subset \mathrm{Ker}\, P_X \cap \mathrm{Ker}\, P_Y$. The reverse inclusion is trivial and hence, (*) is proved. □

Exercise 7.3.1. Suppose that $k+r \geq n$ and $k \neq n$, $r \neq n$.
Show: $E(n,k,r)$ is an open, dense subset of $G(n,k,r)$,
$B(n,k,r)$ is nonempty, compact. □

Let $f \in C^\infty(\mathbb{R}^n,\mathbb{R}^m)$ and $M \subset \mathbb{R}^{n+m}$ be a manifold. With the aid of the function δ in (7.3.43) we can introduce the following continuous function $\psi_{f,M}$ on M which can be seen as a substitute for the function $\phi_{f,M}$:

$$\left.\begin{array}{l} \psi_{f,M}(z) = \|y-f(x)\| + \delta[(T_zM, T_{(x,f(x))}\mathrm{Graph}(f))], \\ \text{where } z \in M \text{ and } \delta: G(n+m, \dim M, n) \rightarrow \mathbb{R} \end{array}\right\} \qquad (7.3.44)$$

The next lemma interrelates the functions $\phi_{f,M}$ and $\psi_{f,M}$ for our purpose.

Lemma 7.3.3. Let $M \subset \mathbb{R}^{n+m}$ be a manifold and $f \in C^\infty(\mathbb{R}^n,\mathbb{R}^m)$. Then, we have:

(i) Graph(f) $\pitchfork M$ iff $\psi_{f,M}(z) > 0$ for all $z \in M$.

(ii) For each $z \in M$: $\psi_{f,M}(z) > 0$ iff $\phi_{f,M}(z) > 0$.

(iii) For compact $K \subset \mathbb{R}^{n+m}$ with $K \cap M \neq \emptyset$:

$$\inf_{z\in K\cap M} \psi_{f,M}(z) > 0 \quad \text{iff} \quad \inf_{z\in K\cap M} \phi_{f,M}(z) > 0. \qquad \square$$

The proof of Lemma 7.3.3 is left as an exercise. (In the proof of (iii) one has to use the "limiting" version of (ii), thereby exploiting the compactedness of $G(n,k,r)$).

Remark 7.3.12. From Lemma 7.3.3 we learn that we may replace - in the statement of Theorem 7.3.7 - the functions ϕ_{f,M_j} by ψ_{f,M_j}.

Proof of Theorem 7.3.7.

Sufficiency-Part. Suppose that Condition $\mathcal{O}$ holds.

We emphasize that this part is a refined version of the "open-part" in the proof of Theorem 7.3.4. A point in $\mathbb{R}^{n+m}$ will again be denoted by z, where $z = (x,y)$, $x \in \mathbb{R}^n$, $y \in \mathbb{R}^m$. Let $B(\bar{x},1)$ be the (Euclidean) ball in $\mathbb{R}^n$ with center $\bar{x}$ and radius 1.

We are done if we can show the existence of an $\varepsilon_{\bar{x}} > 0$ such that $g \in C^\infty(\mathbb{R}^n,\mathbb{R}^m)$ and $\max_{x\in B(\bar{x},1)} \{\|g(x)-f(x)\| + |||Dg(x)-Df(x)|||\} \le \varepsilon_{\bar{x}}$ (with $|||\cdot|||$ as in the proof of Theorem 7.3.4) implies:

$$\psi_{g,M_j}(z) > 0 \text{ for all } z \in (B(\bar{x},1) \times \mathbb{R}^m) \cap M_j,\ j \in J.$$

To this aim we firstly make a reduction step (Step 1):

Step 1. Choose $\delta_{\bar{x}}$ such that for $g \in C^\infty(\mathbb{R}^n,\mathbb{R}^m)$ with $\max_{x\in B(\bar{x},1)} \|g(x)-f(x)\| \le 1$ we have:

$$\text{Graph}(g) \cap (B(\bar{x},1) \times \mathbb{R}^m) \subset B(\bar{x},1) \times B(0,\delta_{\bar{x}}),$$

where $B(0,\delta_{\bar{x}})$ is the ball in $\mathbb{R}^m$ with center $o \in \mathbb{R}^m$ and radius $\delta_{\bar{x}}$. As an abbreviation we denote $K := B(\bar{x},1) \times B(0,\delta_{\bar{x}})$. Since the family $\{M_j, j \in J\}$

is locally finite, it follows that the family of closures $\{\bar{M}_j, j \in J\}$ is locally finite, too. Consequently, only a finite number of sets $\bar{M}_j$ meet the compact set K. Hence, if we take $\varepsilon_{\bar{x}}$ less or equal than one, we can restrict ourselves to the compact set K. But then (as it will be appearant in the sequel) we can assume without loss of generality that we are dealing with only one manifold, say M.

Step 2. If $K \cap \bar{M} = \emptyset$, we put $\varepsilon_{\bar{x}} = 1$.

Step 3. Now, suppose that $K \cap \bar{M} \neq \emptyset$. For $\hat{z} \in K \cap \bar{M}$ we distinguish whether $\hat{z}$ belongs to Graph(f) or not, where $\hat{z} = (\hat{x}, \hat{y})$.

Case 1: $\hat{z} \in K \cap \bar{M}$, but $\hat{z} \notin$ Graph(f).

Obviously, there exist $\rho_1, \rho_2 > 0$ such that $(B(\hat{x},\rho_1) \times B(\hat{y},\rho_2)) \cap \text{Graph}(f) = \emptyset$. But then, an $\varepsilon_{\hat{x}} > 0$ exists such that $g \in C^\infty(\mathbb{R}^n, \mathbb{R}^m)$ and

$$\max_{x \in B(\hat{x},\rho_1)} \|g(x)-f(x)\| \le \varepsilon_{\hat{x}} \quad \text{imply:} \quad (B(\hat{x},\rho_1) \times B(\hat{y},\rho_2)) \cap \text{Graph}(g) = \emptyset.$$

To the point $\hat{z}$ we associate the open set $\mathring{B}(\hat{x},\rho_1) \times \mathring{B}(\hat{y},\rho_2)$ and the number $\varepsilon_{\hat{x}}$.

Case 2: $\hat{z} \in K \cap \bar{M} \cap$ Graph(f).

Let p denote the dimension of M. In view of the validity of Condition $\mathcal{O}$ we must have: $p \ge m$. For a neighborhood U of $\hat{z}$ we define the set $G(U)$ as follows:

$$G(U) = \{(T_z M, T_{(x,f(x))}\text{Graph}(f)) \in \mathcal{G}(n+m,p,n) \,|\, z = (x,y) \in M \cap U\}.$$

Again from Condition $\mathcal{O}$ (cf. also Remark 7.3.12 and (7.3.44)) we obtain the existence of a neighborhood $\hat{U}$ of $\hat{z}$ with $\overline{G(\hat{U})} \cap \mathcal{B}(n+m,p,n) = \emptyset$. Since $\mathcal{G}(n+m,p,n)$ is compact, it follows that $\overline{G(\hat{U})}$ is compact as well. Recall that $\mathcal{G}(n+m,p,n)$ can be (topologically) represented as a subset of $\mathbb{R}^\alpha \times \mathbb{R}^\alpha$, $\alpha = \frac{1}{2}(n+m)(n+m+1)$ (cf. Theorem 7.3.8 and (7.3.40)). So, we are dealing with compact subsets in $\mathbb{R}^\alpha \times \mathbb{R}^\alpha$, namely $\overline{G(\hat{U})}$ and $\mathcal{B}(n+m,p,n)$. Hence, we may perturb the second $\mathbb{R}^\alpha$-factor of every point of $\overline{G(\hat{U})}$ up to an $\varepsilon > 0$ (ε only depending on $\overline{G(\hat{U})}$ and $\mathcal{B}(n+m,p,n)$) without reaching the set $\mathcal{B}(n+m,p,n)$. Note that the second $\mathbb{R}^\alpha$-factor of a point of $G(\hat{U})$ is related to a tangent space of Graph(f) and hence, to the derivative Df. Choose $\rho_1, \rho_2 > 0$ such that $B(\hat{x},\rho_1) \times B(\hat{y},\rho_2)$ is contained in $\hat{U}$. Altogether, we see that there exists an $\varepsilon_{\hat{x}} > 0$ such that $g \in C^\infty(\mathbb{R}^n, \mathbb{R}^m)$ and $\max_{x \in B(\hat{x}, \rho_1)} |||Dg(x) - Df(x)||| \le \varepsilon_{\hat{x}}$

($|||\cdot|||$ as in the proof of Theorem 7.3.4) imply:

$$\psi_{g,M}(z) > 0 \text{ for all } z \in M \cap B(\hat{x}, \rho_1) \times B(\hat{y}, \rho_2).$$

To the point $\hat{z}$ we associate the open set $\mathring{B}(\hat{x}, \rho_1) \times \mathring{B}(\hat{y}, \rho_2)$ and the above number $\varepsilon_{\hat{x}}$.

Next, we cover the compact set $K \cap \overline{M}$ by means of the above mentioned associated open sets and extract a finite subcovering, say generated by $\hat{z}_i$, $i = 1, \ldots, r$. Finally, we put $\varepsilon_{\bar{x}} = \min\{1, \varepsilon_{\hat{x}_1}, \ldots, \varepsilon_{\hat{x}_r}\}$. This number $\varepsilon_{\bar{x}}$ is the number we are looking for.

Necessity-Part. Suppose that Condition $\mathcal{O}$ is violated.
Then, there exists a compact subset $K \subset \mathbb{R}^{n+m}$ and an index $j \in J$ with $K \cap M_j \neq \emptyset$ and $\inf_{z \in K \cap M_j} \psi_{f,M_j}(z) = 0$. Consequently, there exists a sequence $(z_i) \subset M_j$, without loss of generality converging to a point $\bar{z} \in K$, such that $\psi_{f,M_j}(z_i) \downarrow 0$. But $\bar{z} \notin M_j$, since the overall-assumption is: $\text{Graph}(f) \pitchfork M_j, j \in J$. Eventually by taking a subsequence of (z_i), it is not difficult to choose $m \times n$ matrices A_i and vectors b_i with $A_i \to 0$, $b_i \to 0$ such that $\psi_{f_i,M_j}(z_i) = 0$, where $f_i(x) = f(x) + A_i(x - x_i) + b_i$ $(z_i = (x_i, y_i))$.
Next, let $\Phi \in C^\infty(\mathbb{R}^n, \mathbb{R})$ be a function with the properties:
$\Phi(x) = 1$, $\|x\| \le \frac{1}{2}$; $\Phi(x) = 0$, $\|x\| \ge 1$.
Finally, put $\tilde{f}_i(x) = f(x) + \Phi(x - \bar{x})[A_i(x - x_i) + b_i]$. Then, $\tilde{f}_i \to f$ in the C^1-Whitney topology and $\text{Graph}(\tilde{f}_i)$ does not meet M_j transversally at $z_i = (x_i, y_i)$ for all i large enough. □

Remark 7.3.13. Let us return to Example 7.3.2. The set S was firstly partitioned into the manifolds M_1 (= x_3-axis) and M_2. For the function $f(x_1, x_2) = x_1^2 + x_2^2$ we have: $\text{Graph}(f) \pitchfork M_j$, $j = 1,2$. However, Condition $\mathcal{O}$ is not satisfied at the origin w.r.t. the manifold M_2.

Suppose, we have a locally finite family of manifolds in $\mathbb{R}^{n+m}$, say $\{M_j, j \in J\}$. Remark 7.3.13 shows that Condition $\mathcal{O}$ need not be satisfied even if the union of the sets M_j is closed and $\text{Graph}(f) \pitchfork M_j$, all j. So, we have to sharpen the contact-conditions between several manifolds. This gives rise to the following two definitions.

Definition 7.3.2. Let S be a subset of $\mathbb{R}^n$. A family Σ whose elements are subsets of S is called a stratification for S if the following four conditions hold:

<partition>
- (Σ1) $\bigcup_{X\in\Sigma} X = S$,
- (Σ2) for $X,Y \in \Sigma$ either $X = Y$ or $X \cap Y = \emptyset$.

<manifold> (Σ3) Every $X \in \Sigma$ is a manifold

<locally finite> (Σ4) The family Σ is locally finite.

The pair (S,Σ) is called a stratified set in $\mathbb{R}^n$. Each $X \in \Sigma$ is called a stratum* of S. The dimension of (S,Σ) is defined to be: $\max_{X\in\Sigma} \dim X$, and its codimension is $n-\dim(S,\Sigma)$. □

Remark 7.3.14. In Definition 3.1.3 we introduced the natural stratification for an MGB and a stratum was assumed to be (path)connected. In Definition 7.3.2 we used the word stratum* in order to indicate that a stratum* is not necessarily connected. This distinction might seem to be overdone at first glance; however, without additional conditions, the partition of a stratified set (S,Σ) by means of the family of connected components of the elements of Σ need not to be locally finite. An illustrative example is the following:

$$S := \{(x_1,x_2) \mid x_2 = \tfrac{1}{n}x_1 \text{ for some } n \in \{1,2,\dots\}\}.$$

Note that S is the union of a countable number of straight lines in $\mathbb{R}^2$ passing through the origin. Put $X_1 = \{0\}$ and $X_2 = S\setminus\{0\}$. Then, $\Sigma := \{X_1,X_2\}$ is a stratification for S, whereas the family of components of the one-dimensional manifold X_2 is not locally finite; see Fig. 7.3.9. (Note that S is not locally closed at the origin).

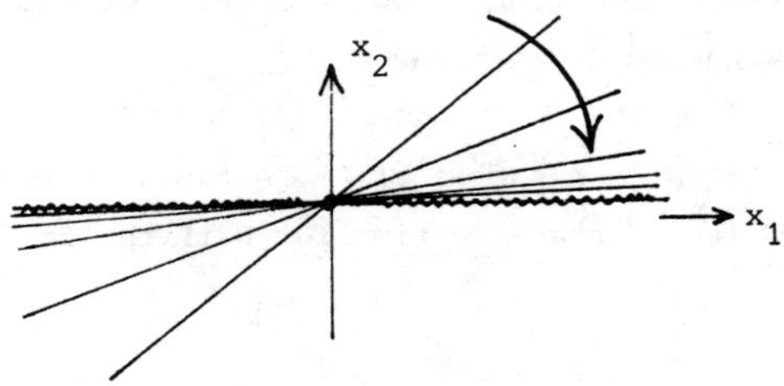

Fig. 7.3.9

Definition 7.3.3. Let (S,Σ) be a stratified subset of $\mathbb{R}^n$. We say that (S,Σ) satisfies Condition $\mathcal{X}$ if the following holds:

$\mathcal{X}1$ For every stratum* X and every point $\bar{x} \in S \cap (\bar{X}\backslash X)$ we have: the stratum* to which $\bar{x}$ belongs has a smaller dimension than dim X.

$\mathcal{X}2$ Let $X,Y \in \Sigma$ and $\bar{x} \in \bar{Y} \cap X$. Let $(y_i) \subset Y$ be any sequence converging to $\bar{x}$ and let T be any accumulation point of the tangent spaces (Ty_i). Then $T_{\bar{x}}X \subset T$.
(Note: Ty_i is considered as a point in the compact Grassmannian $\mathcal{G}(n,\dim Y)$).

Lemma 7.3.4. Let (S,Σ) be a stratified subset of $\mathbb{R}^{n+m}$ and $f \in C^\infty(\mathbb{R}^n,\mathbb{R}^m)$. Suppose, in addition, that S is closed and that S satisfies Condition $\mathcal{X}$. Then, Condition $\mathcal{O}$ is implied by the property Graph(f) $\pitchfork$ X, $X \in \Sigma$, recalling:

Condition $\mathcal{O}$: For every $X \in \Sigma$ and compact $K \subset \mathbb{R}^{n+m}$ with

$$K \cap X \neq \emptyset : \inf_{z \in K \cap X} \phi_{f,X}(z) > 0.$$

Proof. In view of Lemma 7.3.3 we may replace the function $\phi_{f,X}$ by means of $\psi_{f,X}$ (cf. (7.3.44)). Now, suppose that Graph(f) $\pitchfork$ X, $X \in \Sigma$, and that Condition $\mathcal{O}$ is violated. Then there is an $X \in \Sigma$, a compact subset $K \subset \mathbb{R}^{n+m}$ with $K \cap X \neq \emptyset$ and a sequence $(x_i) \subset X$ converging to a point $\bar{x} \in K\backslash X$ with $\psi_{f,X}(x_i) \downarrow 0$. The point $\bar{x}$ belongs to S since S is closed. Hence, $\bar{x} \in Y$, some $Y \in \Sigma$. Since each Grassmannian is compact, we may assume that the tangent spaces $T_{x_i}X$ converge to T. From Condition $\mathcal{X}2$ we have

$T \supset T_{\bar{x}}Y$. Since Graph(f) $\bar{\pitchfork}$ Y, we see that $T_{\bar{x}}$ Graph(f) + $T_{\bar{x}}Y = \mathbb{R}^{n+m}$. Hence, $T_{\bar{x}}$ Graph(f) + T = $\mathbb{R}^{n+m}$. But the latter equality contradicts the fact that $\psi_{f,X}(x_i) \nrightarrow 0$. □

Example 7.3.6. Let $S \subset \mathbb{R}^n$ be an MGB of class C^∞ and Σ its stratification according to Definition 3.1.3 (so each stratum* is connected in this case). Then, (S,Σ) satisfies Condition X. The proof is obvious by considering S in local coordinates.

Corollary 7.3.2. Let $S \subset \mathbb{R}^{n+m}$ be an MGB and Σ its stratification according to Definition 3.1.3. Denote by F the set of all $f \in C^\infty(\mathbb{R}^n,\mathbb{R}^m)$ with Graph(f) $\bar{\pitchfork}$ X, $X \in \Sigma$.

Then, F is C^k-dense for all k; moreover, if S is closed as a subset of $\mathbb{R}^{n+m}$, then F is C^k-open for all $k \geq 1$. □

In Section 7.5 we will return to the subject of stratified sets.

7.4. Jet-Extension, Jet-Transversality.

Throughout Section 7.4, the word manifold will again refer to C^∞-manifold.

Let $f \in C^\infty(\mathbb{R}^n,\mathbb{R}^m)$, $f(x) = (f_1(x),\ldots,f_m(x))^T$. In Section 7.3 we introduced the concept Graph(f) (cf. (7.3.1)). Analogously, we may take the partial derivatives up to order ℓ into account, thus obtaining an "extended graph". Here, we have to pay attention to the fact that the order of forming partial derivatives is irrelevant; for example, in case m = 1, $\frac{\partial^2 f}{\partial x_i \partial x_j} = \frac{\partial^2 f}{\partial x_j \partial x_i}$. Thus, there is a natural symmetry w.r.t. partial differentiation. Therefore, we obtain the actual needed amount of information about f_i and its partial derivatives up to order ℓ, if we cancel out the order of forming partial derivatives.
Given $n \geq 1$, $m \geq 1$, $\ell \geq 0$, we make a list of all partial derivatives of the functions f_i, say by means of the scheme in (7.4.1), thereby cancelling symmetry. Consider the string obtained by writing the rows of (7.4.1) successively after each other; we call this the "order-convention". (Note that this particular order has no intrinsic meaning; in fact, any other order would be fine).

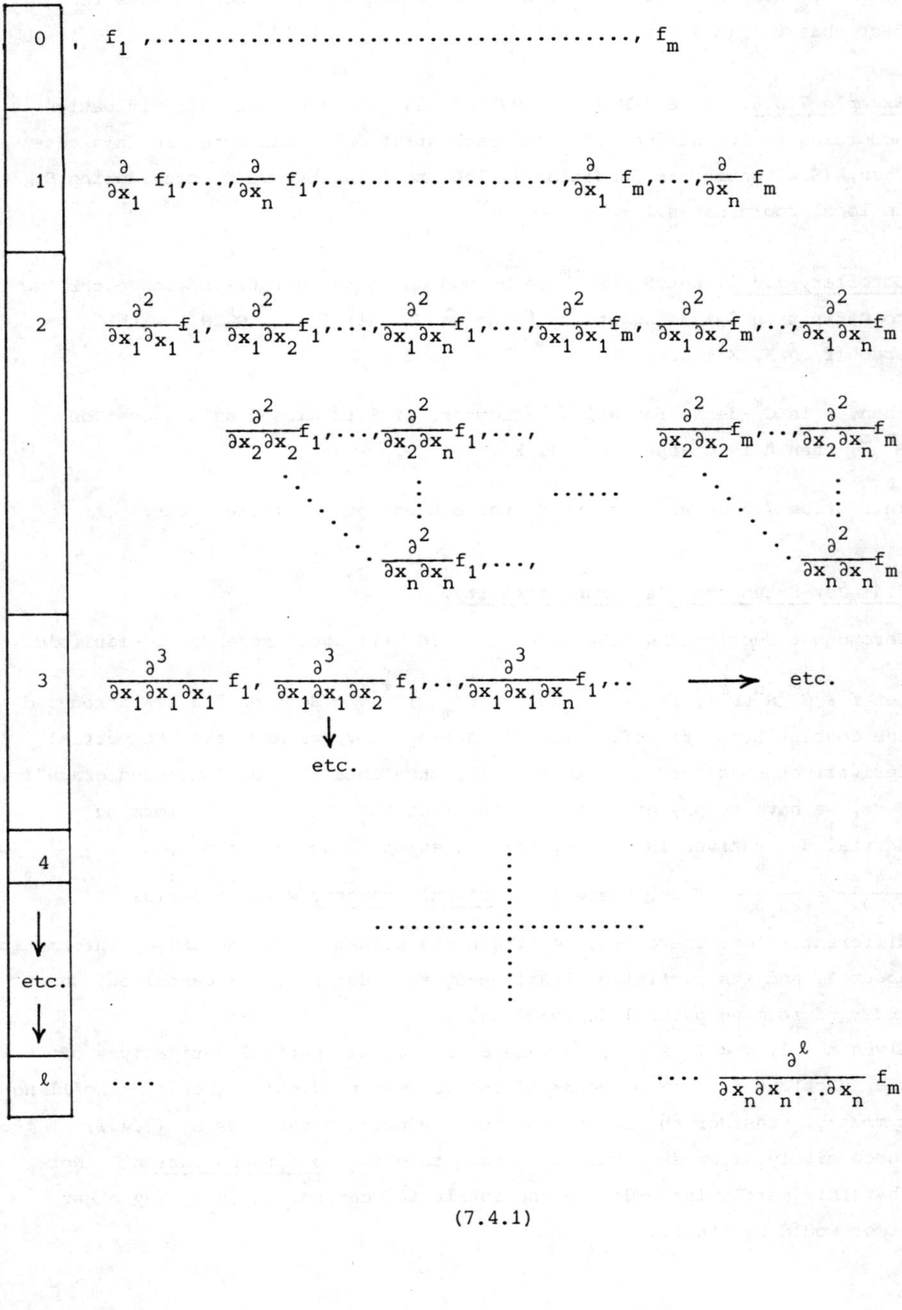
0
$f_1, \ldots, f_m$
1
$\frac{\partial}{\partial x_1} f_1, \ldots, \frac{\partial}{\partial x_n} f_1, \ldots, \frac{\partial}{\partial x_1} f_m, \ldots, \frac{\partial}{\partial x_n} f_m$
2
$\frac{\partial^2}{\partial x_1 \partial x_1} f_1, \frac{\partial^2}{\partial x_1 \partial x_2} f_1, \ldots, \frac{\partial^2}{\partial x_1 \partial x_n} f_1, \ldots, \frac{\partial^2}{\partial x_1 \partial x_1} f_m, \frac{\partial^2}{\partial x_1 \partial x_2} f_m, \ldots, \frac{\partial^2}{\partial x_1 \partial x_n} f_m$
$\frac{\partial^2}{\partial x_2 \partial x_2} f_1, \ldots, \frac{\partial^2}{\partial x_2 \partial x_n} f_1, \ldots, \frac{\partial^2}{\partial x_2 \partial x_2} f_m, \ldots, \frac{\partial^2}{\partial x_2 \partial x_n} f_m$
$\frac{\partial^2}{\partial x_n \partial x_n} f_1, \ldots, \frac{\partial^2}{\partial x_n \partial x_n} f_m$
3
$\frac{\partial^3}{\partial x_1 \partial x_1 \partial x_1} f_1, \frac{\partial^3}{\partial x_1 \partial x_1 \partial x_2} f_1, \ldots, \frac{\partial^3}{\partial x_1 \partial x_1 \partial x_n} f_1, \ldots$
etc.
etc.
4
etc.
ℓ
$\ldots \frac{\partial^\ell}{\partial x_n \partial x_n \ldots \partial x_n} f_m$

(7.4.1)

Definition 7.4.1. Let $f \in C^\infty(\mathbb{R}^n,\mathbb{R}^m)$. The ℓ-jet-extension $j^\ell f$ of f is defined to be the mapping:

$$j^\ell f\colon x \mapsto [x, f_1(x),..,f_m(x), \frac{\partial}{\partial x_1} f_1(x),..,\frac{\partial}{\partial x_n} f_m(x),...,\frac{\partial^\ell}{\partial x_n \partial x_n .. \partial x_n} f_m(x)], \tag{7.4.2}$$

where the partial derivatives are listed according to the "order-convention". Let $N(n,m,\ell)$ denote the length of the vector $j^\ell f(x)$ in (7.4.2).
Then, $j^\ell\colon \mathbb{R}^n \to \mathbb{R}^{N(n,m,\ell)}$.
The jet-space $J(n,m,\ell)$ is defined to be the space $\mathbb{R}^{N(n,m,\ell)}$.

Example 7.4.1. Put m = 1 and consider $j^2 f$ for an $f \in C^\infty(\mathbb{R}^n,\mathbb{R})$.

$$j^2 f(x) = [\underset{\underset{n}{\uparrow}}{x}, \underset{\underset{1}{\uparrow}}{f(x)}, \underbrace{\frac{\partial}{\partial x_1}f(x),..,\frac{\partial}{\partial x_n}f(x)}_{n}, \underbrace{\frac{\partial^2}{\partial x_1 \partial x_1}f(x), \frac{\partial^2}{\partial x_1 \partial x_2}f(x),..,\frac{\partial^2}{\partial x_n \partial x_n}f(x)}_{\frac{1}{2}n(n+1)}]$$

It follows that $\dim. J(n,1,2) = N(n,1,2) = 2n + \frac{1}{2}n(n+1) + 1$.

Example 7.4.2. For $f \in C^\infty(\mathbb{R}^n,\mathbb{R}^m)$ we have $j^o f(x) = (x, f(x))$.
It follows, that $\dim. J(n,m,o) = N(n,m,o) = n+m$.

Remark 7.4.1. Note that $j^\ell f[\mathbb{R}^n]$, where $f \in C^\infty(\mathbb{R}^n,\mathbb{R}^m)$, is an n-dimensional manifold in $J(n,m,\ell)$ (compare the mapping G_f in (7.3.2)). In particular, $j^o f[\mathbb{R}^n] = \text{Graph}(f)$, so $j^\ell f[\mathbb{R}^n]$ is an "extended graph".

The following two theorems generalize Theorems 7.3.4, 7.3.5.

Theorem 7.4.1. (Jet-Transversality Theorem).
Let $n \geq 1$, $m \geq 1$, $\ell \geq 0$ be given. Let $M \subset J(n,m,\ell)$ be a manifold and put $\pitchfork^\ell M = \{f \in C^\infty(\mathbb{R}^n,\mathbb{R}^m) \mid j^\ell f \pitchfork M\}$.
Then, $\pitchfork^\ell M$ is C^k-dense for all k; moreover, if M is closed as a subset of $J(n,m,\ell)$, then $\pitchfork^\ell M$ is C^k-open for all $k \geq \ell+1$. □

Theorem 7.4.2. Let $n \geq 1$, $m \geq 1$, $\ell \geq 0$ be given. Let $N \subset \mathbb{R}^n$ be a manifold, $M \subset \mathbb{R}^{N(n,m,\ell)-n}$ a manifold and put $\mathcal{M} = \mathbb{R}^n \times M$. Moreover, let $\pitchfork^\ell \mathcal{M} = \{f \in C^\infty(\mathbb{R}^n,\mathbb{R}^m) \mid j^\ell f \pitchfork \mathcal{M}\}$ and define $\mathcal{F} \subset \pitchfork^\ell \mathcal{M}$ as follows: an element $f \in \pitchfork^\ell \mathcal{M}$ lies in $\mathcal{F}$ iff $(j^\ell f)^{-1}(\mathcal{M}) \pitchfork N$.
Then, $\mathcal{F}$ is C^k-dense in $C^\infty(\mathbb{R}^n,\mathbb{R}^m)$ for all k; moreover, if both N is closed as a subset of $\mathbb{R}^n$ and M is closed as a subset of $\mathbb{R}^{N(n,m,\ell)-n}$, then $\mathcal{F}$ is C^k-open for all $k \geq \ell+1$.

Proof of Theorem 7.4.1 and Theorem 7.4.2.
Put $j^\ell f(x) = (x,F(x))$, $p = N(n,m,\ell)-n$. Then $F \in C^\infty(\mathbb{R}^n,\mathbb{R}^p)$ and we say that F is induced by f. Note that $j^\ell(f) \pitchfork \mathcal{M}$ iff Graph(F) $\pitchfork$ $\mathcal{M}$. From this and Theorems 7.3.4, 7.3.5 the constraint $k \geq \ell+1$ in the open part of Theorems 7.4.1, 7.4.2 becomes obvious, since the entries in DF are partial derivatives of f up to order $\ell+1$.
The open part itself of Theorems 7.4.1, 7.4.2 follows also immediately from Theorems 7.3.4, 7.3.5. In fact, note that the induced mapping F is just a special element of the function space $C^\infty(\mathbb{R}^n,\mathbb{R}^p)$ and that the set $(j^\ell f)^{-1}(\mathcal{M})$ in Theorem 7.4.2 is nothing else but the set $F^{-1}(M)$.
Now we treat the dense-part. In view of the proof of Theorems 7.3.4, 7.3.5 it is obvious that we merely have to develop a suitable framework in which Sard's Theorem is applicable.
Let us denote by $\mathcal{A}$ the linear subspace of $C^\infty(\mathbb{R}^n,\mathbb{R}^p)$ consisting of all mappings $F \in C^\infty(\mathbb{R}^n,\mathbb{R}^p)$ which are induced by an $f \in C^\infty(\mathbb{R}^n,\mathbb{R}^m)$. Let $\bar{x} \in \mathbb{R}^n$. It suffices to show that we can suitably parametrize a whole $J(n,m,\ell)$-neighborhood of $j^\ell f(\bar{x})$ with manifolds of the type Graph(F), $F \in \mathcal{A}$.
It will be clear that we may assume, without loss of generality, that $\bar{x} = 0$.
Let us write a vector $c \in \mathbb{R}^p$ as follows:

$$c = (c^0_1,\ldots,c^0_m,c^1_{1,1},\ldots,c^1_{n,1},c^1_{1,2},\ldots,c^1_{n,2},\ldots,c^1_{n,m},\ldots,c^\ell_{\underbrace{n,n,\ldots,n}_{\ell},m})^T. \tag{7.4.3}$$

The ordering of the components of c in (7.4.3) corresponds to the "order convention" (cf. (7.4.1)).
As an example, $c^3_{i,j,k,r}$ corresponds to $\frac{\partial^3}{\partial x_i \partial x_j \partial x_k} f_r$.

By P_c^i we mean the following <u>polynomial</u> in n variables:

$$P_c^i(x) = c_i^0 + c_{1,i}^1 x_1 + .. + c_{n,i}^1 x_n + c_{1,1,i}^2 x_1 x_1 + c_{1,2,i}^2 x_1 x_2 + \ldots$$

$$\ldots + c_{n,n,..,n,i}^{\ell} x_n x_n .. x_n \tag{7.4.4}$$

Next we put

$$g_c = (f_1 + P_c^1, f_2 + P_c^2, \ldots, f_m + P_c^m)^T := (g_c^1, g_c^2, \ldots, g_c^m)^T. \tag{7.4.5}$$

Let G_c denote the mapping induced by g_c, i.e.

$$j^{\ell} g_c(x) = (x, G_c(x)). \tag{7.4.6}$$

Note that $G_0 = F$. The mappings $G_c \in A$, $c \in \mathbb{R}^p$, will serve as a p-parameter family of functions in order to "move" Graph(F) locally around $j^{\ell}f(0)$ as c varies in a neighborhood of the origin $o \in \mathbb{R}^p$.
It remains to show that there exist an open $J(n,m,\ell)$-neighborhood $\mathcal{O}$ of $j^{\ell}f(0)$ and a mapping $\psi \in C^{\infty}(\mathcal{O},\mathbb{R}^p)$ such that for every $z \in \mathcal{O}$, $z = (x,y)$, we have: $j^{\ell}g_{\psi(z)}(x) = z$.
To this aim we make the Implicit Function Theorem applicable. Let the functions $h_1, \ldots, h_p \in C^{\infty}(\mathbb{R}^n \times \mathbb{R}^p \times \mathbb{R}^p, \mathbb{R})$ be defined as follows (taking the "order-convention" into account):

$$\left.\begin{aligned}
h_1(x,y,c) &= y_1 - g_c^1(x) \quad , \quad \underline{x \in \mathbb{R}^n \text{ and } y,c \in \mathbb{R}^p,}\\
h_2(x,y,c) &= y_2 - g_c^2(x) \quad ,\\
&\vdots\\
h_m(x,y,c) &= y_m - g_c^m(x) \quad ,\\
h_{m+1}(x,y,c) &= y_{m+1} - \frac{\partial}{\partial x_1} g_c^1(x),\\
&\vdots\\
h_p(x,y,c) &= y_p - \frac{\partial^{\ell}}{\partial x_n .. \partial x_n} g_c^m(x).
\end{aligned}\right\} \tag{7.4.7}$$

As an abbreviation we put $h = (h_1, \ldots, h_p)^T$. Obviously, for fixed c, the manifold Graph(G_c) is represented by the equation $h(x,y,c) = 0$.
In particular, Graph(F) is given by $h(x,y,0)$ and thus $h(j^{\ell}f(0),0) = 0$.
Note that $D_c h(j^{\ell}f(0),0) = \text{diag}(\alpha_1, .., \alpha_p)$, where α_i is a negative integer, $i = 1,..,p$.

Thus, $D_c h(j^\ell f(0),0)$ is nonsingular. Consequently, by the Implicit Function Theorem, there exists an open $J(n,m,\ell)$-neighborhood $\mathcal{O}$ of $j^\ell f(0)$ and a mapping $\psi \in C^\infty(\mathcal{O},\mathbb{R}^p)$ such that

$$h(z,\psi(z)) \equiv 0. \tag{7.4.8}$$

(In view of $D\psi(0) = -(D_c h(j^\ell f(0),0))^{-1} D_z h(j^\ell f(0),0)$, $z = (x,y)$, the matrix $D\psi(0)$ has rank p, since $D_z h(j^\ell f(0),0)$ has rank p (cf. (7.4.7)). Therefore, ψ takes all values in a neighborhood of $0 \in \mathbb{R}^p$ as z varies in a $J(n,m,\ell)$-neighborhood of $j^\ell f(0)$).

By construction, we obtain from (7.4.8) that $z \in \text{Graph}(G_{\psi(z)})$, i.e. $j^\ell g_{\psi(z)}(x) = z$. The mapping ψ gives us exactly the parameter-distribution which we need in order to apply Sard's Theorem. □

Remark 7.4.2. In the same spirit as Remarks 7.3.9, 7.3.11, we have the following slight generalization of Theorems 7.4.1, 7.4.2.
Let $C^\infty(\mathbb{R}^n,\mathbb{R}^m)$ be endowed with the C^∞-topology.
Generalization of Theorem 7.4.1: Let $\{M_j, j \in J\}$ be a countable family of manifolds in $\mathbb{R}^{n+m}$. Then, $\bigcap_{j\in J} \pitchfork^\ell M_j$ is a generic subset of $C^\infty(\mathbb{R}^n,\mathbb{R}^m)$.
Generalization of Theorem 7.4.2: Let $\{M_j, j \in J\}$, resp. $\{N_i, i \in I\}$ be a countable family of manifolds in $\mathbb{R}^m$, resp. $\mathbb{R}^n$ and put $\mathcal{M}_j = \mathbb{R}^n \times M_j$, $j \in J$. Define the subset F of $C^\infty(\mathbb{R}^n,\mathbb{R}^m)$ as follows:

$$F = \{f \in \bigcap_{j\in J} \pitchfork^\ell \mathcal{M}_j \mid (j^\ell f)^{-1}(\mathcal{M}_j) \pitchfork N_i \quad , \quad \text{all } i \in I, j \in J\}.$$

Then, F is a generic subset.

Example 7.4.3. Let $F \subset C^\infty(\mathbb{R}^n,\mathbb{R})$ denote the set of all nondegenerate functions. Then, F is C^k-dense for all k and C^k-open for $k \geq 2$ (compare Theorem 7.1.3). This result follows easily from Theorem 7.4.1. In fact, consider $j^1 f$:

$$j^1 f(x) = [x, f(x), \frac{\partial}{\partial x_1} f(x), \ldots, \frac{\partial}{\partial x_n} f(x)].$$

Define $M \subset J(n,1,1)$ to be the manifold $\mathbb{R}^n \times \mathbb{R} \times \{0\}$, $0 \in \mathbb{R}^n$.
Then, $j^1 f(\bar{x}) \in M$ iff $\frac{\partial}{\partial x_i} f(\bar{x}) = 0$, $i = 1,\ldots,n$, i.e. iff $\bar{x}$ is a critical point for f.

Suppose that $j^1f \pitchfork M$ and that $j^1f(\bar{x}) \in M$. Then, we have:

$$Dj^1f(\bar{x})[\mathbb{R}^n] + T_{j^1f(\bar{x})}M = J(n,1,1) \; . \qquad (7.4.9)$$

Note that $\dim Dj^1f(\bar{x})[\mathbb{R}^n] = n = \operatorname{codim} M$. Formula (7.4.9) holds iff the $(2n+1)\times(2n+1)$ matrix in (7.4.10) is nonsingular.

$$\left(\begin{array}{c|c} \begin{array}{c} I_n \\ \hline Df(\bar{x}) \end{array} & I_{n+1} \\ \hline D^2f(\bar{x}) & 0 \end{array}\right), \quad I_k = k\times k \text{ identity matrix} \qquad (7.4.10)$$

$$\underbrace{\qquad\qquad\qquad}_{\substack{\text{column span}\\ Dj^1f(\bar{x})[\mathbb{R}^n]}}\qquad\underbrace{\qquad\qquad\qquad}_{\substack{\text{columns span}\\ T_{j^1f(\bar{x})}M}}$$

Obviously, the matrix in (7.4.10) is nonsingular iff $D^2f(\bar{x})$ is nonsingular (i.e. $\bar{x}$ is a nondegenerate critical point for f).
Consequently, $f \in F$ iff $j^1f \pitchfork M$.
If $j^1f \pitchfork M$ and $(j^1f)^{-1}(M) \neq \emptyset$, then $(j^1f)^{-1}(M)$, being exactly the set of critical points for f, is a manifold in $\mathbb{R}^n$ having the same codimension as the codimension of M in $J(n,1,1)$. Since $\operatorname{codim} M = n$, we have $\operatorname{codim}(j^1f)^{-1}(M) = n$ and thus $\dim(j^1f)^{-1}(M) = 0$, i.e. nondegenerate critical points are isolated critical points.

The following lemma shows how in-essential information can be split off in transversality considerations. Its application to jet-transversality will be discussed in Remark 7.4.3.

<u>Lemma 7.4.1</u>. Let $F \in C^\infty(\mathbb{R}^n,\mathbb{R}^k)$ and $M \subset \mathbb{R}^k$ a manifold. Suppose that $\mathbb{R}^k = \mathbb{R}^p \times \mathbb{R}^q$ and that $M = \tilde{M} \times \mathbb{R}^q$, where $\tilde{M}$ is a manifold in $\mathbb{R}^p$.
Let $\Pi\colon \mathbb{R}^k \to \mathbb{R}^p$ be the projection $(\xi,\eta) \mapsto \xi$ and define $\tilde{F} = \Pi\circ F$. Then, we have:

$$F \pitchfork M \quad \text{iff} \quad \tilde{F} \pitchfork \tilde{M}.$$

Proof. The proof is an easy exercise, using the following observations: $F(x) \in M$ iff $\tilde{F}(x) \in \tilde{M}$; $D\tilde{F} = D(\Pi \circ F) = \Pi \cdot DF$; $T_{F(x)}M = \{0\} \times \mathbb{R}^q + T_{\tilde{F}(x)}\tilde{M} \times \{0\}$ and hence, $T_{\tilde{F}(x)}\tilde{M} = \Pi[T_{F(x)}M]$. □

Remark 7.4.3. The idea of Lemma 7.4.1 can be translated in terms of jet-transversality by means of the substitution:

$$F = j^{\ell} f, \text{ where } f \in C^{\infty}(\mathbb{R}^n, \mathbb{R}^m) \ ; \ \mathbb{R}^k = J(n,m,\ell).$$

The mapping $\tilde{F}$ in Lemma 7.4.1 now becomes a "reduced" jet-extension, say $\widetilde{j^{\ell}}f$. The Jet-Transversality Theorem (Theorem 7.4.1) then remains valid in the obviously formulated "reduced" form.

As a simple example let us consider Example 7.4.3 from this new point of view. The manifold $M = \mathbb{R}^n \times \mathbb{R} \times \{0\}$ can be interpreted, in relation with the jet-extension $j^1 f(x) = [x, f(x), Df(x)]$, as follows: it does not matter at which place a critical point appears ("$\mathbb{R}^n$") and it does not matter what the actual functional value at the critical point is ("$\mathbb{R}$"). So, if we are only interested in the set of nondegenerate functions without further specification, then the "$\mathbb{R}^n \times \mathbb{R}$"-factor in M is superfluous. Therefore, we split $J(n,1,1)$ as follows:

$$J(n,1,1) = (\mathbb{R}^n \times \mathbb{R}) \times (\mathbb{R}^n)$$

The reduced 1-jet-extension $\widetilde{j^1}f\colon \mathbb{R}^n \to \mathbb{R}^n$ then becomes the mapping $x \mapsto Df(x)$, whereas the manifold $\tilde{M}$ is the 0-dimensional manifold $\{0\} \subset \mathbb{R}^n$. In view of Lemma 7.4.1 we see that $j^1 f \pitchfork M$ iff $\widetilde{j^1}f \pitchfork \tilde{M}$. Note that the condition $\widetilde{j^1}f \pitchfork \tilde{M}$ is nothing else but the condition $Df \pitchfork \{0\}$; so, we reduced the $(2n+1)\times(2n+1)$ matrix in (7.4.10) to the essential lower left part $(D^2 f)$.

Example 7.4.4. Let $N \subset \mathbb{R}^n$ be a manifold of dimension $k < n$.

Problem: Given an $f \in C^{\infty}(\mathbb{R}^n, \mathbb{R})$. Can we approximate f arbitrarily well in the C^k-sense by an $\tilde{f} \in C^{\infty}(\mathbb{R}^n, \mathbb{R})$ such that: $\tilde{f}$ is nondegenerate and no critical points of $\tilde{f}$ lie on N ?

Solution: Consider the manifold $M \subset J(n,1,1)$ defined by $M = N \times \mathbb{R} \times \{0\}$, $0 \in \mathbb{R}^n$ and apply Theorem 7.4.1. In fact, suppose that $g \in C^{\infty}(\mathbb{R}^n, \mathbb{R})$ has the property that $j^1 g \pitchfork M$. Then $(j^1 g)^{-1}(M)$ is a manifold in $\mathbb{R}^n$ having the same codimension as M has. Thus, in order that $(j^1 g)^{-1}(M) \neq \emptyset$ we must have codim(M) $\leq$ n. However, codim(M) = n + codim N = n+(n-k) > n, since k < n.

Remark: Note that we cannot avoid that $f_{|N}$ has critical points. In fact, if N is compact, then $f_{|N}$ attains (by the Weierstrass-Theorem) its minimum and maximum on N.

The Jet-Transversality Theorem deals with the set of functions $f \in C^\infty(\mathbb{R}^n,\mathbb{R}^m)$ for which $j^\ell f$ meets a given manifold $M \subset J(n,m,\ell)$ transversally. We emphasize that the condition "$j^\ell f \pitchfork M$" is a condition which gives information about the relation between f and M w.r.t. every single point $x \in \mathbb{R}^n$. We obtain a generalization of the jet-transversality idea if we study a certain regular behaviour of f with respect to every subset of p points $\{x^1,\dots,x^p\} \subset \mathbb{R}^n$. For example, consider the set F_s of all $f \in C^\infty(\mathbb{R}^n,\mathbb{R})$ having the property: distinct critical points of f have distinct functional values. Note that the fact "$f \in F_s$" is a condition on every subset of two distinct points in $\mathbb{R}^n$. Let $u,v \in \mathbb{R}^n$ be distinct points. Then, the set $\{u,v\}$ can be represented by the points (u,v) and (v,u) in $\mathbb{R}^n \times \mathbb{R}^n \setminus \Delta$, where Δ is the diagonal set $\{(x,x) \mid x \in \mathbb{R}^n\}$. This gives rise to the following definition.

Definition 7.4.2. Let $n \geq 1$, $m \geq 1$, $p > 1$, $\ell \geq 0$ be given.

Define:

$$\mathbb{R}^n_p = \{(x^1,\dots,x^p) \in \mathop{\times}_{i=1}^{p} \mathbb{R}^n \mid x^i \neq x^j,\ 1 \leq i < j \leq p\}. \tag{7.4.11}$$

The set $\mathop{\times}_{i=1}^{p} \mathbb{R}^n \setminus \mathbb{R}^n_p$ is called the generalized diagonal.

Let Π: $J(n,m,\ell) \to \mathbb{R}^n$ be the projection on the first $\mathbb{R}^n$-factor, $\Pi(x,u) = x$.

Analogously, let Π_p: $\mathop{\times}_{i=1}^{p} J(n,m,\ell) \to \mathop{\times}_{i=1}^{p} \mathbb{R}^n$ be the induced product-projection.

The multi-jet space $J_p(n,m,\ell)$ is defined to be $(\Pi_p)^{-1}[\mathbb{R}^n_p]$.

Finally, the multi-jet extension $j^\ell_p(f)$ is the following natural mapping:

$$\left.\begin{aligned} &j^\ell_p f : \mathbb{R}^n_p \to J_p(n,m,\ell) \\ &j^\ell_p f(x^1,\dots,x^p) = [j^\ell f(x^1),\dots,j^\ell f(x^p)] \end{aligned}\right\} \tag{7.4.12}$$

□

Remark 7.4.4. Note that $\mathbb{R}^n_p$ is an open (and dense) subset of $\overset{p}{\underset{i=1}{X}} \mathbb{R}^n$; further, $J_p(n,m,\ell)$ is an open (and dense) subset of $\overset{p}{\underset{i=1}{X}} J(n,m,\ell)$. Hence, the concepts "$M$ is a manifold in $J_p(n,m,\ell)$" and "$j^\ell_p f \pitchfork M$" are obviously defined.

Theorem 7.4.3. (Multi-jet Transversality Theorem). Let M be a manifold in $J_p(n,m,\ell)$ and endow $C^\infty(\mathbb{R}^n,\mathbb{R}^m)$ with the C^∞-topology. Put:

$$F = \{f \in C^\infty(\mathbb{R}^n,\mathbb{R}^m) \mid j^\ell_p f \pitchfork M\}.$$

Then, the set F is generic. □

We will not prove Theorem 7.4.3 here in detail (cf. e.g. [ ]). The proof depends on the following observation. Let $(x^1,\dots,x^p) \in \mathbb{R}^n_p$. Then, there exist pairwise disjoint balls $B(x^i,\rho_i) \subset \mathbb{R}^n$, $i = 1,\dots,p$ ($B(x^i,\rho_i)$ being the ball with center x^i and radius ρ_i). Now, we can perturb f on the balls $B(x^i,\frac{1}{2}\rho_i)$ simultaneously and independently, thereby keeping f unchanged outside $\bigcup_{i=1}^{p} B(x^i,\rho_i)$. The question whether the set F in Theorem 7.4.3 is open, is more delicate and we will not go into these details; we just mention (and this is easily shown) that F is open if $M \subset J_p(n,m,\ell)$ is compact (cf. also [18]).

Example 7.4.5. Let F_s be the set of all $f \in C^\infty(\mathbb{R}^n,\mathbb{R})$ having the property: distinct critical points of f have distinct functional values.
Contention: the set F_s is generic (w.r.t. C^∞-topology).

We prove the contention by using the Multi-jet Transversality Theorem. With $j^1_2 f(x^1,x^2) = [x^1,f(x^1),Df(x^1), x^2,f(x^2),Df(x^2)]$ we see that $f \in F_s$ iff $j^1_2 f[\mathbb{R}^n_2] \cap M = \emptyset$, where M is the following manifold in $J_2(n,1,1)$:

$$M = \left\{ \begin{array}{c} [\mathbb{R}^n \times \underset{\alpha}{\underset{|}{\mathbb{R}}} \times \{0\} \times \mathbb{R}^n \times \underset{\beta}{\underset{|}{\mathbb{R}}} \times \{0\}] \cap J_2(n,1,1) \\ \alpha = \beta \ \underline{\text{additional constraint}} \end{array} \right\} \quad (0 \in \mathbb{R}^n)$$

In fact, $f \in F_s$ iff for $x^1 \neq x^2$ the set of equations $Df(x^1) = 0$, $Df(x^2) = 0$, $f(x^1) = f(x^2)$ is <u>not simultaneously</u> fulfilled.

Note that $\dim j_2^1 f[\mathbb{R}_2^n] = 2n$ and $\operatorname{codim} M = 2n+1$. But then, $j_2^1 f \pitchfork M$ iff $j_2^1 f[\mathbb{R}_2^n] \cap M = \emptyset$. Application of Theorem 7.4.3 yields the desired result.

<u>Note 1</u>. We emphasize that the representation of the set of two distinct points $\{u,v\} \subset \mathbb{R}^n$ by means of two points in $\mathbb{R}_2^n$, namely (u,v) and (v,u), does not cause any troubles.

<u>Note 2</u>. The set F_s is not open (cf. also Remark 7.3.9).

<u>Exercise 7.4.1</u>. Let $z \in \mathbb{R}^{n+1}$ be partitioned as $z = (x,y)$, $x \in \mathbb{R}^n$, $y \in \mathbb{R}$. Define F as follows:

$$F = \left\{ f \in C^\infty(\mathbb{R}^{n+1},\mathbb{R}) \;\middle|\; \begin{array}{l} \underline{\text{for every}}\ \bar{x} \in \mathbb{R}^n \text{ we have: the equations} \\ f(\bar{x},y) = 0,\ \frac{\partial}{\partial y} f(\bar{x},y) = 0 \text{ are simultaneously} \\ \text{fulfilled for at most } n \text{ points } y. \end{array} \right\}$$

Endow $C^\infty(\mathbb{R}^{n+1},\mathbb{R})$ with the C^∞-topology and show, using the Multi-jet Transversality Theorem, that F is generic.

<u>Hint</u>: Take $p = n+1$ and exploit the additional constraints $x^1 = x^2, \ldots, x^{p-1} = x^p$. □

Loosely speaking, up to now we considered a certain manifold M and we showed that <u>generically</u> a function (or its jet-extension) meets this manifold transversally; generic, here, is related to the space of <u>all</u> smooth functions. This means: if we restrict ourselves to a subfamily of functions, e.g. described by a finite number of parameters, then it need not be true that for almost all parameters the corresponding function belongs to the desired generic class. In fact, the whole subfamily might belong to the complement of the generic class. So, we have to impose an additional condition on the <u>family as a whole</u> in order that for many parameters the corresponding function belongs to the desired class. This leads to the following theorem.

<u>Theorem 7.4.4.</u> (<u>Parametric Transversality Theorem</u>).
Let $\mathbb{R}^n = \mathbb{R}^p \times \mathbb{R}^q$ and partition $x \in \mathbb{R}^n$ as $x = (u,v)$, $u \in \mathbb{R}^p$, $v \in \mathbb{R}^q$.
The role of the parameter is played by u.
Let $f \in C^\infty(\mathbb{R}^n,\mathbb{R}^m)$ and $M \subset \mathbb{R}^m$ a manifold.
If $f \pitchfork M$, then $f(u,\cdot) \pitchfork M$ for <u>almost all u</u>.

<u>Proof</u>. If $f^{-1}(M) = \emptyset$, there is nothing to prove. So, suppose $f^{-1}(M) \neq \emptyset$. From Theorem 7.3.1 we know that $f^{-1}(M)$ is a manifold and, moreover, for $\bar{x} \in f^{-1}(M)$ we have the formula:

$$T_{\bar{x}}f^{-1}(M) = Df(\bar{x})^{-1}T_{f(\bar{x})}M. \tag{7.4.13}$$

Note that $f(\bar{u},\cdot) \pitchfork M$ iff for all v with $f(\bar{u},v) \in M$:

$$D_vf(\bar{u},v)[\mathbb{R}^q] + T_{f(\bar{u},v)}M = \mathbb{R}^m. \tag{7.4.14}$$

Obviously, we have:

$$D_vf(\bar{u},v)[\mathbb{R}^q] = Df(\bar{u},v)[\{0\} \times \mathbb{R}^q]. \tag{7.4.15}$$

Since $\mathbb{R}^n = Df(\bar{u},v)^{-1}[\mathbb{R}^m]$, it follows from (7.4.14), taking (7.4.13) and (7.4.15) into account, that

$$\{0\} \times \mathbb{R}^q + T_{(\bar{u},v)}f^{-1}(M) = \mathbb{R}^n. \tag{7.4.16}$$

Since (7.4.16) holds for all v with $f(\bar{u},v) \in M$, it follows:

$$\{\bar{u}\} \times \mathbb{R}^q \pitchfork f^{-1}(M). \tag{7.4.17}$$

On the other hand, (7.4.17) obviously implies (7.4.14) for all v with $f(\bar{u},v) \in M$. Further, (7.4.17) is equivalent with the fact that $\bar{u}$ is a regular value for $\Pi_{|f^{-1}(M)}$, where $\Pi: \mathbb{R}^p \times \mathbb{R}^q \to \mathbb{R}^p$ is the projection $\Pi(u,v) = u$. Finally, the set $\{u | u \text{ is a regular value for } \Pi_{|f^{-1}(M)}\}$ has full measure in view of Sard's Theorem on manifolds (cf. Theorem 7.1.1 and Remark 7.1.5). This proves the theorem. □

Finally, let us return once again to Example 7.4.3. We may look at this example from two points of view.

<u>Point of view 1</u>. (ℓ-jet-extension; $\ell = 1$). We <u>cannot avoid</u> that a function $f \in C^\infty(\mathbb{R}^n,\mathbb{R})$ has critical points, i.e. that $Df(x)$ meets 0 at some points in $\mathbb{R}^n$. However, if $Df(x)$ meets 0, then - in general - $Df(x)$ will meet 0 transversally.

<u>Point of view 2</u>. ((ℓ+1)-jet-extension; $\ell = 1$). In general we <u>can avoid</u> that the equations $Df(x) = 0$, $\det D^2f(x) = 0$ are simultaneously satisfied in $\mathbb{R}^n$.

Let us consider the second point of view a little closer in case $n = 2$. As an abbreviation we put $\partial_i f = \frac{\partial}{\partial x_i} f$, $\partial_{ij} f = \frac{\partial^2}{\partial x_i \partial x_j} f$. Then,

$$j^2 f(x) = [x, f(x), \partial_1 f(x), \partial_2 f(x), \partial_{11} f(x), \partial_{12} f(x), \partial_{22} f(x)].$$

In the jet-space $J(2,1,2)$ we consider the set M:

$$M = \mathbb{R}^2 \times \mathbb{R} \times \{0\} \times \{0\} \times \Sigma,$$

where $\Sigma = \{(\xi_1,\xi_2,\xi_3) \mid \xi_1\xi_3 - \xi_2^2 = 0\}$.
The set Σ represents exactly the set of points (ξ_1,ξ_2,ξ_3) satisfying $\det\begin{pmatrix} \xi_1 & \xi_2 \\ \xi_2 & \xi_3 \end{pmatrix} = 0$, i.e. the zero-set of the polynomial $p(\xi_1,\xi_2,\xi_3) = \xi_1\xi_3 - \xi_2^2$.

Unfortunately, M is not a manifold since Σ is not a manifold. In fact, consider a neighborhood of $0 \in \Sigma$ (Fig. 7.4.1).

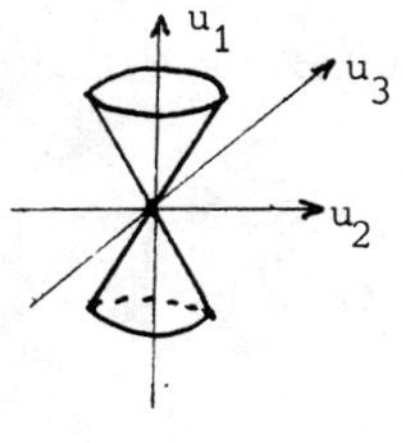

Fig. 7.4.1

Σ after a linear coordinate transformation

$$\begin{pmatrix} \xi_1 \\ \xi_2 \\ \xi_3 \end{pmatrix} = \underbrace{\begin{pmatrix} 1 & 1 & 0 \\ 0 & 0 & 1 \\ 1 & -1 & 0 \end{pmatrix}}_{\det \neq 0} \begin{pmatrix} u_1 \\ u_2 \\ u_3 \end{pmatrix}$$

$$\xi_1\xi_3 - \xi_2^2 = 0 \iff u_1^2 - u_2^2 - u_3^2 = 0$$

However, $\Sigma\setminus\{0\}$ is a manifold of dimension 2, since $Dp = (\xi_3, -2\xi_2, \xi_1)$ and thus $Dp \neq 0$ outside the origin. Consequently we may view at Σ as been built up by the two manifolds $\Sigma\setminus\{0\}$ and $\{0\}$, i.e. $\{\{0\}, \Sigma\setminus\{0\}\}$ is a stratification for Σ. Put

$$M_1 = \mathbb{R}^2 \times \mathbb{R} \times \{0\} \times \{0\} \times (\Sigma\setminus\{0\})$$

$$M_2 = \mathbb{R}^2 \times \mathbb{R} \times \{0\} \times \{0\} \times \underbrace{\{0\}}_{(0,0,0)}$$

Then, codim $M_1 = 3$, codim $M_2 = 5$. Define

$$F = \{f \in C^\infty(\mathbb{R}^2, \mathbb{R}) \mid j^2 f \pitchfork M_i,\ i = 1,2\}.$$

By Remark 7.4.2, F is C^k-dense for all k. Since codim $M_i > 2$, $i = 1,2$, for an $f \in F$ we have $(j^2 f)^{-1}(M_i) = \emptyset$, $i = 1,2$, i.e. $f \in F$ implies that the equations $Df(x) = 0$, $\det D^2 f(x) = 0$ are not satisfied simultaneously, i.e. f is nondegenerate.

7.5. Whitney-Regularity, Final Remarks.

In Section 7.3 we introduced the concept of a stratified set. Now we consider an often used and important regularity condition for stratified sets, namely Whitney-Regularity. For convenience, we use the definition as in Gibson et al. [16]. We recall that the set of linear k-dimensional subspaces in $\mathbb{R}^n$, denoted by $G(n,k)$, constitutes a compact manifold (Grassmann-manifold); cf. Theorem 7.3.8.

Definition 7.5.1. (Whitney-Regularity). Let (S,Σ) be a stratified set in $\mathbb{R}^n$. Further, let $X, Y \in \Sigma$, $X \neq Y$ and $\bar{x} \in X$.
Then, Y is called Whitney-regular over X at the point $\bar{x}$, if for every sequence $((x_i, y_i)) \subset X \times Y$ with the properties (1), (2), (3),

(1) $x_i \to \bar{x}$, $y_i \to \bar{x}$

(2) $Ty_i \to T$ in $G(n, \dim Y)$

(3) $L_i \to L$ in $G(n,1)$, $L_i := \text{span}\{y_i - x_i\}$,

the following inclusion holds: $L \subset T$.

A stratum* $Y \in \Sigma$ is said to be Whitney-regular over $X \in \Sigma$, $X \neq Y$, if Y is Whitney-regular over X at every point $x \in X$. The stratification Σ for S is called Whitney-regular (or: (S,Σ) is Whitney-regular) if every stratum* $Y \in \Sigma$ is Whitney-regular over every stratum* $X \in \Sigma$, $X \neq Y$.

Remark 7.5.1. In order that property (1) in Definition 7.5.1 is satisfied, the point $\bar{x}$ must lie in the closure $\bar{Y}$ of Y.

Remark 7.5.2. Let (S,Σ) be a Whitney-regular stratification and let $\Phi \in C^\infty(\mathbb{R}^n,\mathbb{R}^n)$ be a diffeomorphism. Then, $(\Phi(S),\Phi(\Sigma))$ is also a Whitney-regular stratification, where $\Phi(\Sigma) := \{\Phi(W) | W \in \Sigma\}$. To see this, note that the assertion is trivial if Φ is (affine) linear. Now, the Whitney-regularity condition is a local condition (we work "im Kleinen") and the assertion follows immediately from the fact that - as a neighborhood $\mathcal{U}_{\bar{x}}$ shrinks to $\bar{x}$ - Φ is approximated on $\mathcal{U}_{\bar{x}}$ arbitrarily well by its linear approximation $x \mapsto \Phi(\bar{x}) + D\Phi(\bar{x})(x-\bar{x})$.

Example 7.5.1. Let $S \subset \mathbb{R}^n$ be an MGB of class C^∞ and Σ its stratification according to Definition 3.1.3. Then, (S,Σ) is Whitney-regular. In fact, consider S in local coordinates (cf. Definition 3.1.1) and the proof becomes obvious. In particular, every Regular Constraint Set (of class C^∞) admits a Whitney-regular stratification.

Example 7.5.2. Let us return to Example 7.3.2. The set $S \subset \mathbb{R}^3$ is the zero-set of the polynomial $p(x_1,x_2,x_3) = x_2^2-x_1^2x_3$. The stratification $\Sigma = \{M_1,M_2\}$, where M_1 = x_3-axis and $M_2 = S \backslash M_1$, is not Whitney-regular. In fact, M_2 is not Whitney-regular over M_1 at the origin. To see this, consider the sequence $\{(\alpha_i,\beta_i),\ i = 1,2,\ldots\}$ in $M_1 \times M_2$ with $\alpha_i = (0,0,-\frac{1}{i})$ and $\beta_i = (\frac{1}{i},0,0)$. Now, $T_{\beta_i}M_2$ is constant and equals the (x_1,x_2)-plane, whereas the direction of the line through α_i, β_i is also constant for all i, being represented by the vector (1,0,1). But (1,0,1) does not lie in the (x_1,x_2)-plane! The origin is the only point at which M_2 fails to be Whitney-regular over M_1. Hence, the refined stratification as shown in Fig. 7.3.6.c is Whitney-regular.

In Section 7.3 we introduced the Condition X for a stratified subset (S,Σ) of $\mathbb{R}^n$. Together with Lemma 7.3.4 we then can apply the "Openess Principle" (Theorem 7.3.7).

Lemma 7.5.1. A Whitney-regular stratified subset (S,Σ) in $\mathbb{R}^n$ satisfies Condition X. □

The proof of Lemma 7.5.1 follows from the following Lemma 7.5.2 and Corollary 7.5.2.

Lemma 7.5.2. Let (S,Σ) be a stratified subset in $\mathbb{R}^n$. Further, let $X,Y \in \Sigma$, $X \neq Y$, and $\bar{x} \in X \cap \bar{Y}$.
Suppose that Y is Whitney-regular over X at $\bar{x}$. Then, for every sequence $(y_i) \subset Y$ with $y_i \to \bar{x}$ and $T_{y_i} \to T$ (in $G(n,\dim Y)$) we have: $T_{\bar{x}}X \subset T$.

Proof. If $\dim X = 0$, there is nothing to prove. So, let $\dim X > 0$. From Remark 7.5.2 it follows that we may look at the problem in suitable local coordinates. Thus, without loss of generality, we may assume that $X = \mathbb{R}^k \times \{0\} \subset \mathbb{R}^n$, $k \geq 1$, and that $\bar{x} = 0$. Take a point $\hat{x} \in \mathbb{R}^k \times \{0\}$ with $\|\hat{x}\| = 1$ and put $L = \operatorname{span}\{\hat{x}\}$. Let $(y_i) \subset Y$ be a sequence converging to the origin such that $T_{y_i}Y \to T$ and define $d(y_i,L) = \inf\limits_{x \in L} \|x-y_i\|$.
Eventually after taking a subsequence of (y_i), we may assume that $d(y_i,L) \leq i^{-2}$, $i = 1,2,\ldots$. Put $x_i = \frac{1}{i}\hat{x}$. Then, $x_i \to 0$ and $\operatorname{span}\{y_i - x_i\} \to L$ (in $G(n,1)$). Finally, the validity of the Whitney-regularity at $\bar{x} = 0$ implies $L \subset T$. This proves the lemma. □

The next corollary follows easily from Lemma 7.5.2 and Lemma 7.3.2, and the proof is omitted.

Corollary 7.5.1. Under the same assumptions as in Lemma 7.5.2 let $M \subset \mathbb{R}^n$ be a manifold intersecting X transversally at $\bar{x}$. Then, there exists an $\mathbb{R}^n$-neighborhood O of $\bar{x}$ such that M intersects Y transversally in O. □

Corollary 7.5.2. Under the same assumptions as in Lemma 7.5.2 we have: $\dim X < \dim Y$.

Proof. Let $(y_i) \subset Y$ be a sequence converging to $\bar{x}$. Without loss of generality we may assume that $T_{y_i}Y \to T$ (in $G(n,\dim Y)$). Since $y_i \to \bar{x}$, for sufficiently large i there exists an $x_i \in X$ which minimizes the Euclidean distance w.r.t. y_i (look a priori at the problem in suitable local coordinates for X). Without loss of generality we may assume that $L_i := \mathrm{span}\{y_i - x_i\}$ converges to L in $G(n,1)$. The Whitney-regularity implies that $L \subset T$. Thus taking Lemma 7.5.2 into account, we have: $T_{\bar{x}}X + L \subset T$. Now, the choice of x_i is such that $L_i \perp T_{x_i}X$ and thus, by continuity, $L \perp T_{\bar{x}}X$. Consequently, we have $\dim T_{\bar{x}}X < \dim(T_{\bar{x}}X+L) \leq \dim T$, and hence the assertion in the corollary follows in virtue of $\dim T_{\bar{x}}X = \dim X$, $\dim T = \dim Y$. □

Corollary 7.5.3. Let (S,Σ) be a Whitney-regular stratified subset of $\mathbb{R}^n$ and suppose that S is a closed set.
Let Σ^i denote the union of all i-dimensional strata* in Σ, $i = 0,1,\ldots,n$. Then, $\bigcup_{i=0}^{k} \Sigma^i$ is closed for all $k \in \{0,1,\ldots,n\}$.

Proof. A point $\bar{x}$ belonging to the closure of $\bigcup_{i=0}^{k} \Sigma^i$, but not to $\bigcup_{i=0}^{k} \Sigma^i$, belongs to a stratum* X of dimension $\ell > k$. On the other hand, $\bar{x}$ belongs to the closure of a stratum* Y of dimension less or equal to k. In virtue of Corollary 7.5.2, we must have: $\dim X < \dim Y$, a contradiction. □

Let (S_1,Σ_1), (S_2,Σ_2) be Whitney-regular stratified subsets of $\mathbb{R}^n$. If $X_1 \pitchfork X_2$ for all $X_1 \in \Sigma_1$, $X_2 \in \Sigma_2$, then it is easily seen that $\Sigma_1 \cap \Sigma_2 := \{X_1 \cap X_2 \mid X_i \in \Sigma_i\}$ is a Whitney-regular stratification for the intersection $S_1 \cap S_2$. This generalizes transversal intersection of manifolds (moreover, it can be extended in a natural way to Whitney-regular stratified sets "in general position"). As a generalization of Theorem 7.4.1 we have:

Theorem 7.5.1. (Jet-Transversality Theorem for Stratified Sets).
Let $n \geq 1$, $m \geq 1$, $\ell \geq 0$ be given. Further, let (S,Σ) be a Whitney-regular stratified set in $J(n,m,\ell)$ and define

$$\pitchfork^{\ell} S = \{f \in C^{\infty}(\mathbb{R}^n,\mathbb{R}^m) \mid j^{\ell}f \pitchfork X \text{ for all } X \in \Sigma\}.$$

Then, $\pitchfork^{\ell}S$ is C^k-dense for all k; moreover, if S is a closed subset then $\pitchfork^{\ell}S$ is C^k-open for all $k \geq \ell+1$.
Furthermore, if $f \in \pitchfork^{\ell}S$, then $\{(j^{\ell}f)^{-1}(X) \mid X \in \Sigma\}$ is a Whitney-regular stratification for $(j^{\ell}f)^{-1}(S)$. □

The non-trivial part in Theorem 7.5.1 is the "open part". However, now the "open part" follows directly from a combination of Lemma 7.5.1, Lemma 7.3.4 and Theorem 7.3.7. A corresponding generalization of Theorem 7.4.2 is left to the reader.

Stratification theory is a very difficult theory and a further study would go beyond of the scope of this presentation. However, we emphasize that it is a natural tool in familiar sets. For example, a subset $X \subset \mathbb{R}^n$ is called semi-algebraic if X is obtained by means of finitely many times taking intersections, unions and set differences from the basic sets: $\{x \in \mathbb{R}^n \mid p(x) > 0\}$ where p is a polynomial. An important theorem is: every semi-algebraic set admits a Whitney-regular stratification (see [16] for a nice (partial) proof). As another example, the decomposition of the set of (symmetric) matrices - cf. Examples 7.3.3 and 7.3.4 - constitutes already a Whitney-regular stratification (see also [16]). Among the nice properties of Whitney-regular stratifications (cf. [16], [57], [61]) we mention the following concerning the (path) components of strata*:

Theorem 7.5.2. ([16]).
Let (S,Σ) be a Whitney-regular stratified subset of $\mathbb{R}^n$ and suppose, in addition, that S is locally closed.
Let Σ^c denote the family of all connected components of the elements of Σ.
Then:

1. Σ^c is locally finite (hence, (S,Σ^c) is also a Whitney-regular stratification).

2. If $X,Y \in \Sigma^c$ and $X \cap \bar{Y} \neq \emptyset$, then $X \subset \bar{Y}$ (i.e. the socalled Frontier Condition holds). □

Remark 7.5.3. The set S in Remark 7.3.14 is Whitney-regular w.r.t. the given stratification. However, S is not locally closed and hence, Theorem 7.5.2 is not applicable. Note on the other hand: for an MGB with its natural stratification the Frontier Condition obviously holds.

For additional reading on transversality theory we refer to Gibson [17], Golubitsky and Guillemin [18], Hirsch[31] and Mather [60]. Moreover, extensions to infinite dimensional spaces are treated in Abraham and Robbin [1] and in the recent beautiful book of Aubin and Ekeland [3]. In particular, in Smale [70] Sard's theorem is generalized.

8. GRADIENT FLOWS.

Descent methods in optimization for finding a local minimum of a function $f \in C^{\infty}(\mathbb{R}^n, \mathbb{R})$ are usually based on gradient techniques. In fact, they can be regarded as discretizations of the (negative) "gradient" differential equation

$$\dot{x} = -F(x),$$

where $F(x) = Q(x).D^T f(x)$, $Q(x)$ being a symmetric positive definite matrix, smoothly depending on x.
If $Q(x)$ = Identity, then $-F(x)$ points into the socalled "steepest descent" direction. The inverse matrix $Q(x)^{-1}$ can be interpreted (as it will be shown) as a metric depending on x ("variable metric" or Riemannian metric).

The aim of this chapter is a description of local and global aspects of the phase portrait of the gradient differential equation "in general position". From this we learn in particular two basic (theoretical) principles for finding - iteratively - all local minima of a function. Note that we already used gradient techniques in the deformation theorems in Morse theory (cf. Chapters 2 and 3). So, it can be expected that the ideas of Morse theory are closely related to those presented in this chapter.

8.1. Flows, hyperbolic singularities, linearization.

Let $U \subset \mathbb{R}^n$ be open and let $F \in C^{\infty}(U, \mathbb{R}^n)$ be a vectorfield. The differential equation

$$\frac{dx}{dt} = F(x) \tag{8.1.1}$$

associated with F is called an <u>autonomous</u> differential equation (i.e. the righthandside of (8.1.1) does not depend on the variable t).

Let $x(t) = \Phi(t,x)$ denote the solution of (8.1.1) with $x(0) = x$ as initial condition. The definition domain of Φ is then a certain open neighborhood of $\{o\} \times U$ in $\mathbb{R}^1 \times U$ (cf. Section 2.3) and Φ is called the <u>flow</u> of the vectorfield F. Recall that $\frac{\partial \Phi}{\partial t}(t,x) = F(\Phi(t,x)) = F \circ \Phi(t,x)$ and $\Phi(0,x) = x$ for all $x \in U$.

<u>Remark 8.1.1</u>. Let us consider vectorfields and flows in <u>new coordinates</u>. To this aim, let $U,V \subset \mathbb{R}^n$ be open and $\Psi \in C^\infty(U,V)$ a diffeomorphism. Put $y = \Psi(x)$.

The vectorfield $F(x)$ is transformed under Ψ into the vectorfield $\tilde F \in C^\infty(V,\mathbb{R}^n)$:

$$\tilde F(y) = [D\Psi \cdot F]\circ\Psi^{-1}(y). \tag{8.1.2}$$

On the other hand, the flow $\Phi(t,x)$ is mapped to $\tilde\Phi(t,y)$:

$$\tilde\Phi(t,y) = \Psi\circ\Phi(t,\Psi^{-1}(y)), \tag{8.1.3}$$

or shortly,

$$\tilde\Phi_t = \Psi\circ\Phi_t\circ\Psi^{-1} \quad \text{(cf. also Section 2.3).}$$

We contend that $\tilde\Phi(t,y)$ is the flow of the vectorfield $\tilde F(y)$. In fact, differentation of (8.1.3) with respect to t yields:

$$\left.\frac{\partial\tilde\Phi}{\partial t}(t,y) = \underbrace{D\Psi[\Phi(t,\Psi^{-1}(y))]}_{\downarrow}\cdot\underbrace{\frac{\partial}{\partial t}\Phi(t,\Psi^{-1}(y))}_{F[\underbrace{\Phi(t,\Psi^{-1}(y))}_{\swarrow}]} \quad \Psi^{-1}[\tilde\Phi(t,y)]\right\} \tag{8.1.4}$$

Consequently, from (8.1.2) and (8.1.4) we obtain

$$\frac{\partial\tilde\Phi}{\partial t} = \tilde F\circ\tilde\Phi \tag{8.1.5}$$

From (8.1.3) we obtain

$$\tilde\Phi(0,y) = \Psi\circ\underbrace{\Phi(0,\Psi^{-1}(y))}_{\Psi^{-1}(y)} = y \tag{8.1.6}$$

Finally, from (8.1.5), (8.1.6) and the uniqueness theorem of differential equations (cf. Theorem 2.3.1) it follows that $\tilde\Phi$ is the flow of $\tilde F$. □

Now, let $\bar{x} \in U$ be a point at which F does not vanish. We show that there exist smooth local coordinates such that the transformed vector field is just the constant vectorfield $(1,0,\ldots,0)^T$. This gives us the simplest normal form of a vectorfield and it is the basic idea of a so called "flow box" (cf. also [30]).

With the aid of an affine coordinate transformation of the form $T(x) = A(x-\bar{x})$, A being a nonsingular n*n matrix, we may assume without loss of generality that $\bar{x} = 0$ and $F(0) = (1,0,\ldots,0)^T$ (use also Remark 8.1.1). The main idea consists of a parametrization of a neighborhood of the origin by means of the linear subspace $\{(y_1,\ldots,y_n) \mid y_1 = 0\}$ of codimension one and the one dimensional integralcurves of F through this subspace. It is crucial to note that the integralcurves of F intersect the subspace $\{y \mid y_1 = 0\}$ transversally in a neighborhood of the origin.
If we integrate the vectorfield F starting at $(0,y_2,\ldots,y_n)$ during the time y_1, we get to the point $\Phi[y_1,(0,y_2,\ldots,y_n)]$ (this is "what we have"). If we do the same with the constant vectorfield $(1,0,\ldots,0)^T$ we get to the point $(y_1,y_2,\ldots,y_n)$ (this is "what we want"). So a reasonable candidate Ψ for the new local coordinates is formed by mapping $(x_1,\ldots,x_n) := \Phi[y_1,(0,y_2,\ldots,y_n)]$ to the point $(y_1,y_2,\ldots,y_n)$ (cf. Fig. 8.1.1).

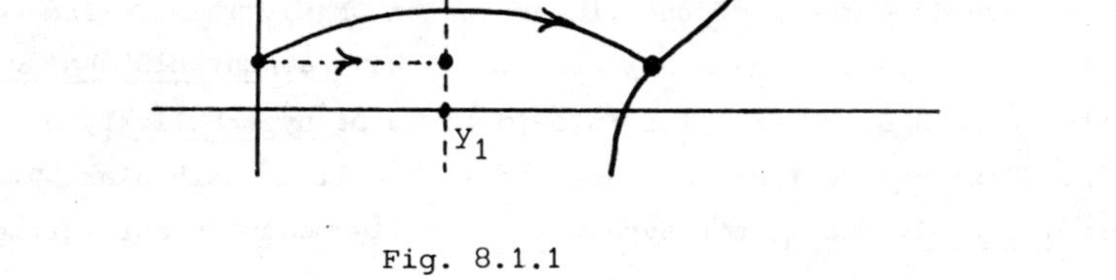

Fig. 8.1.1

It remains to point out two details. Firstly, we have to show that Ψ is well defined as a local diffeomorphism. Secondly, we have to verify that Ψ

transforms the vectorfield F into the constant vectorfield $(1,0,\ldots,0)^T$.

First detail. The mapping Ψ is nothing else but the (local) inverse of the mapping η: $y \mapsto \Phi[y_1,(0,y_2,\ldots,y_n)]$. Note that the restriction of η to the linear subspace $\{y|y_1 = 0\}$ is the identity and $\frac{\partial}{\partial y_1}\eta(0) = F(0) = (1,0,..,0)^T$. Hence, $D\eta(0) = I_n$ is nonsingular and the inverse function theorem then implies that η is locally invertible.

Second detail. Let $\tilde{\Phi}$ denote the flow corresponding to the constant vectorfield $(1,0,\ldots,0)^T$. Then we obviously have:

$$\tilde{\Phi}[t,(y_1,y_2,\ldots,y_n)] = (t+y_1,y_2,\ldots,y_n). \tag{8.1.7}$$

In view of Remark 8.1.1 we are done if relation (8.1.3) is satisfied. But this follows from the following calculation ($|t|$ and $\|y\|$ sufficiently small):

$$\Psi\circ\Phi(t,\Psi^{-1}(y)) = \Psi\circ\Phi(t,\Phi[y_1,(0,y_2,\ldots,y_n)]) =$$

$$= \Psi\circ\Phi[t+y_1,(0,y_2,\ldots,y_n)] = \Psi\circ\Psi^{-1}(t+y_1,y_2,\ldots,y_n) =$$

$$= (t+y_1,y_2,\ldots,y_n) = \tilde{\Phi}(t,y).$$

From the foregoing we see that the phase portrait (= family of integral curves) of (8.1.1) can be linearized by means of a smooth coordinate transformation in a neighborhood of any point where the vector field F does not vanish.

A point $\bar{x} \in U$ at which F vanishes is called a singular point of (8.1.1). A singular point $\bar{x}$ is a rest point (equilibrium) for (8.1.1) since $\Phi(t,\bar{x}) = \bar{x}$ for all t. From a Taylor expansion of F around a singular point $\bar{x}$ it might be expected that the local behaviour of the phase portrait of (8.1.1) is governed by means of the linearization $DF(\bar{x})$. This is indeed true if all eigenvalues of the $n\times n$ matrix $DF(\bar{x})$ have a nonvanishing real part. In the latter case a singular point is said to be of hyperbolic type. If, in addition, $DF(\bar{x})$ has real eigenvalues, we will call $\bar{x}$ a singular point of real hyperbolic type. The latter type will actually occur in our further considerations.

Let us start with the linear case of a real hyperbolic singularity. So, we look at the system

$$\frac{dx}{dt} = Ax \quad , \tag{8.1.8}$$

where A is a <u>nonsingular</u> matrix with <u>real</u> eigenvalues. Replacing A by -A in (8.1.8) is equivalent with replacing the integration variable t by -t (i.e. integrating (8.1.8) in the negative sense). Up to this sign we have in case n = 2 the following three typical situations for A (apart from multiples of the identity):

(i) $A = \begin{pmatrix} \lambda_1 & 0 \\ 0 & \lambda_2 \end{pmatrix}$ $0 > \lambda_1 > \lambda_2$ (eigenvalues have same sign),

(ii) $A = \begin{pmatrix} \lambda_1 & 0 \\ 0 & \lambda_2 \end{pmatrix}$ $\lambda_1 < 0 < \lambda_2$ (eigenvalues have different sign),

(iii) $A = \begin{pmatrix} \lambda & 0 \\ 1 & \lambda \end{pmatrix}$ $\lambda < 0$.

The matrix A in case (iii) is defect, i.e. the algebraic multiplicity of the eigenvalue λ (= multiplicity of λ as a root of the characteristic polynomial $p(\lambda) := \det(\lambda I-A)$) differs from the geometric multiplicity of λ (= dimension of the eigenspace belonging to λ). The phase portraits corresponding to (i), (ii) and (iii) are depicted in Fig. 8.1.2.

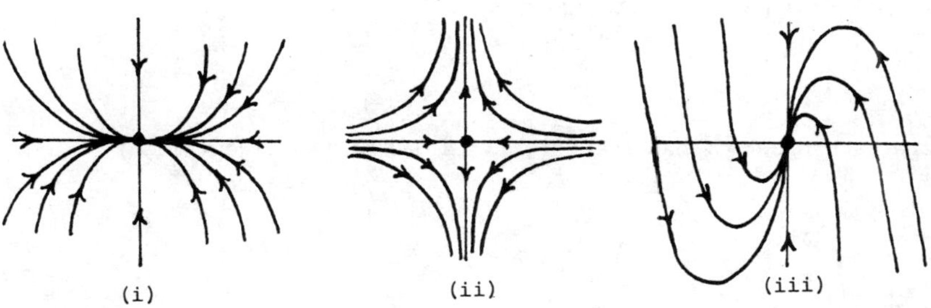

Fig. 8.1.2

The word "hyperbolic" comes from Fig. 8.1.2 (ii). In fact, the solution of (8.1.8) in case (ii) with $(\bar{x}_1,\bar{x}_2)$ as initial point has the form $(x_1(t),x_2(t)) = (\bar{x}_1 e^{\lambda_1 t}, \bar{x}_2 e^{\lambda_2 t})$. If in particular $\lambda_1 = -\lambda_2$, then $x_1(t).x_2(t) = \bar{x}_1.\bar{x}_2$ and so, the integral curve through $(\bar{x}_1,\bar{x}_2)$ lies on the <u>hyperbola</u> defined by the equation $x_1 \cdot x_2 = \bar{x}_1 \cdot \bar{x}_2$.

Next we turn to the n-dimensional case. In order to study the behaviour of the phase portrait of (8.1.8) in detail we uncouple the system (8.1.8) into smaller independent subsystems. To this aim we consider a linear coordinate transformation

$$y = Bx \quad , \tag{8.1.9}$$

where B is a nonsingular n×n matrix.
According to Remark 8.1.1 the system (8.1.8) in the y-coordinates becomes

$$\frac{dy}{dt} = BAB^{-1}y. \tag{8.1.10}$$

Note that the eigenvalues of A and BAB^{-1} coincide. Since the eigenvalues of A are <u>real</u>, we can choose (the <u>real</u> matrix) B such that BAB^{-1} has the so called Jordan normal form (cf. [12], [30])

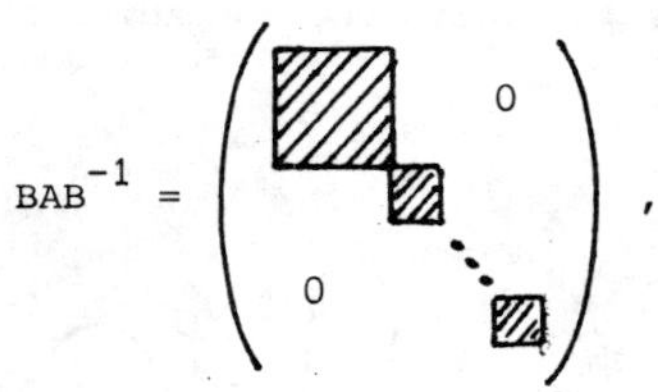

(8.1.11)

where each block  has the form (λ is one of the eigenvalues)

$$= \begin{pmatrix} \lambda & & & 0 \\ 1 & \lambda & & \\ & 1 & \lambda & \\ 0 & & 1 & \lambda \end{pmatrix}, \tag{8.1.12}$$

the number of blocks associated to λ being equal to the geometric multiplicity of λ; so the total number of blocks is equal to the number of linearly independent eigenvectors. In this way the system (8.1.8) is decomposed into smaller independent subsystems whose matrices have the form (8.1.12). Now, suppose that the $n\times n$ matrix A has the form (8.1.12) and let $p_i(t)$ denote a polynomial of degree i. Then the solution of (8.1.8) starting at some $\bar{x} \in \mathbb{R}^n$ has the form:

$$x_i(t) = p_{i-1}(t)e^{\lambda t}, \quad i = 1,\ldots,n . \tag{8.1.13}$$

In particular, if $\lambda < 0$ we see that $x_i(t)$ tends to zero as t tends to $+\infty$. By means of a permutation of the blocks in (8.1.11) we may assume that BAB^{-1} has the following blockstructure:

$$BAB^{-1} = \left(\begin{array}{c|c} A_1 & 0 \\ \hline 0 & A_2 \end{array}\right) \qquad \begin{array}{l} A_1: \; m\times m \\ A_2: \; (n-m)\times(n-m) \end{array} \tag{8.1.14}$$

where the matrix A_1 (resp. A_2) has negative (resp. positive) eigenvalues. Let us decompose an $x \in \mathbb{R}^n$ in accordance with (8.1.14) as $x = (x_1,x_2)$, with $x_1 \in \mathbb{R}^m$ and $x_2 \in \mathbb{R}^{n-m}$. The solution of (8.1.10) with $\bar{x} = (\bar{x}_1,\bar{x}_2)$ as initial point, now denoted by $x(t)$, becomes:

$$(x_1(t),x_2(t)) = (\exp(tA_1)\cdot \bar{x}_1, \exp(tA_2)\cdot \bar{x}_2). \tag{8.1.15}$$

In (8.1.15) the matrix exponential $\exp(tA)$ stands for the solution of the matrix differential equation $\dot{X} = AX$ with $X(0) = \text{Identity}$. From (8.1.13) and (8.1.15) we learn:

If $\bar{x}_1 \neq 0$ then, $\|x_1(t)\| \to 0(\infty)$ if $t \to \infty(-\infty)$ and
if $\bar{x}_2 \neq 0$ then, $\|x_2(t)\| \to \infty(0)$ if $t \to \infty(-\infty)$.

The matrix $\exp(tA_1)$ serves as a <u>contraction</u> part whereas $\exp(tA_2)$ represents an <u>expansion</u> part as $t \to \infty$. The linear subspaces $\mathbb{R}^m \times \{o\}$ and $\{o\} \times \mathbb{R}^{n-m}$ play a special role in the following sense: the solution of (8.1.10) with $\bar{x} \in \mathbb{R}^n$ as initial point tends to the singular point (the origin in this case) as $t \to \infty$ (resp. $-\infty$) iff $\bar{x} \in \mathbb{R}^m \times \{o\}$ (resp. $\bar{x} \in \{o\} \times \mathbb{R}^{n-m}$). Because of this property $\mathbb{R}^m \times \{o\}$ (resp. $\{o\} \times \mathbb{R}^{n-m}$) is called the <u>stable</u> (resp. <u>unstable</u>) <u>manifold</u> at the singular point. The stable (resp. unstable) manifold of the original system (8.1.8) at the singular point obviously becomes the linear subspace $\{B^{-1}x \mid x_2 = 0\}$ (resp. $\{B^{-1}x \mid x_1 = 0\}$). Note that the stable and unstable manifolds intersect <u>transversally</u> in $\mathbb{R}^n$ at the singular point (of course they need not be orthogonal to each other unless the transformation matrix B is orthogonal).

Let us turn back to the nonlinear case. We consider a singularity $\bar{x}$ for (8.1.1) of <u>real hyperbolic type</u>, i.e. $F(\bar{x}) = 0$ and $DF(\bar{x})$ has nonvanishing real eigenvalues.

Let $\Psi(x) := B(x-\bar{x})$ be an affine coordinate transformation and let $\tilde{F}$ denote the transformed vector field according to (8.1.2). Then, using the fact that $F(\bar{x}) = 0$, differentiation of (8.1.2) gives:

$D\tilde{F}(0) = D\Psi \cdot DF(\bar{x}) \cdot D\Psi^{-1} = B \cdot DF(\bar{x}) \cdot B^{-1}$. Now we can choose B in such a way that $D\tilde{F}(0)$ becomes:

$$D\tilde{F}(0) = \left(\begin{array}{c|c} A_1 & 0 \\ \hline 0 & A_2 \end{array}\right), \qquad \begin{array}{l} A_1 : m\times m \\ \\ A_2 : (n-m)\times(n-m) \end{array} \tag{8.1.16}$$

where A_1 (resp. A_2) has negative (resp. positive) eigenvalues. From the latter calculation we see that we may assume, without loss of generality, that $DF(\bar{x})$ takes the form (8.1.16) and that $\bar{x} = 0$.

Next, we take a sufficiently small open neighborhood $\mathcal{O}$ of the origin ($\mathcal{O}$ to be adjusted later on) and try to give a precise description of the <u>(local) stable manifold</u> W^s of (8.1.1) to the singular point $\bar{x} = 0$, where

$$W^s = \left\{ x \in \mathcal{O} \;\middle|\; \begin{array}{l} \Phi(t,x) \text{ is defined in } \mathcal{O} \\ \text{for all } t \geq 0 \text{ and} \\ \lim\limits_{t\to\infty} \Phi(t,x) = \bar{x} \end{array} \right\}. \tag{8.1.17}$$

As it might be expected, the manifold W^s is indeed a smooth manifold which is tangent at $\bar{x} = 0$ to the stable manifold corresponding to the linearized differential equation $\dot{x} = DF(0)x$; cf. Fig. 8.1.3.

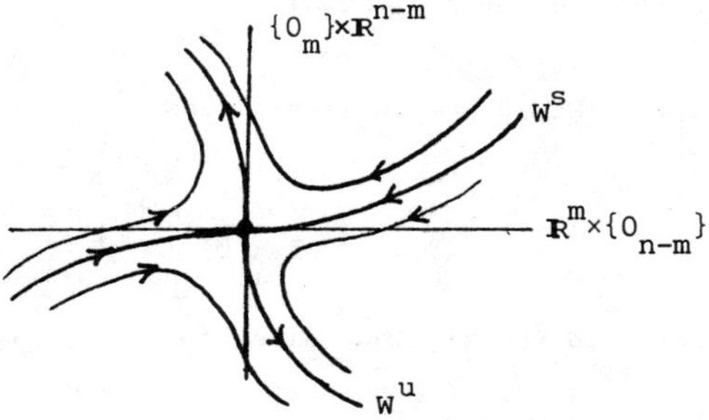

Fig. 8.1.3

We will restrict to a short outline of the equations which actually define W^s and for the omitted proofs we refer to the excellent presentation in [8 , p. 330 ff.].

With $G(x) = F(x) - DF(0)x$ we obtain for (8.1.1):

$$\frac{dx}{dt} = F(x) = DF(0)x + G(x). \tag{8.1.18}$$

Note that $G(0) = 0$ and $DG(0) = 0$.

Following [8], we put:

$$U_1(t) = \left(\begin{array}{c|c} \exp(tA_1) & 0 \\ \hline 0 & 0 \end{array}\right) \quad \underline{\text{linear contraction part}}$$

$$U_2(t) = \left(\begin{array}{c|c} 0 & 0 \\ \hline 0 & \exp(tA_2) \end{array}\right) \quad \underline{\text{linear expansion part}}$$

The stable manifold W^s is obtained by means of an <u>iterative</u> process, starting with the linear case (cf. Fig. 8.1.4).

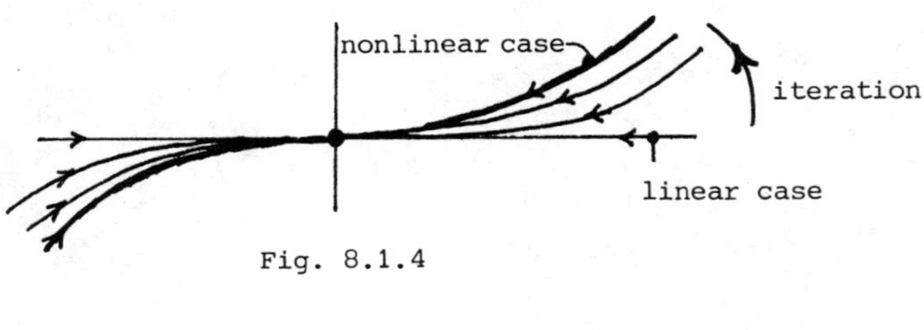

Fig. 8.1.4

To this aim consider the following integral equation ([8]):

$$\left.\begin{aligned} \Theta(t,x) &= \underbrace{U_1(t)x}_{\text{linear part}} + \int_0^t U_1(t-s)G(*)ds - \int_t^\infty U_2(t-s)G(*)ds \\ &* = \Theta(s,x) \quad (t \geq 0) \end{aligned}\right\} \tag{8.1.19}$$

The iteration runs as follows: in the righthandside of (8.1.19) we substitute $\Theta^{(i)}(t,x)$ and obtain $\Theta^{(i+1)}(t,x)$ as a result. As starting function the zero function $\Theta^{(0)}(t,x) = 0$ is substituted.
From $G(0) = 0$ and the fact that in the product $U_1(t)x$ the last $(n-m)$ components of x are annihilated, we see that the last $(n-m)$ components of x have no influence on the first iteration function $\Theta^{(1)}(t,x)$ and hence, they do not enter into a solution $\Theta(t,x)$. Consequently, for a solution of (8.1.19) we have:

$$\Theta(t,(x_1,\ldots,x_n)) = \Theta(t,(x_1,\ldots,x_m,0,\ldots,0)). \tag{8.1.20}$$

From (8.1.20) we see that $\Theta(0,x)$ is completely determined by the first m components of x (hence by the linear subspace $\mathbb{R}^m \times \{0\}$, which in fact is the stable manifold of the linearized system!).

A calculation (cf. [8]) shows that for $x \in \mathcal{O}$, $\mathcal{O}$ suitably chosen, equation (8.1.19) has indeed a well-defined solution which tends exponentially to zero for $t \to \infty$; in fact, it can be shown that $\|\Theta(t,x)\| \leq c\|x\|e^{-\gamma t}$, where c and γ are positive constants.

Furthermore, differentiation of (8.1.19) with respect to t shows that $\Theta(t,x)$ is also a solution of the differential equation (8.1.18).

For the initial point $\Theta(0,x)$ we obtain, using (8.1.19) and the special blockstructure of $U_1(t)$, $U_2(t)$, with $\Theta = (\Theta_1,\ldots,\Theta_n)$:

$$\left.\begin{aligned} \Theta_i(0,x) &= x_i, \quad i = 1,\ldots,m \\ \Theta_j(0,x) &= \left(-\int_0^\infty U_2(-s)G(\Theta(s,x))ds\right)_j, \quad j = m+1,\ldots,n \end{aligned}\right\}, \tag{8.1.21}$$

where $(\cdot)_j$ stands for the j-th component.

Finally, let us put

$$h_j(x_1,\ldots,x_m) = \Theta_j(0,x_1,\ldots,x_m,0,\ldots,0), \; j = m+1,\ldots,n. \tag{8.1.22}$$

The functions h_j in (8.1.22) play the crucial role in the sense that they determine the stable manifold W^s. In fact, it can be shown ([8]):

(i) h_j are of class C^∞, $j = m+1,\ldots,n$. (More precisely, if F is of class C^k, $k \geq 1$, then the same is true for the functions h_j).
Moreover, the derivatives $\frac{\partial}{\partial x_i} h_j$ vanish at the origin for $i = 1,\ldots,m$ and $j = m+1,\ldots,n$.

(ii) Let x(t) be the solution of (8.1.18) with initial point $x(0) = \tilde{x}$.
Then x(t) tends within $\mathcal{O}$ to 0 (the singular point) as $t \to \infty$ iff $\tilde{x}_j = h_j(\tilde{x}_1,\ldots,\tilde{x}_m)$, $j = m+1,\ldots,n$.

From (ii) above it follows that the stable manifold W^s is defined by the equations $x_j = h_j(x_1,\ldots,x_m)$, $j = m+1,\ldots,n$; together with property (i) we see that W^s is a smooth manifold of dimension m and its tangent space at the origin is $\mathbb{R}^m \times \{0\}$, the stable manifold of the linearized system.

For the unstable manifold W^u the reasoning is essentially similar since W^u is the stable manifold to the singular point corresponding to the negative vectorfield $-F(x)$.

Although our consideration appealed to <u>real</u> hyperbolic singularities, the theory for stable/unstable manifolds for general hyperbolic singularities runs basically along the same lines.

<u>Remark 8.1.2</u>. In the foregoing discussion we considered (un)stable manifolds in a <u>neighborhood</u> of a hyperbolic singularity. Globally, it can even happen that the stable and unstable manifolds coincide. In fact, consider a function $f \in C^\infty(\mathbb{R}^2,\mathbb{R})$ with two local minima and one saddlepoint $\bar{x}$. Some level lines are sketched in Fig. 8.1.5.a.

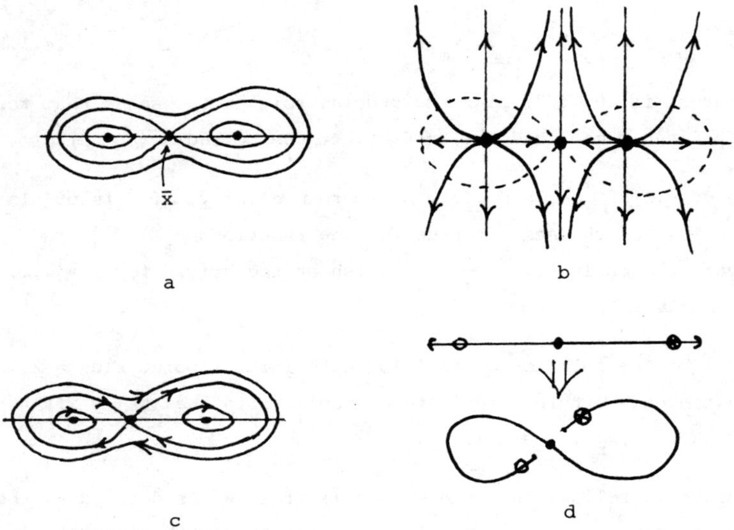

Fig. 8.1.5

Consider the vectorfields F_1, F_2:

$$F_1(x) = \begin{pmatrix} 1 & 0 \\ 0 & 1 \end{pmatrix} D^T f(x) \ , \quad F_2(x) = \begin{pmatrix} 0 & 1 \\ -1 & 0 \end{pmatrix} D^T f(x).$$

Note that $Df(x)\cdot F_2(x) \equiv 0$. Hence, the integral curves of F_2 lie on level lines of f. The phase portraits corresponding to F_1, resp. F_2 are depicted in Fig. 8.1.5.b, resp. c. We see that the stable and unstable manifolds to the saddlepoint in Fig. 8.1.5.c "coincide" and that they have the form of the figure "eight"; so, they are not manifolds. In fact, the stable manifold (as well as the unstable one) can be parametrized by means of an open interval (i.e. we are dealing with an <u>injective immersion of a manifold</u>); see Fig. 8.1.5.d.
Note that the local minima of f are hyperbolic singularities for the "gradient" vector field F_1 but non-hyperbolic for F_2. □

In a neighborhood of a point where the vector field F does not vanish, we found local smooth coordinates such that the vector field in the new coordinates is transformed into a constant vector field; so, the phase portrait in the new coordinates consists of parallel straight lines. Now suppose that F vanishes at $\bar{x}$ and that $\bar{x}$ is of hyperbolic type. In order to find a normal form for the phase portrait in a neighborhood of $\bar{x}$, we ask for a local coordinate transformation H sending the trajectories of the original differential equation (8.1.1) onto the trajectories corresponding to a "simple" vector field, preserving the orientation on the trajectories. In general one cannot expect such a coordinate transformation H to be smooth. We will explain this phenomenon in the next two examples concerning real hyperbolic singularities in dimension two.

Example 8.1.1. Consider the vector fields F_1, F_2 on $\mathbb{R}^2$:

$$F_1(x) = -\begin{pmatrix} 1 & 0 \\ 0 & 1 \end{pmatrix} x \,, \quad F_2(x) = \begin{pmatrix} \lambda_1 & 0 \\ 0 & \lambda_2 \end{pmatrix} x \text{ with } 0 > \lambda_1 > \lambda_2.$$

The phase portraits corresponding to F_1, F_2 are sketched in Fig. 8.1.6.a,b.

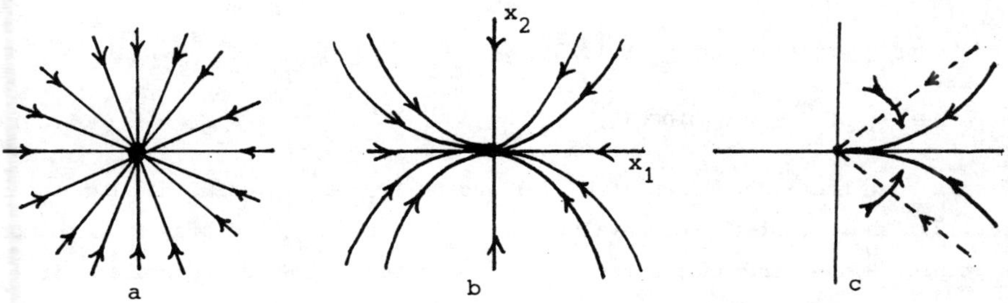

Fig. 8.1.6

If a coordinate transformation H of class C^1 would exist, sending trajectories of F_1 onto those of F_2, then most of the (straight line) trajectories corresponding to F_1 are pinched (Fig. 8.1.6.c). However, this is in contradistinction with the fact that the Jacobian matrix DH(0) is nonsingular. □

Example 8.1.2. At first glance, one might expect that the situation of Fig. 8.1.6.a is exceptional and that between phase portraits of the type as in Fig. 8.1.6.b a differentiable coordinate transformation exists. However, this is not true and the argument is more subtle:

Let F_1, F_2 be smooth vector fields on $\mathbb{R}^2$ both having the origin as a real hyperbolic singularity. Suppose that H is a local smooth coordinate transformation sending the origin onto itself and mapping - thereby preserving the orientation - trajectories of F_1 onto trajectories of F_2.
Using (8.1.2), the vector field F_1 is transformed (locally) under H to the vector field $\tilde{F}_1$, where

$$\tilde{F}_1(y) = (DH \cdot F_1) \circ H^{-1}(y). \tag{8.1.23}$$

Since H maps trajectories onto trajectories, we conclude that $\tilde{F}_1(y)$ is tangent to the integral curve of F_2 through y. Hence, $\tilde{F}_1(y) = \lambda(y) \cdot F_2(y)$, where λ is a real valued function, smooth outside the origin.
Suppose that λ can be extended continuously to the origin, then we obtain by taking derivatives at 0, using the fact that $F_2(0) = 0$:

$$D\tilde{F}_1(0) = \lambda(0) DF_2(0) \tag{8.1.24}$$

and, consequently by using (8.1.23):

$$DH \cdot DF_1 \cdot DH^{-1}|_0 = \lambda(0) DF_2(0) \tag{8.1.25}$$

Recall that the eigenvalues of $DF_1(0)$ and $DH \cdot DF_1 \cdot DH^{-1}|_0$ coincide. But then, (8.1.25) shows that the eigenvalues of $DF_1(0)$ and $DF_2(0)$ differ by a common multiple. Hence, the quotient of the eigenvalues of $DF_1(0)$ coincides with the quotient of those of $DF_2(0)$. We conclude that the quotient of the eigenvalues is invariant under smooth (even under C^1-) coordinate transformations.

In order to obtain (8.1.24) we assumed that λ can be extended continuously to the origin. Independently from this extension question, Formula (8.1.24) can be derived as follows: Put $\tilde{F}_1 = (\tilde{f}_1, \tilde{f}_2)$, $F_2 = (g_1, g_2)$ and let $\partial_i \phi$ denote the derivative $\frac{\partial}{\partial x_i} \phi$ evaluated at the origin, ϕ being a real valued C^1-function. Since $\tilde{F}_1(y)$ and $F_2(y)$ are linearly dependent at every y, we obtain, $|\cdot|$ standing for determinant:

$$\begin{vmatrix} \tilde{f}_1 & g_1 \\ \tilde{f}_2 & g_2 \end{vmatrix} \equiv 0 \tag{8.1.26}$$

Using the fact that $\tilde{F}_1$ and F_2 vanish at the origin, we can take (even if F is of class C^2) the <u>second</u> derivative of the l.h.s. in (8.1.26) <u>at the origin</u> and we obtain:

$$\begin{vmatrix} \partial_1\tilde{f}_1 & \partial_1 g_1 \\ \partial_1\tilde{f}_2 & \partial_1 g_2 \end{vmatrix} = \begin{vmatrix} \partial_2\tilde{f}_1 & \partial_2 g_1 \\ \partial_2\tilde{f}_2 & \partial_2 g_2 \end{vmatrix} = 0, \tag{8.1.27}$$

$$\begin{vmatrix} \partial_1\tilde{f}_1 & \partial_2 g_1 \\ \partial_1\tilde{f}_2 & \partial_2 g_2 \end{vmatrix} + \begin{vmatrix} \partial_2\tilde{f}_1 & \partial_1 g_1 \\ \partial_2\tilde{f}_2 & \partial_1 g_2 \end{vmatrix} = 0. \tag{8.1.28}$$

From the fact that det $DF_2(0) \neq 0$ we see that both $\begin{pmatrix} \partial_1 g_1 \\ \partial_1 g_2 \end{pmatrix}$ and $\begin{pmatrix} \partial_2 g_1 \\ \partial_2 g_2 \end{pmatrix}$ do not vanish. Hence, (8.1.27) yields:

$$\begin{pmatrix} \partial_1\tilde{f}_1 \\ \partial_1\tilde{f}_2 \end{pmatrix} = \alpha_1 \begin{pmatrix} \partial_1 g_1 \\ \partial_1 g_2 \end{pmatrix}, \quad \begin{pmatrix} \partial_2\tilde{f}_1 \\ \partial_2\tilde{f}_2 \end{pmatrix} = \alpha_2 \begin{pmatrix} \partial_2 g_1 \\ \partial_2 g_2 \end{pmatrix}. \tag{8.1.29}$$

Substitution of (8.1.29) into (8.1.28) gives:

$$\alpha_1 \det DF_2(0) - \alpha_2 \det DF_2(0) = 0.$$

Finally, we obtain $\alpha_1 = \alpha_2 =: \alpha$ and then, (8.1.29) implies $D\tilde{F}_1(0) = \alpha DF_2(0)$ which is again Formula (8.1.24) with $\lambda(0)$ replaced by α. □

The Examples 8.1.1, 8.1.2 suggest the introduction of topological equivalence in order to obtain simple normal forms:
Let $U \subset \mathbb{R}^n$ be open and let $F_1, F_2 \in C^\infty(U, \mathbb{R}^n)$ be vector fields. Then, F_1 and F_2 are said to be <u>topologically equivalent</u> if there exists a <u>homeomorphism</u> $H: U \to U$ mapping trajectories of F_1 onto trajectories of F_2, preserving the <u>orientation</u> of integration (but not necessarily preserving the parametrizations of the integral curves by means of the integration time t).

The following basic result shows that in the case of hyperbolic singularities, it suffices to restrict to the linearization in order to obtain normal forms:

Theorem 8.1.1. (Grobman [20]/Hartman [25])
Let $U \subset \mathbb{R}^n$ be an open set containing the origin, and suppose that the origin is a singularity of hyperbolic type for the vector field $F \in C^\infty(U,\mathbb{R}^n)$. Then: F and its linearization $\tilde{F}(x) := DF(0)x$ are topologically equivalent on a neighborhood of the origin. □

Next we study the linear case corresponding to a real hyperbolic singularity. To this aim we firstly consider the following useful lemma.

Lemma 8.1.1. Let A be a nonsingular real $n\times n$ matrix with real eigenvalues, exactly m of them being negative. Then, there exists a nonsingular real $n\times n$ matrix B such that for $\tilde{A} := BAB^{-1}$ the following properties hold:

(i) (blockstructure) $$\tilde{A} = \left(\begin{array}{c|c} A_1 & 0 \\ \hline 0 & A_2 \end{array}\right), \quad \begin{array}{l} A_1: m\times m \\ A_2: (n-m)\times(n-m) \end{array}, \tag{8.1.30}$$

(ii) (definiteness) the symmetric matrix $\frac{1}{2}(A_1+A_1^T)$, resp. $\frac{1}{2}(A_2+A_2^T)$ is negative, resp. positive definite.

Proof.
Step 1. In this step we consider the special case that A is a Jordan block, where λ is a nonvanishing real number:

$$A = \begin{pmatrix} \lambda & & & 0 \\ 1 & \lambda & & \\ & 1 & \ddots & \\ 0 & & 1 & \lambda \end{pmatrix}, \quad A: n\times n \quad . \tag{8.1.31}$$

Let p be a nonnegative integer, $\varepsilon \neq 0$, and let D_ε be the following $n\times n$ diagonal matrix:

$$D_\varepsilon = \mathrm{diag}(\varepsilon^{p+1}, \varepsilon^{p+2}, \ldots, \varepsilon^{p+n}) \; . \qquad (8.1.32)$$

A short calculation shows:

$$A_\varepsilon := D_\varepsilon A D_\varepsilon^{-1} = \begin{pmatrix} \lambda & & & \\ \varepsilon & \lambda & & \\ & \varepsilon & \ddots & \\ 0 & & \varepsilon & \lambda \end{pmatrix} . \qquad (8.1.33)$$

For the symmetrization of A we get a tridiagonal matrix:

$$\hat{A}_\varepsilon := \frac{1}{2}(A_\varepsilon + A_\varepsilon^T) = \begin{pmatrix} \lambda & \frac{1}{2}\varepsilon & & 0 \\ \frac{1}{2}\varepsilon & \lambda & \ddots & \\ & \ddots & \ddots & \frac{1}{2}\varepsilon \\ 0 & & \frac{1}{2}\varepsilon & \lambda \end{pmatrix} \qquad (8.1.34)$$

From the Gershgorin theorem (cf. [12]), applied to the matrix $\hat{A}_\varepsilon$ in (8.1.34), we obtain the following estimate for any eigenvalue $\hat{\lambda}_\varepsilon$ of $\hat{A}_\varepsilon$:

$$|\lambda - \hat{\lambda}_\varepsilon| < |\varepsilon| \; . \qquad (8.1.35)$$

Consequently, if $|\varepsilon| < |\lambda|$, then $\hat{A}_\varepsilon$ is positive definite, resp. negative definite, according to $\lambda > 0$, resp. $\lambda < 0$. This establishes Statement (ii) of Lemma 8.1.1 for matrices A of the type (8.1.31) with $\lambda \neq 0$.

Step 2. Now we turn over to the general case. As a preparation-step we choose a nonsingular real matrix $\tilde{B}$ such that $\tilde{B}A\tilde{B}^{-1}$ has the Jordan normal form as given in (8.1.11), (8.1.12), such that the upper (resp. lower) blocks correspond to negative (resp. positive) eigenvalues. This, in particular, gives us already a blockstructure as meant in Statement (i) of the lemma.

For $\varepsilon \neq 0$, let D_ε be the $n \times n$ diagonal matrix:

$$D_\varepsilon = (\varepsilon, \varepsilon^2, \ldots, \varepsilon^n) \; , \qquad (8.1.36)$$

and put

$$A_\varepsilon = (D_\varepsilon \tilde{B}) A (D_\varepsilon \tilde{B})^{-1} . \qquad (8.1.37)$$

The matrix A_ε in (8.1.37) then also has the blockstructure as in Statement (i). Let λ_i, $i = 1,\ldots,n$, denote the eigenvalues of A and choose ε such that $0 < |\varepsilon| < \min_{i=1,\ldots,n} |\lambda_i|$. Then, using Step 1, we see that with $B := D_\varepsilon \tilde{B}$, the corresponding matrix BAB^{-1} satisfies also the properties in (ii). □

Example 8.1.3. Let $\alpha \in \mathbb{R}$, $\alpha > 0$, and put $A = \begin{pmatrix} \alpha & 1 \\ 0 & \alpha \end{pmatrix}$.
Then, $\tilde{A} := \frac{1}{2}(A+A^T) = \begin{pmatrix} \alpha & \frac{1}{2} \\ \frac{1}{2} & \alpha \end{pmatrix}$. Note that $\det \tilde{A} = \alpha^2 - \frac{1}{4}$. Consequently, for $0 < \alpha < \frac{1}{2}$ we see that $\tilde{A}$ is not positive definite, although the eigenvalues of A, both equal to α, are positive.

Theorem 8.1.2. Let A be a nonsingular real n×n matrix with real eigenvalues, exactly m of them being negative. Then, the following two vector fields F_1 and F_2 are topologically equivalent:

$$F_1(x) = Ax \quad , \quad F_2(x) = \left(\begin{array}{c|c} -I_m & 0 \\ \hline 0 & I_{n-m} \end{array}\right) x. \tag{8.1.38}$$

Proof. Firstly, we consider the case m = n. A combination of Lemma 8.1.1 and Remark 8.1.1 shows that we may assume, eventually by means of a preceding linear coordinate transformation $y = Bx$, that the matrix $\frac{1}{2}(A+A^T)$ is negative definite.
Let γ denote the maximal eigenvalue of $\frac{1}{2}(A+A^T)$. Then, we have:

$$x^T Ax = \frac{1}{2} x^T(A+A^T)x \le \gamma\|x\|^2 \quad , \; \gamma < 0. \tag{8.1.39}$$

Let $x(t)$ be the solution of the differential equation $\dot{x} = Ax$ with initial point $x(0) \neq 0$. Then,

$$\frac{d}{dt} \ln\|x(t)\| = \frac{1}{2}\frac{d}{dt} \ln x(t)^T x(t) =$$

$$= \frac{1}{2}\cdot\frac{1}{\|x(t)\|^2}\cdot 2\cdot x(t)^T\cdot\frac{dx(t)}{dt} = \frac{x(t)^T Ax(t)}{\|x(t)\|^2} \le \gamma.$$

Consequently,

$$\ln\|x(t)\| - \ln\|x(0)\| \le \gamma t \text{ (resp. } \ge \gamma t) \text{ for } t \ge 0 \text{ (resp. } t \le 0),$$

and, hence:

$$\|x(t)\| \le \|x(0)\|e^{\gamma t},\ t \ge 0 \quad\text{and}\quad \|x(t)\| \ge \|x(0)\|e^{\gamma t},\ t \le 0 \qquad (8.1.40)$$

Since $\gamma < 0$ we see from (8.1.40) that $x(t) \to 0$ (resp. ∞) as $t \to \infty$ (resp. $-\infty$) (compare also (8.1.13)).

Let D^n denote the unit ball $\{x \in \mathbb{R}^n \mid \|x\| \le 1\}$ and ∂D^n its boundary. Note that the <u>vector Ax is transversal to ∂D^n</u> at every point $x \in \partial D^n$, since $x^T Ax \ne 0$ for $x \in \partial D^n$.
Let $\Phi_1(t,x)$ denote the flow corresponding to F_1, so $\Phi_1(t,x) = \exp(tA)x$.

Together with the above reasoning we see that there exists a unique C^∞-map $T: \mathbb{R}^n\backslash\{o\} \to \mathbb{R}$ with the property that $\Phi_1(T(x),x) \in \partial D^n$.

Now we "connect" the integral curves of the vector fields $F_1(x) := Ax$ and $F_2(x) := -x$.

Choose $x \in \mathbb{R}^n\backslash\{o\}$. In order to reach x, starting at the point $\frac{x}{\|x\|} \in \partial D^n$ via an integral curve of $F_2(x) := -x$, we need the integration time t determined by $e^{-t} = \|x\|$. Hence, $t = -\ln\|x\|$. This observation gives rise to the following mapping $H: \mathbb{R}^n \to \mathbb{R}^n$ (cf. Fig. 8.1.7):

$$H(x) = \begin{cases} 0, & x = 0 \\ \Phi_1(-\ln\|x\|, \frac{x}{\|x\|}), & x \ne 0. \end{cases} \qquad (8.1.41)$$

The mapping H is invertible with inverse H^{-1}:

$$H^{-1}(y) = \begin{cases} 0, & y = 0 \\ \exp(T(y))\Phi_1(T(y),y), & y \ne 0 . \end{cases} \qquad (8.1.42)$$

From (8.1.41), (8.1.42) we see that H and H^{-1} are smooth outside the origin.
From (8.1.40), (8.1.41) we obtain the following implication ($\varepsilon > 0$):

$$\|x\| \le \exp\left(\frac{-\ln\varepsilon}{\gamma}\right) \Rightarrow \|H(x)\| \le \varepsilon ,$$

which shows, in particular, that H is continuous at the origin. Now, let $(y^k) \subset \mathbb{R}^n \backslash \{o\}$ be a sequence tending to the origin. Then, it is not difficult to see that $T(y^k)$ tends to $-\infty$ and hence, from (8.1.42) we see that $\|H^{-1}(y^k)\|$ tends to zero since $\|H^{-1}(y^k)\| = \exp(T(y^k))$. This establishes the continuity of H^{-1} at the origin.

Altogether we see that H is a homeomorphism from $\mathbb{R}^n$ onto $\mathbb{R}^n$. From the very construction it follows immediately that H maps trajectories of $F_2(x) := -x$ onto trajectories of $F_1(x) := Ax$, preserving the orientation of integration. Hence, H (or H^{-1}) serves as a topological equivalence.

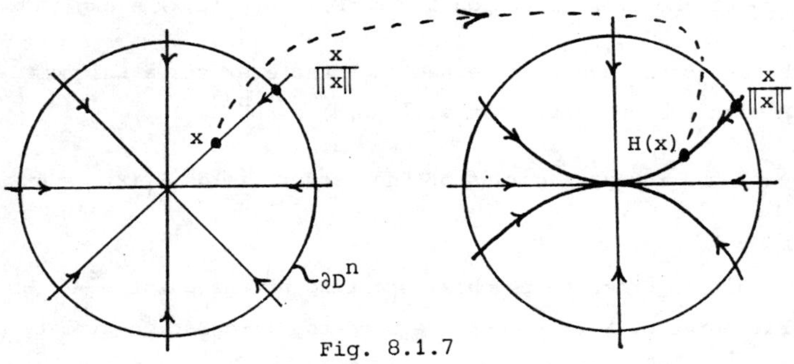

Fig. 8.1.7

The case m = 0 is accomplished by defining

$$H(x) = \Phi_1(\ln\|x\|, \frac{x}{\|x\|})$$

for $x \neq 0$.

Finally, we arrive to the case $0 < m < n$. In view of Lemma 8.1.1 we may assume (using a preceding linear coordinate transformation $y = Bx$) that A has the blockstructure as in the r.h.s. of (8.1.30). According to that blockstructure we decompose a vector $x \in \mathbb{R}^n$ as $x = (x_1, x_2)$ where $x_1 \in \mathbb{R}^m$ and $x_2 \in \mathbb{R}^{n-m}$. Let Φ_1 (resp. Φ_2) denote the flow of the systems $\dot{x}_1 = A_1x_1$ (resp. $\dot{x}_2 = A_2x_2$). The required homeomorphism H then takes the form:

$$\left.\begin{aligned} H(x) &= (H_1(x_1), H_2(x_2)) , \\ H_1(0) &= 0 , H_2(0) = 0 , \\ H_1(x_1) &= \Phi_1(-\ln\|x_1\|, \frac{x_1}{\|x_1\|}) \text{ if } x_1 \neq 0 \\ H_2(x_2) &= \Phi_2(\ln\|x_2\|, \frac{x_2}{\|x_2\|}) \text{ if } x_2 \neq 0 \end{aligned}\right\} \quad (8.1.43)$$

□

Remark 8.1.3. Consider the vector field $F_2^k(x) := \left(\begin{array}{c|c} -I_k & 0 \\ \hline 0 & I_{n-k} \end{array}\right) x.$

The vector field F_2 in (8.1.38) is precisely F_2^m. One might ask whether the vector field F_1 in Theorem 8.1.2 can be topologically equivalent to F_2^k with $k \neq m$. Or, otherwise stated: can F_2^k be topologically equivalent to F_2^m, $k \neq m$?

This, however, cannot be true in view of the following reasoning. If F_2^k is topologically equivalent to F_2^m, then the stable manifold of F_2^k (corresponding to the origin) is mapped homeomorphically under the topological equivalence, onto the stable manifold of F_2^m. But the stable manifold of $F_2^k(F_2^m)$ is homeomorphic with $\mathbb{R}^k(\mathbb{R}^m)$ and the linear spaces $\mathbb{R}^k$ and $\mathbb{R}^m$ are homeomorphic only if k = m, the latter fact being a consequence of the famous theorem of Brouwer on the invariance of domain (cf. [73]).

Remark 8.1.4. In Theorem 8.1.2 we restricted ourselves to the case of a real n×n matrix with nonvanishing real eigenvalues. However, with slight modifications, the statement of Theorem 8.1.2 remains true in the case that A is a real n×n matrix with eigenvalues having a nonvanishing real part (m denoting the number of them with negative real part); i.e. the origin is a hyperbolic singularity for the linear system $\dot{x} = Ax$ rather than a real hyperbolic singularity (see also [30, page 129]).

8.2. Variable metric and gradient systems.

We recall that an inner product $\langle\langle\cdot,\cdot\rangle\rangle: \mathbb{R}^n \times \mathbb{R}^n \to \mathbb{R}$ is a symmetric, bilinear and positive definite mapping. The associated norm $|||\cdot|||$ is defined as $|||x||| = \langle\langle x,x\rangle\rangle^{\frac{1}{2}}$. The next obvious lemma connects inner products and symmetric positive definite matrices.

Lemma 8.2.1. Let $\langle\langle\cdot,\cdot\rangle\rangle$ be an inner product on $\mathbb{R}^n$. Then the n×n matrix $A = (a_{ij})$, defined by $a_{ij} = \langle\langle e_i, e_j\rangle\rangle$ with e_i as the i-th unit vector, is a symmetric positive definite matrix. Moreover, we have: $\langle\langle x,y\rangle\rangle = x^T Ay$ (A is called the generating matrix for $\langle\langle\cdot,\cdot\rangle\rangle$).
On the other hand, if A is a symmetric positive definite n×n matrix, then $\langle\langle x,y\rangle\rangle := x^T Ay$ defines an inner product on $\mathbb{R}^n$. □

With the aid of an inner product we can define the gradient of a diffentiable function.

Definition 8.2.1. Let $f \in C^\infty(\mathbb{R}^n, \mathbb{R})$ and let $\langle\langle\cdot,\cdot\rangle\rangle$ be an inner product on $\mathbb{R}^n$. The gradient grad f(x) of f with respect to $\langle\langle\cdot,\cdot\rangle\rangle$ at a point $x \in \mathbb{R}^n$ is the unique solution vector of the following linear equation:

$$\langle\langle \text{grad } f(x), \xi\rangle\rangle = Df(x)\xi \quad , \quad \xi \in \mathbb{R}^n. \tag{8.2.1}$$

In terms of the generating matrix A for an inner product $\langle\langle\cdot,\cdot\rangle\rangle$ we can give an explicit expression for grad f(x). In fact, noting that (8.2.1) is satisfied for all $\xi \in \mathbb{R}^n$ iff it holds for the unit vectors e_i, i = 1,...,n, we obtain form (8.2.1)

$$A.\text{grad } f(x) = D^T f(x), \tag{8.2.2.a}$$

or, equivalently,

$$\text{grad } f(x) = A^{-1} D^T f(x). \tag{8.2.2.b}$$

Exercise 8.2.1. Suppose that $Df(\bar{x}) \neq 0$. Then show that grad $f(\bar{x})$ is a positive multiple of the solution of the following optimization problem:

$$\text{maximize } Df(\bar{x})\xi \quad \text{on} \quad \{\xi \,\big|\, |||\xi||| = 1\} . \tag{8.2.3}$$

Remark 8.2.1. If $Df(\bar{x}) \neq 0$, then the level set $\{x \in \mathbb{R}^n \,|\, f(x) = f(\bar{x})\}$ is a C^∞-manifold in a neighborhood of $\bar{x}$ and its tangent space T at $\bar{x}$ becomes $T = \{\xi \in \mathbb{R}^n \,|\, Df(\bar{x})\xi = 0\}$. From (8.2.1) we see that grad $f(\bar{x})$ is orthogonal to T with respect to the inner product $\langle\langle\cdot,\cdot\rangle\rangle$. As a consequence, the direction of grad $f(\bar{x})$ is completely determined by both the latter orthogonality and the orientation $Df(\bar{x})\cdot\text{grad } f(\bar{x}) > 0$.

In optimization algorithms using gradient techniques, the orthogonality with respect to <<•,•>> is often connected with the generating matrix A. In this context, two vectors $\xi,\eta \in \mathbb{R}^n$ are called <u>conjugate</u> with respect to A, or <u>A-orthogonal</u>, if $\xi^T A\eta = 0$ (cf. [26]). Conjugacy has a nice geometric interpretation. In fact, consider the quadratic function $g(x) = \frac{1}{2}x^T Ax$, whose level sets are ellipsoids, since A is positive definite.

Let us fix $k \in \{1,\ldots,n-1\}$ and let $\xi_1,\ldots,\xi_k \in \mathbb{R}^n$ be linearly independent. For two points $\bar{x}_1,\bar{x}_2 \in \mathbb{R}^n$, the sets

$$L_i = \{\bar{x}_i + \sum_{j=1}^{k} \alpha_j\xi_j \mid \alpha_j \in \mathbb{R},\ j = 1,\ldots,k\},\ i = 1,2,$$

are parallel planes of dimension k.

Let $\tilde{x}_i \in L_i$ be the global minimum for $g_{|L_i}$, $i = 1,2$.

<u>Proposition</u>: The vector $\tilde{x}_2 - \tilde{x}_1$ is conjugate to every vector ξ_j, $j = 1,\ldots,k$, with respect to A (cf. Fig. 8.2.1.a).

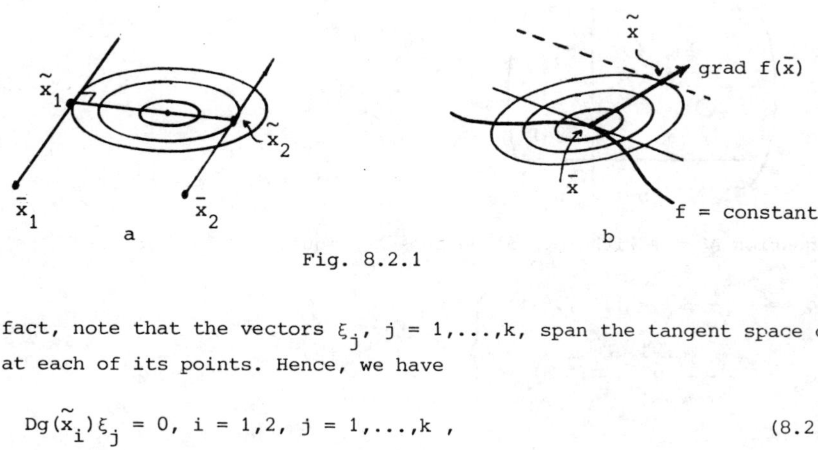

Fig. 8.2.1

In fact, note that the vectors ξ_j, $j = 1,\ldots,k$, span the tangent space of L_i at each of its points. Hence, we have

$$Dg(\tilde{x}_i)\xi_j = 0,\ i = 1,2,\ j = 1,\ldots,k\ , \tag{8.2.4}$$

since $\tilde{x}_i$ is in particular a critical point for $g_{|L_i}$, $i = 1,2$.

But $Dg(\tilde{x}_i) = \tilde{x}_i^T A$. Substituting this in (8.2.4) and taking the difference yields:

$$0 = (Dg(\tilde{x}_2)-Dg(\tilde{x}_1))\xi_j = (\tilde{x}_2-\tilde{x}_1)^T A\xi_j,\ j = 1,\ldots,k\ ,$$

which proves the proposition.

From the above observation the direction of grad $f(\bar{x})$ at a point $\bar{x}$ with $Df(\bar{x}) \neq 0$ can be obtained as follows. Take a plane V of dimension $(n-1)$ not passing through $\bar{x}$ but parallel to the tangent space of the level set of f at $\bar{x}$. Let $\tilde{x}$ be the global minimum of the quadratic function $\frac{1}{2}(x-\bar{x})^T A(x-\bar{x})$ on V. Then the direction of grad $f(\bar{x})$ coincides with the direction of either $(\tilde{x}-\bar{x})$ or $(\bar{x}-\tilde{x})$, the choice of which is unique since $Df(\bar{x}).\text{grad } f(\bar{x}) > 0$ (cf. Fig. 8.2.1.b). □

Remark 8.2.2. Different inner products may produce the same gradient at some point $\bar{x}$: in fact, the equation $A\xi = \eta$ has, for fixed nonvanishing ξ and η, many solutions for the $n\times n$ matrix A.
We will now actually construct, for given ξ and η with $\xi^T\eta > 0$, a symmetric positive definite $n\times n$ matrix A with $A\xi = \eta$.

Case 1. We consider the special case: $\eta = (0,\dots,0,\alpha)$, $\alpha \neq 0$. So, $\xi^T\eta = \xi_n\alpha > 0$. We take a special structure for A which makes it unique:

$$A = \left(\begin{array}{cccc|c} 1 & & & & \alpha_1 \\ & 1 & & 0 & \alpha_2 \\ & & \ddots & & \vdots \\ & 0 & & 1 & \alpha_{n-1} \\ \hline \alpha_1 & \alpha_2 & \dots & \alpha_{n-1} & \alpha_n \end{array}\right) \tag{8.2.5}$$

The equation $A\xi = \eta$ with (8.2.5) defines A uniquely:

$$\left.\begin{aligned} \alpha_i &= -\frac{\xi_i}{\xi_n}, \quad i = 1,\dots,n-1 \\ \alpha_n &= \frac{\alpha}{\xi_n} + \sum_{i=1}^{n-1} \left(\frac{\xi_i}{\xi_n}\right)^2 \end{aligned}\right\} \tag{8.2.6}$$

So, it remains to show that A is positive definite.
Let e_i be the i-th unit vector. Since $\xi_n \neq 0$ we see that the vectors $e_1,\dots,e_{n-1},\xi$ form a basis for $\mathbb{R}^n$ and we write a vector $x \in \mathbb{R}^n$ with respect to this basis as follows:

$$x = \lambda_1 e_1 + \dots + \lambda_{n-1}e_{n-1} + \lambda_n\xi$$

It follows that $x^T Ax = \sum_{i=1}^{n-1} \lambda_i^2 + \lambda_n^2 \cdot \xi_n \cdot \alpha$. Since $\xi_n \cdot \alpha > 0$ we see that A is positive definite.

Case 2. $\xi, \eta \in \mathbb{R}^n$ arbitrary, but $\xi^T \eta > 0$.
Choose an orthogonal matrix Q with $\eta/\|\eta\|$ as its last column. Hence, $Q^T \eta = (0,\ldots,0,\|\eta\|)$. Put $\tilde{\eta} = Q^T \eta$, $\tilde{\xi} = Q^T \xi$. Since Q is orthogonal we have in particular $\tilde{\xi}^T \tilde{\eta} = \xi^T \eta$ (> 0). Let $\tilde{A}$ be the symmetric positive definite matrix of the form (8.2.5) solving $\tilde{A} \cdot \tilde{\xi} = \tilde{\eta}$. Then, $A := Q\tilde{A}Q^T$ solves $A\xi = \eta$ and moreover, A is symmetric and positive definite as well. □

Instead of considering a fixed inner product $\langle\langle \cdot, \cdot \rangle\rangle$ on $\mathbb{R}^n$, it is also convenient to admit inner products which depend on the points of $\mathbb{R}^n$. This gives rise to the following definition.

Definition 8.2.2. Let $U \subset \mathbb{R}^n$ be an open subset and $\mathcal{A}$ the space of symmetric n×n matrices. A C^∞-variable metric or C^∞-Riemannian metric on U is a C^∞-mapping $\mathcal{R}: U \to \mathcal{A}$ with $\mathcal{R}(x)$ positive definite for all $x \in U$.

If $f \in C^\infty(U, \mathbb{R}^n)$ and $\mathcal{R}$ a C^∞-Riemannian metric on U, then the C^∞-gradient vector field $\mathrm{grad}_{\mathcal{R}} f(x)$ is defined, according to (8.2.2.b), as:

$$\mathrm{grad}_{\mathcal{R}} f(x) = \mathcal{R}(x)^{-1} \cdot D^T f(x) \quad , \; x \in U. \tag{8.2.7}$$

The gradient vector field $\mathrm{grad}_{\mathcal{R}} f(x)$ defines the gradient differential equation (resp. negative gradient differential equation):

$$\frac{dx}{dt} = \mathrm{grad}_{\mathcal{R}} f(x) \quad \left(\text{resp. } \frac{dx}{dt} = -\mathrm{grad}_{\mathcal{R}} f(x)\right) \tag{8.2.8}$$

Example 8.2.1. Let $\bar{x} \in \mathbb{R}^n$ be a nondegenerate local minimum for $f \in C^\infty(\mathbb{R}^n, \mathbb{R})$; so, $Df(\bar{x}) = 0$ and $D^2 f(\bar{x})$ positive definite. Let U be an open neighborhood of $\bar{x}$ where $D^2 f(x)$ positive definite. On U we consider the special Riemannian metric $\mathcal{R}: x \mapsto D^2 f(x)$. The corresponding negative gradient differential equation then becomes:

$$\frac{dx}{dt} = -\mathrm{grad}_{\mathcal{R}} f(x) = -D^2 f(x)^{-1} D^T f(x) \tag{8.2.9}$$

Since $Df(\bar{x})$ vanishes, the point $\bar{x}$ is a singular point for (8.2.9). For the Jacobian matrix at $\bar{x}$ we obtain:

$$D(-\mathrm{grad}_R f(x))_{|\bar{x}} = -I \tag{8.2.10}$$

Consequently, $\bar{x}$ is a singularity of real hyperbolic type and the local stable manifold is an open subset containing $\bar{x}$; so $\bar{x}$ is an attractor for (8.2.9) as $t \to \infty$.

The (simplest) Euler discretization of (8.2.9) becomes:

$$x_{i+1} - x_i = -\mathrm{grad}_R f(x_i) \quad , \text{ i.e.}$$

$$x_{i+1} = x_i - D^2 f(x_i)^{-1} D^T f(x_i). \tag{8.2.11}$$

Formula (8.2.11) is the well-known <u>Newton iteration formula</u> for the search of zeros of the mapping $x \mapsto D^T f(x)$. □

The next theorem gives a sufficient condition for a vector field to be the gradient vector field of a given function in <u>some</u> Riemannian metric.

<u>Theorem 8.2.1</u>. Let $U \subset \mathbb{R}^n$ be open and $f \in C^\infty(\mathbb{R}^n, \mathbb{R})$. Let $F \in C^\infty(U, \mathbb{R}^n)$ be a vector field with the property:

$$Df(x) \cdot F(x) > 0 \quad \text{for all} \quad x \in U. \tag{8.2.12}$$

Then, there exists a C^∞-Riemannian metric R on U such that $F(x) = \mathrm{grad}_R f(x)$.

<u>Proof</u>. Suppose that we have proved the following local part:

<u>Local part</u>: for each $\bar{x} \in U$ there exists an open neighborhood $U_{\bar{x}} \subset U$ of $\bar{x}$ together with a C^∞-Riemannian metric $R_{\bar{x}}$ on $U_{\bar{x}}$ such that F is the gradient vector field for f with respect to $R_{\bar{x}}$.

Then the globalization is obtained by using the technique of a C^∞-partition of unity of U subordinate to the covering $\{U_{\bar{x}}, \bar{x} \in U\}$ (cf. Section 2.2). In fact, note that <u>convex combinations</u> of positive definite matrices are positive definite, and that the equations $A_i \xi = \eta$, $i = 1,\dots,k$ imply that $\sum_{i=1}^{k} \lambda_i A_i \xi = \eta$ if $\sum_{i=1}^{k} \lambda_i = 1$.

So, let us return to the local part. In view of the construction made in Remark 8.2.2 we only need to define <u>locally</u> around $\bar{x} \in U$ a C^∞-mapping $x \mapsto Q(x)$, $Q(x)$ being an <u>orthogonal</u> $n \times n$ matrix with $D^T f(x)/\|Df(x)\|$ as its last column (recall that the matrix A in (8.2.5) is uniquely determined). But this can be achieved by means of a Gram-Schmidt orthogonalization technique (cf. also [26]):
Start by choosing $Q(\bar{x})$ as an orthogonal matrix with columns $\bar{\xi}_1, \ldots, \bar{\xi}_{n-1}$, $\bar{\xi}_n := D^T f(\bar{x})/\|Df(\bar{x})\|$.

Then, in a neighborhood of $\bar{x}$ the matrix $Q(x)$ with columns $\xi_1(x), \ldots, \xi_n(x)$ is well-defined and smooth if we put:

$$\xi_n(x) = D^T f(x)/\|Df(x)\|$$

$$\xi_{n-1}(x) = \underbrace{(\bar{\xi}_{n-1} - \langle \bar{\xi}_{n-1}, \xi_n(x) \rangle \xi_n(x))}_{\ast}/\|\ast\|$$

$$\xi_{n-2}(x) = \underbrace{(\bar{\xi}_{n-2} - \langle \bar{\xi}_{n-2}, \xi_{n-1}(x) \rangle \xi_{n-1}(x) - \langle \bar{\xi}_{n-2}, \xi_n(x) \rangle \xi_n(x))}_{\ast}/\|\ast\|$$

etc. ...

This completes the proof of the theorem. □

Let $U \subset \mathbb{R}^n$ be open, $f \in C^\infty(U, \mathbb{R})$ and $\mathcal{R}(x)$ a C^∞-Riemannian metric on U. From (8.2.7) we see that the <u>singular points</u> of the gradient differential equation are exactly the <u>critical points</u> of f. Let $\Phi(t,x)$ denote the flow of $\mathrm{grad}_{\mathcal{R}} f(x)$. For $\bar{x} \in U$ we consider the behaviour of f along an integral curve of $\mathrm{grad}_{\mathcal{R}} f(x)$. So we regard the composite function $\psi(t)$:

$$\psi(t) = f \circ \Phi(t, \bar{x}) \tag{8.2.13}$$

Then, recalling (8.2.7) we obtain for the derivative at $t = 0$:

$$\frac{d\psi}{dt}(0) = Df(\bar{x}) \cdot \mathcal{R}(\bar{x})^{-1} \cdot D^T f(\bar{x}). \tag{8.2.14}$$

Consequently, if $Df(\bar{x}) \neq 0$, then $\frac{d\psi}{dt}(0) > 0$ since $\mathcal{R}(\bar{x})^{-1}$ is positive definite. So, $\psi(t)$ increases strictly for increasing t if the starting point $\bar{x}$ is not a critical point. Hence, there are no periodic solutions of the gradient differential equation with positive period.

Let $\bar{x} \in U$ be a nondegenerate critical point of f. Then, the point $\bar{x}$, regarded as a singular point for the gradient differential equation, is of real hyperbolic type. This follows in particular from the following lemma, thereby noting that the Jacobian matrix of $\text{grad}_R f(x)$ at the critical point $\bar{x}$ becomes:

$$D(\text{grad}_R f(x))|_{\bar{x}} = R(\bar{x})^{-1} \cdot D^2 f(\bar{x}) \tag{8.2.15}$$

Lemma 8.2.2. Let A, B be nonsingular symmetric n×n matrices with A positive definite. Then the following statements hold:

a. The eigenvalues of AB are real and there exists a system of n linearly independent eigenvectors.

b. The number of negative (resp. positive) eigenvalues of AB equals the index (resp. coindex) of B.

c. Let L^- (resp. L^+) denote the subspace of $\mathbb{R}^n$ spanned by those eigenvectors of AB corresponding to negative (resp. positive) eigenvalues. Then, $B_{|L^-}$ (resp. $B_{|L^+}$) is negative (resp. positive) definite.

d. Let $\xi \in L^-$ and $\eta \in L^+$. Then $\xi A^{-1}\eta = 0$, i.e. ξ and η are A^{-1}-orthogonal.

Proof.

a). Let P be a (nonsingular) symmetric n×n matrix with A = PP. Let $\lambda_1, \ldots, \lambda_n$ be the eigenvalues of the symmetric matrix PBP with $w_1, \ldots, w_n$ pairwise orthogonal eigenvectors. From $PBPw_i = \lambda_i w_i$ we obtain $AB(Pw_i) = P(PBP)w_i = \lambda_i Pw_i$. Hence, λ_i, Pw_i, $i = 1, \ldots, n$ are eigenvalues, resp. eigenvectors for AB. Obviously Pw_i, $i = 1, \ldots, n$, are linearly independent since P is nonsingular.

It remains to construct the matrix P. Let Q be an orthogonal matrix whose columns are eigenvectors for the symmetric matrix A. So, $Q^T AQ = \Lambda := \text{diag}(\tilde{\lambda}_1, \ldots, \tilde{\lambda}_n)$. Since A is positive definite, its eigenvalues $\tilde{\lambda}_i$ are positive. Put $P = Q \cdot \text{diag}(\tilde{\lambda}_1^{\frac{1}{2}}, \ldots, \tilde{\lambda}_n^{\frac{1}{2}}) \cdot Q^T$. Then, $P = P^T$ and $PP = A$.

b). With the matrix P as above and Lemma 2.5.1, we see that (co)index of PBP = (co)index B. Now, Statement b follows from the fact that the eigenvalues of PBP and AB coincide.

c). Let $v \in L^-\backslash\{o\}$. Then, $w := P^{-1}v$ lies in the linear space spanned by the eigenvectors of PBP corresponding to negative eigenvalues. Statement c follows from the inequality:

$$0 > w^T(PBP)w = v^TBv.$$

d). If Statement d holds in the special case that ξ and η are eigenvectors of the matrix AB, then it obviously holds in the general case. So, let $AB\xi = \lambda\xi$, $AB\eta = \mu\eta$ with $\lambda < 0$, $\mu > 0$. Put $v = P^{-1}\xi$, $w = P^{-1}\eta$. Then, v, w are <u>eigenvectors</u> of the <u>symmetric</u> matrix PBP corresponding to the <u>different</u> <u>eigenvalues</u> λ,μ, which implies that $v^Tw = 0$.
But, $v^Tw = (P^{-1}\xi)^T(P^{-1}\eta) = \xi^TP^{-2}\eta = \xi^TA^{-1}\eta$, and hence, $\xi^TA^{-1}\eta = 0$. □

Let k be the <u>(quadratic) index</u> of the nondegenerate critical point $\bar{x}$, i.e. k = number of negative eigenvalues of $D^2f(\bar{x})$. From Lemma 8.2.2 we see that $R(\bar{x})^{-1}D^2f(\bar{x})$ has k negative, resp. (n-k) positive eigenvalues.
Let V^-, resp. V^+, denote the k, resp. (n-k) dimensional linear subspace spanned by the eigenvectors of $R(\bar{x})^{-1}D^2f(\bar{x})$ corresponding to negative, resp. positive eigenvalues.
From Lemma 8.2.2.d we see that V^- and V^+ are $R(\bar{x})$-orthogonal. Let W^s, resp. W^u be, <u>locally</u> around $\bar{x}$, the stable, resp. unstable manifold for the gradient vector field (8.2.7). From Section 8.1 it follows that the tangent space of W^s (W^u) at $\bar{x}$ just equals V^- (V^+).
Since $Df(\bar{x}) = 0$, the point $\bar{x}$ is also a critical point for $f_{|W^s}$ and $f_{|W^u}$.
In any local representation of W^s by means of a <u>defining system</u> of (n-k) C^∞-functions, the <u>Lagrange parameters</u> at $\bar{x}$ <u>vanish</u>. Hence, the quadratic (co)index of $\bar{x}$ as a critical point for $f_{|W^s}$ is equal to the (co)index of $D^2f(\bar{x})_{|V^-}$ (see also Example 2.5.3).
Consequently, $\bar{x}$ is a <u>nondegenerate local maximum</u> for $f_{|W^s}$. But then, for sufficiently small $\varepsilon > 0$ the set $W^s \cap \{x \in \mathbb{R}^n | f(x) = f(\bar{x})-\varepsilon\}$ is C^∞-diffeomorphic to the sphere S^{k-1} of dimension k-1 (cf. also the Morse-lemma).
So, locally, the stable manifold W^s (having dimension k) is generated by means of a sphere S^{k-1}, embedded in a natural way via f in W^s, for $k \geq 1$.
(The case k = 0 makes no real sense within this context, since then $W^s = \{\bar{x}\}$ and the corresponding sphere S^{-1} is -formally- the empty set).
Similarly, W^u is generated by means of a sphere S^{n-k-1} (cf. Fig. 8.2.2).

The spheres, S^{k-1}/S^{n-k-1} generating W^s/W^u actually characterize the singularity $\bar{x}$.

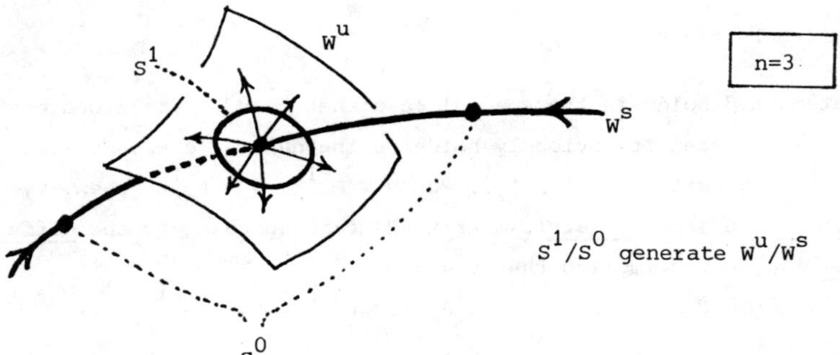

Fig. 8.2.2

Remark 8.2.3. Let $U, V \subset \mathbb{R}^n$ be open and $\psi: U \to V$ a C^∞-diffeomorphism. If $f \in C^\infty(U,\mathbb{R})$ and $\mathcal{R}$ a C^∞-Riemannian metric on U, then $\text{grad}_{\mathcal{R}} f(x)$ transforms under ψ to a gradient vector field on V in the following natural way. Consider the composed function $g \in C^\infty(V,\mathbb{R})$:

$$g(y) = f \circ \psi^{-1}(y). \tag{8.2.16}$$

According to Remark 8.1.1 the (gradient) vector field $\mathcal{R}(x)^{-1} D^T f(x)$ transforms under ψ into the vector field

$$(D\psi \cdot \mathcal{R}^{-1} \cdot D^T f) \circ \psi^{-1}(y). \tag{8.2.17}$$

From (8.2.16) we obtain $D^T g(y) = D^T \psi^{-1}(y) \cdot D^T f(\psi^{-1}(y))$, and substitution into (8.2.17) produces the vector field:

$$(D\psi \cdot \mathcal{R}^{-1} \cdot D^T \psi)_{|\psi^{-1}(y)} \cdot D^T g(y). \tag{8.2.18}$$

Note that $(D\psi(\psi^{-1}(y)))^{-1} = D\psi^{-1}(y)$ and put

$$\tilde{\mathcal{R}}(y) = D^T \psi^{-1}_{|y} \cdot \mathcal{R}_{|\psi^{-1}(y)} \cdot D\psi^{-1}_{|y} . \tag{8.2.19}$$

Then, $\tilde{\mathcal{R}}(y)$ in (8.2.19) is also symmetric, positive definite and hence, a Riemannian metric. As a consequence, the vector field in (8.2.18) is nothing else but $\mathrm{grad}_{\tilde{R}}\, g(y)$. □

The concept of a gradient vector field can be easily extended to manifolds in $\mathbb{R}^n$.
In fact, let $M \subset \mathbb{R}^n$ be a C^∞-manifold, U an open ($\mathbb{R}^n$-)neighborhood of M and R a C^∞-Riemannian metric on U. For $f \in C^\infty(U,\mathbb{R})$ the <u>gradient vector</u> $\mathrm{grad}_R f_{|M}$ at $x \in M$ now becomes the <u>unique vector in the tangent space</u> T_xM satisfying:

$$(\mathrm{grad}_R f_{|M}(x))^T \cdot R(x) \cdot \xi = Df(x) \cdot \xi, \; \xi \in T_xM \tag{8.2.20}$$

In particular, we obtain that $\mathrm{grad}_R f_{|M} \in C^\infty(M,\mathbb{R}^n)$ (cf. Definition 3.1.4).

<u>Exercise 8.2.2</u>. Let M be a C^∞-manifold in $\mathbb{R}^n$ of codimension m. Put $R(x) \equiv I_n$; so $R(x)$ produces the standard inner product on $\mathbb{R}^n$.

a). Show that $\mathrm{grad}_R f_{|M}(x)$ is the <u>orthogonal projection</u> of $D^Tf(x)$ on T_xM.

Let $\bar{x} \in M$, $U_{\bar{x}}$ an open $\mathbb{R}^n$-neighborhood of $\bar{x}$ and $h_i \in C^\infty(U_{\bar{x}},\mathbb{R})$, $i = 1,\ldots,m$ a defining system of functions for $M \cap U_{\bar{x}}$.
Choose $\xi_j \in \mathbb{R}^n$, $j = m+1,\ldots,n$, such that $D^Th_i(\bar{x}), \xi_j, i = 1,..,m, j = m+1,..,n$, form a basis for $\mathbb{R}^n$.
Consider the local coordinate transformation $y := \Phi(x)$ sending $\bar{x}$ onto the origin, defined by

$$y_i = h_i(x),\; i = 1,\ldots,m\;,\; y_j = \xi_j^T(x-\bar{x})\;,\; j = m+1,\ldots,n.$$

Put $h = (h_1,\ldots,h_m)^T$, $A = (\xi_{m+1}|\ldots|\xi_n)$ and
$B(x) = A^T[I_n - D^Th(Dh\cdot D^Th)^{-1}Dh_{|x}]A$.

b). Show that B(x) is positive definite for x in a neighborhood of $\bar{x}$ and that (for $x \in M$) the gradient vector field $\mathrm{grad}_R f_{|M}(x)$ transforms under Φ into the vector field

$$\left(\begin{array}{c|c} 0 & 0 \\ \hline 0 & B(x) \end{array}\right) D^T g_{|\Phi(x)}\;, \text{ where } g = f \circ \Phi^{-1}.$$

□

8.3. Gradient systems in general position.

In this section we proceed with the discussion on stable and unstable manifolds corresponding to singularities of gradient vector fields. The original treatment of S. Smale on this subject (cf. [69]) is very instructive and is to be seen as the basis for this section.

In the preceding two sections we considered (gradient) vector fields mainly defined on an open subset of $\mathbb{R}^n$. However, we also paid attention to changes of coordinates and so, we may equally well treat (gradient) vector fields on manifolds (cf. also Definition 3.2.5); the verification of omitted details are left to the reader.

The main assumptions in this section are:

- a. $M \subset \mathbb{R}^n$ is a compact C^∞-manifold of dimension m.
- b. $\mathcal{R}$ is a C^∞-Riemannian metric on some open $\mathbb{R}^n$-neighborhood of M.
- c. $f \in C^\infty(\mathbb{R}^n,\mathbb{R})$ and $f_{|M}$ is nondegenerate.

Recall that the set of functions $f \in C^\infty(\mathbb{R}^n,\mathbb{R})$ with $f_{|M}$ nondegenerate is C^k-open and dense, for all $k \geq 2$ (cf. Chapter 7). Consequently, from the point of view of "general position", the nondegeneracy of $f_{|M}$ is not a severe restriction.

We are interested in the gradient vector field $\mathrm{grad}_{\mathcal{R}} f_{|M}$ (cf. (8.2.20)).

The singularities of $\mathrm{grad}_{\mathcal{R}} f_{|M}$ are exactly the critical points for $f_{|M}$ and they are of real hyperbolic type since $f_{|M}$ is nondegenerate. From the compactedness of M we conclude that any C^∞-vector field on M, and thus in particular $\mathrm{grad}_{\mathcal{R}} f_{|M}$, is completely integrable. Hence, the flow $\Phi(t,x)$ of the gradient vector field is a smooth map from $\mathbb{R} \times M$ to M.

Let $\bar{x} \in M$ be a critical point for $f_{|M}$ of (quadratic) index k. Then, the stable (resp. unstable) "manifold" $W^s_{\bar{x}}$ (resp. $W^u_{\bar{x}}$) of grad $f_{|M}$ corresponding to $\bar{x}$ is defined to be the set of points x with $\lim_{t\to+\infty} \Phi(t,x) = \bar{x}$ (resp. $\lim_{t\to-\infty} \Phi(t,x) = \bar{x}$). If $x \in M$ is not a critical point for $f_{|M}$, then $f\circ\Phi(t,x)$ is strictly monotone increasing. Consequently, we have the following result:

$$W^s_{\bar{x}} \cap W^u_{\bar{x}} = \{\bar{x}\}. \tag{8.3.1}$$

In particular, we see that $W^s_{\bar{x}}$ (resp. $W^u_{\bar{x}}$) is indeed a C^∞-manifold of dimension k (resp. m-k, where m = dim M); compare also Remark 8.1.2.

Moreover, if $\bar{x},\bar{y} \in M$ are <u>different</u> critical points for $f_{\lceil M}$, then:

$$W^s_{\bar{x}} \cap W^s_{\bar{y}} = \emptyset \quad \text{and} \quad W^u_{\bar{x}} \cap W^u_{\bar{y}} = \emptyset , \tag{8.3.2}$$

and

$$W^s_{\bar{x}} \cap W^u_{\bar{y}} = \emptyset \quad \text{if} \quad f(\bar{x}) \le f(\bar{y}). \tag{8.3.3}$$

On the other hand it is not exceptional (as it will become clear) that $W^s_{\bar{x}} \cap W^u_{\bar{y}} \neq \emptyset$ in case $f(\bar{x}) > f(\bar{y})$. In that case the intersection $W^s_{\bar{x}} \cap W^u_{\bar{y}}$ consists of a union of (1-dimensional) integral curves of $\text{grad}_R f_{|M}$. To be more precise, let $S^\alpha \subset M$, resp. $S^\beta \subset M$, be a generating sphere for $W^s_{\bar{x}}$, resp. $W^u_{\bar{y}}$ (cf. Section 8.2). Then $W^s_{\bar{x}} \cap W^u_{\bar{y}}$ can be parametrized by means of the subset of S^α consisting of those points which are connected by means of an integral curve with some point of S^β. Moreover, if $W^s_{\bar{x}}$ and $W^u_{\bar{x}}$ <u>intersect transversally</u> at some $\bar{z} \in M$, then they <u>intersect transversally at all points</u> $\Phi(t,\bar{z})$, $t \in \mathbb{R}$ (recall that $\Phi_t(\cdot) := \Phi(t,\cdot)$ is a C^∞-diffeomorphism on M).

Before we state the transversality theorem of this section, we need to introduce the concept of C^k-topology of the linear space of C^∞-vector fields on the compact C^∞- manifold M. To this aim we cover M by means of a <u>finite</u> number of <u>bounded</u>, ($\mathbb{R}^n$-)open subsets V_i, $i = 1,\dots,r$, with $V_i \subset \bar{V}_i \subset \tilde{V}_i$ where $\bar{V}_i$ is the closure of V_i and $\tilde{V}_i$ is open, such that $\tilde{V}_i$ is a local coordinate neighborhood for M; i.e. there exists an open neighborhood $\tilde{U}_i$ of the origin in $\mathbb{R}^n$ and a C^∞-diffeomorphism $\psi_i: \tilde{V}_i \to \tilde{U}_i$ with $\psi_i(\tilde{V}_i \cap M) = \tilde{U}_i \cap (\{0_{n-m}\} \times \mathbb{R}^m)$; see also Theorem 3.1.1.
In this way, a smooth vector field on M is transformed for each i, under the Jacobian matrix of ψ_i, to a smooth vector field on $\tilde{U}_i \cap (\{0_{n-m}\} \times \mathbb{R}^m)$. Taking the derivatives up to order k of the transformed vector field and restricting them to the <u>compact</u> set $\psi_i(\bar{V}_i)$, <u>$i = 1,\dots,r$</u>, the C^k-topology for C^∞-vector fields on M is obviously defined.

<u>Exercise 8.3.1</u>. Fill in the details for the definition of the C^k-topology of C^∞-vector fields on M, and show that it is independent from the chosen finite covering of M by means of the sets V_i, $i = 1,\dots,r$.

The next (transversality) theorem is the main goal of this section.

<u>Theorem 8.3.1. (S. Smale)</u>. Let f, M, $\mathcal{R}$ satisfy the main assumptions of this section.

Then, for every k, there exists a C^∞-vector field χ on M satisfying:

(i) χ arbitrarily C^k-close to $\mathrm{grad}_{\mathcal{R}} f_{|M}$.

(ii) χ coincides with $\mathrm{grad}_{\mathcal{R}} f_{|M}$ in a neighborhood of the critical points for $f_{|M}$.

(iii) <u>All stable and unstable manifolds</u> corresponding to singular points of χ intersect <u>transversally</u>.

<u>Remark 8.3.1</u>. From Theorem 8.3.1 (i) and (ii) it follows that χ can be chosen in such a way that the singular points of χ coincide with the critical points for $f_{|M}$ and that $Df(x)\cdot\chi(x) > 0$ on M <u>outside</u> the critical point set of $f_{|M}$. But then, it <u>essentially</u> follows from Theorem 8.2.1 that a new Riemannian metric $\mathcal{R}(\chi)$ on some neighborhood of M exists with the property:

$$\chi = \mathrm{grad}_{\mathcal{R}(\chi)} f_{|M} \ . \qquad \square$$

<u>The proof of Theorem 8.3.1.</u>

Let $c_1 < c_2 < \ldots < c_p$ be the critical values of $f_{|M}$ and choose numbers $\gamma_i \in \mathbb{R}$ satisfying

$$\gamma_0 < c_1 < \gamma_1 < \ldots < \gamma_{p-1} < c_p < \gamma_p \ . \tag{8.3.4}$$

For fixed $i \in \{0,1,\ldots,p\}$, consider the set $M(\gamma_i)$ defined by

$$M(\gamma_i) = \{x \in M \mid f(x) > \gamma_i\} \ . \tag{8.3.5}$$

<u>Step 1</u>. Suppose that χ is a C^k-approximation of $\mathrm{grad}_{\mathcal{R}} f_{|M}$ satisfying (i) and (ii) in Theorem 8.3.1, and, moreover, also satisfying the following relaxed condition ($W^s_{\bar{x}}$, $W^u_{\bar{y}}$ corresponding to χ):

(iii)* $W^s_{\bar{x}}$ and $W^u_{\bar{y}}$ intersect transversally for every pair of critical points $\bar{x},\bar{y}$ of $f_{|M}$ lying in $M(\gamma_i)$.

Then, Condition (iii)* remains true if we replace χ by another approximation $\tilde{\chi}$ satisfying $\tilde{\chi}_{|M(\gamma_i)} = \chi_{|M(\gamma_i)}$.

Note. If i = p, then (iii)* is obvious; if i = 0, then (iii)* coincides with condition (iii) in Theorem 8.3.1.

Step 2. Let χ satisfy (i), (ii) in Theorem 8.3.1 and condition (iii)* above. Let $\bar{x}$ be a critical point for $f_{|M}$ with $f(\bar{x}) = c_i$.

Then, given an arbitrarily small neighborhood $\mathcal{O}_{\bar{x}}$ of $\bar{x}$ and a positive integer k, there exists a vector field $\tilde{\chi}$ on M satisfying:

1. $\tilde{\chi}$ coincides with χ outside $\mathcal{O}_{\bar{x}}$ and on some open neighborhood $V_{\bar{x}} \subset \mathcal{O}_{\bar{x}}$ of $\bar{x}$.
2. $\tilde{\chi}$ is arbitrarily C^k-close to χ.
3. $W^u_{\bar{x}}$ intersects $W^s_{\bar{y}}$ transversally for all critical points $\bar{y}$ of $f_{|M}$ lying in $M(\gamma_i)$.
   (Here, $W^u_{\bar{x}}$ and $W^s_{\bar{y}}$ correspond to $\tilde{\chi}$).

Step 3. Suppose that Step 2 has been carried out successively for all critical points of $f_{|M}$ with functional value equal to c_i, and denote the resulting vector field by $\approx\!\!\chi$. Then, with (8.3.1) - (8.3.3) it is easily seen that $\approx\!\!\chi$ satisfies the assumption in Step 1 with χ replaced by $\approx\!\!\chi$ and γ_i replaced by γ_{i-1}.

Now, from the final note in Step 1 it follows that the proof is complete, provided that Step 2 is verified. □

The verification of Step 2 above will firstly be illustrated by showing how to break a saddle-connection in $\mathbb{R}^2$:
To this aim let $U \subset \mathbb{R}^2$ be an open subset, $f \in C^\infty(U,\mathbb{R})$ a nondegenerate function and $\mathcal{R}$ a smooth Riemannian metric on U. Moreover, let $\bar{x},\bar{y} \in U$ be distinct critical points for f of (quadratic) index 1, and suppose that there is a trajectory of $\mathrm{grad}_{\mathcal{R}} f$ connecting $\bar{x}$ and $\bar{y}$ (cf. Fig. 8.3.1).
Choose a point $\bar{z}$ on this trajectory. In view of the "flow-box" construction in Section 8.1, we can find a compact neighborhood $C_{\bar{z}}$ of $\bar{z}$, positive numbers

ε, δ, and a C^∞-diffeomorphism ψ: $C_{\bar{z}} \to [-\varepsilon,\varepsilon] \times [-\delta,\delta]$ with the property that $\psi(\bar{z}) = 0$ and that the transformed vector field $\mathrm{grad}_R f$ becomes the constant vector field $(1,0)^T$.

The new local coordinates are denoted by (y_1,y_2) and as an abbreviation we put

$$Q = [-\varepsilon,\varepsilon] \times [-\delta,\delta] . \tag{8.3.6}$$

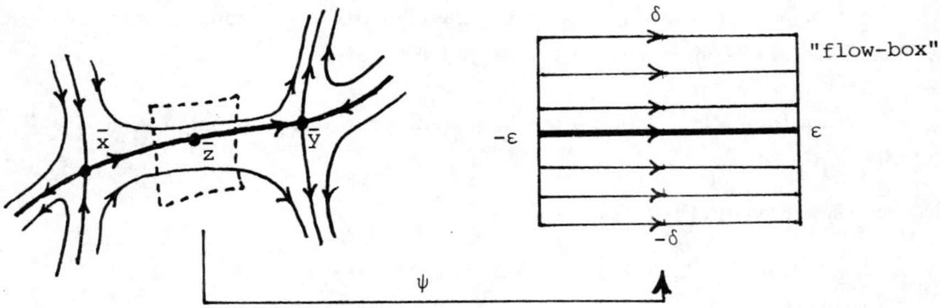

Fig. 8.3.1

The idea for breaking the saddle-connection consists of a perturbation of the constant vector field $(1,0)^T$ inside the flow-box Q in such a way that after the perturbation the trajectory coming from $\bar{x}$ enters the flow-box at the same point, but leaves it at a different y_2-level (cf. Fig. 8.3.2).

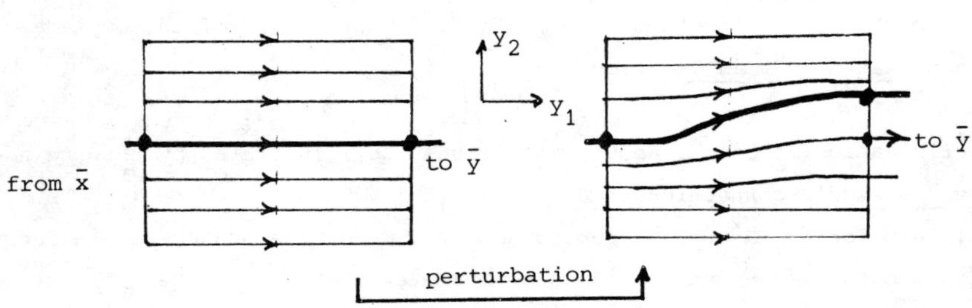

Fig. 8.3.2

This perturbation can be achieved by moving the trajectories up (or down) by means of a local addition of a certain "amount" of the complementary direction $(0,1)^T$ to the constant vector field $(1,0)^T$ (cf. Fig. 8.3.3.a). The global effect will be as depicted in Fig. 8.3.3.b.

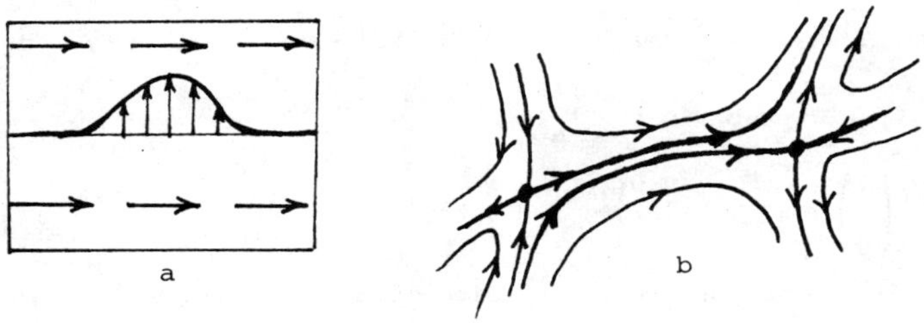

Fig. 8.3.3

For the mathematical realization of the mentioned perturbation we introduce two functions $\xi, \eta \in C^\infty(\mathbb{R},\mathbb{R})$ with the properties:

$$\left.\begin{array}{l} \xi(x) \geq 0 \quad \text{for all } x \quad ; \quad \xi(x) = 0 \text{ outside } [-\frac{1}{2}\varepsilon, \frac{1}{2}\varepsilon] \\ \int_{-\varepsilon}^{\varepsilon} \xi(x)dx = 1 \end{array}\right\} \quad (8.3.7)$$

$$\left.\begin{array}{l} 0 \leq \eta(x) \leq 1 \quad \text{for all } x \\ \eta(x) = 1 \text{ on } [-\frac{1}{2}\delta, \frac{1}{2}\delta] \; ; \; \eta(x) = 0 \text{ outside } [-\frac{3}{4}\delta, \frac{3}{4}\delta] \end{array}\right\} \quad (8.3.8)$$

The function $\xi(y_1)$ will be used in order to move trajectories of the constant vector field $(1,0)^T$ in the y_2-direction, whereas the function $\eta(|y_2|)$ takes care that the perturbation is also tapered off in the y_2-direction.

Since the functions ξ and η have compact support, each of their derivatives is uniformly bounded. Hence, given k, the function $\alpha \cdot \xi(y_1) \cdot \eta(|y_2|)$ is arbitrarily C^k-close to the zero-function for sufficiently small $|\alpha|$. This allows us to produce a perturbed vector field which is arbitrarily C^k-close to the original vector field $\text{grad}_R f$.

On the flow-box Q we consider the vector field F_α, depending on the additional real parameter α:

$$F_\alpha(y_1,y_2) = \begin{pmatrix}1\\0\end{pmatrix} + \alpha\cdot\xi(y_1)\cdot\eta(|y_2|)\begin{pmatrix}0\\1\end{pmatrix} \qquad (8.3.9)$$

Define the subsets $\tilde{Q}$, $\tilde{\tilde{Q}}$ of Q as follows:

$$\tilde{Q} = \{y \in Q|\ |y_2| \le \tfrac{1}{2}\delta\}\ ;\ \tilde{\tilde{Q}} = \{y \in Q|\ |y_2| \le \tfrac{1}{4}\delta\} \qquad (8.3.10)$$

As a consequence, we obtain for F_α:

$$F_\alpha(y) = \begin{pmatrix}1\\0\end{pmatrix} + \alpha\cdot\xi(y_1)\begin{pmatrix}0\\1\end{pmatrix},\ \text{for } y \in \tilde{Q}\ . \qquad (8.3.11)$$

Integration of the vector field F_α, using (8.3.7) yields:

If $|\alpha| \le \frac{1}{4}\delta$, then the trajectory of F_α through any point of $\tilde{\tilde{Q}}$ remains in $\tilde{Q}$.

Let Φ^α denote the flow of F_α. If follows:

$$\Phi^\alpha(2\varepsilon,(-\varepsilon,0)) = (\varepsilon,\alpha) \quad \text{for } |\alpha| \le \tfrac{1}{4}\delta \qquad (8.3.12)$$

So, in order to break the saddle-connection of $\text{grad}_R f$ we can take any α satisfying $0 < |\alpha| \le \frac{1}{4}\delta$.

After this intermezzo we proceed with the verification of Step 2 in the proof of Theorem 8.3.1 (This will be done in $\mathbb{R}^n$ rather than in M, since the construction is done essentially in local coordinates).
To this aim let $U \subset \mathbb{R}^n$ be an open subset, $f \in C^\infty(U,\mathbb{R})$ a nondegenerate function, R a smooth Riemannian metric on U and $\bar{x} \in U$ a critical point for f with (quadratic) index k. (It will become clear from the sequel that the only relevant case is: $0 < k < n$; this will tacitly be assumed).
The unstable manifold $W^u_{\bar{x}}$ for $\text{grad}_R f$ then has dimension n-k. We will construct a (generalized) "flow-box" in which $W^u_{\bar{x}}$ takes a simple form. Without loss of generality we assume:

$$\bar{x} = 0\ ,\ f(\bar{x}) = 0\ ,\ T_{\bar{x}}W^u_{\bar{x}} = \mathbb{R}^{n-k}\times\{0_k\} \qquad (8.3.13)$$

Obviously, there exists an open neighborhood $\mathcal{O}_{\bar{x}}$ of $\bar{x}$ with:

1. $\bar{x}$ is the only critical point in the closure of $\mathcal{O}_{\bar{x}}$
2. The manifolds $W^u_{\bar{x}}$ and $(x + \{0_{n-k}\} \times \mathbb{R}^k)$ intersect transversally at all points $x \in \mathcal{O}_{\bar{x}}$.

Next, we take $c > 0$ so small that the generating sphere $f^{-1}(c) \cap W^u_{\bar{x}}$ for $W^u_{\bar{x}}$ is completely contained in $\mathcal{O}_{\bar{x}}$ (cf. Fig. 8.3.4), and we put

$$S = f^{-1}(c) \cap W^u_{\bar{x}}. \tag{8.3.14}$$

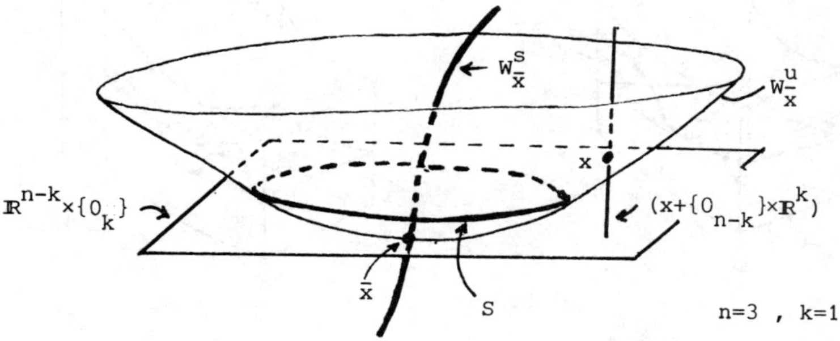

Fig. 8.3.4

Let Φ denote the flow of $\text{grad}_R f$. Next, let s denote a point of S and let v denote an $\mathbb{R}^k$-vector. Finally put

$$D_\delta = \{v \in \mathbb{R}^k \mid \|v\| \leq \delta\}, \ \mathring{D}_\delta = \text{Interior}(D_\delta), \tag{8.3.15}$$

and consider the map H:

$$H: (t,s,v) \mapsto \Phi(t,(s+(0,v))) \ , \quad 0 \in \mathbb{R}^{n-k}. \tag{8.3.16}$$

It is not difficult to see that there exist positive numbers ε, δ, such that the restriction of H to Q, where now

$$Q = [-\varepsilon,\varepsilon] \times S \times D_\delta, \tag{8.3.17}$$

is a diffeomorphism onto a neighborhood of S in $\mathbb{R}^n$.
Analogously as in the construction of a flow-box in Section 8.1, it follows from the very construction that the inverse map H^{-1} transforms $\text{grad}_R f$ to the constant vector field (1,0,0) on the manifold $(-\varepsilon,\varepsilon) \times S \times \mathring{D}_\delta$, where the second zero in (1,0,0) is to be understood as the zero in the tangent space $T_s S$ at $s \in S$.

Note that $W^u_{\bar{x}}$ in the flow-box Q is the set $[-\varepsilon,\varepsilon] \times S \times \{\underset{\uparrow}{0}\}$.

From now on we concentrate on the flow-box Q as defined in (8.3.17). Let $\mathcal{W}$ be a (countable number of) submanifold(s) of $\{\varepsilon\} \times S \times \mathring{D}_\delta$ (cf. Fig. 8.3.5.a).

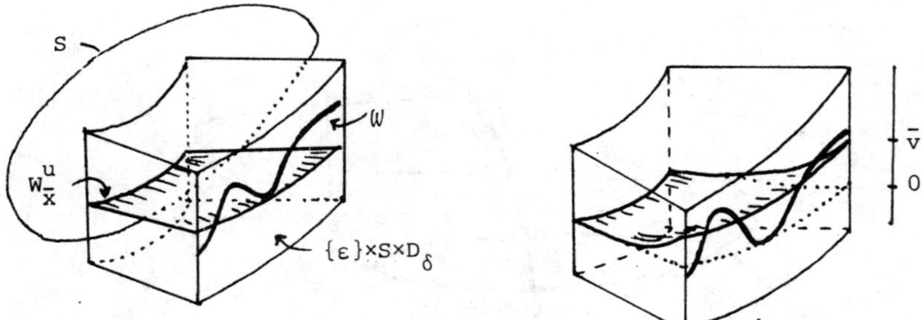

a b

Fig. 8.3.5

It might be that $\{\varepsilon\} \times S \times \{0\}$ (= intersection of $W^u_{\bar{x}}$ and $\{\varepsilon\} \times S \times \mathring{D}_\delta$) and $\mathcal{W}$ do not intersect transversally in $\{\varepsilon\} \times S \times \mathring{D}_\delta$. But then as will be shown, we can move $\{\varepsilon\} \times S \times \{0\}$ slightly in order that it intersects $\mathcal{W}$ transversally, with the additional aid of Sard's theorem.

We can parametrize the whole set $\{\varepsilon\} \times S \times \mathring{D}_\delta$ with sets of the form $\{\varepsilon\} \times S \times \{v\}$, $v \in \mathbb{R}^k$. The parameter v is just the value of the projection $\Pi: \{\varepsilon\} \times S \times \mathbb{R}^k \to \mathbb{R}^k$ on the last component, i.e. $\Pi(\varepsilon,s,v) = v$. Now we restrict this map Π to the manifold $\mathcal{W}$ and we can take a regular value of $\Pi_{|\mathcal{W}}$ (so, here Sard's theorem comes into play!).

Note that $\{\varepsilon\} \times S \times \{v\}$ intersects $\mathcal{W}$ transversally iff v is a regular value of $\Pi_{|\mathcal{W}}$ (exercise; see also Remark 7.3.3).

It is important to note that the role of $\mathcal{W}$ can be played by means of the intersection of the stable manifolds $W^s_{\bar{y}}$ with $\{\varepsilon\} \times S \times \mathring{D}_\delta$ corresponding to all critical points $\bar{y}$ with $f(\bar{y}) > f(\bar{x})$.

(This essentially yields the verification of Step 2 in the proof of Theorem 8.3.1).

Choose a regular value $\bar{v}$ of $\Pi_{|\mathcal{W}}$. Using the functions ξ and η in (8.3.7), (8.3.8), we define the vector field $F_{\bar{v}}$.

$$F_{\bar{v}}(t,s,v) = (1,0,0) + \xi(t)\cdot\eta(\|v\|)\cdot\bar{v}. \tag{8.3.18}$$

Since the regular value $\bar{v} \in \mathbb{R}^k$ can be taken arbitrarily close to $0 \in \mathbb{R}^k$, the vector field $F_{\bar{v}}$ approximates $\mathrm{grad}_{\mathcal{R}} f$ arbitrarily well and brings $W^u_{\bar{x}}$ in transversal position to $W^s_{\bar{y}}$ for all critical points $\bar{y}$ with $f(\bar{y}) > f(\bar{x})$ (cf. Fig. 8.3.5).

In fact, note that for all critical points $\bar{y}$ with $f(\bar{y}) > f(\bar{x})$ we have:

a. The intersection of $W^s_{\bar{y}}$ with $\{\varepsilon\} \times S \times \mathring{D}_\delta$ remains unchanged under the perturbation.
b. Each intersection point of $W^s_{\bar{y}}$ with $W^u_{\bar{x}}$ (the manifolds considered with respect to the perturbed vector field) can be connected via an integral curve with a point in $\{\varepsilon\} \times S \times \mathring{D}_\delta$.
c. Transversal intersection of the <u>restriction</u> of $W^s_{\bar{y}}$ and $W^u_{\bar{x}}$ to $\{\varepsilon\} \times S \times \mathring{D}_\delta$ at a point $\bar{z} \in \{\varepsilon\} \times S \times \mathring{D}_\delta$ implies transversal intersection of $W^s_{\bar{y}}$ and $W^u_{\bar{x}}$ at $\bar{z}$.
d. Transversal intersection of $W^s_{\bar{y}}$ and $W^u_{\bar{x}}$ at some point $\bar{z}$ implies transversal intersection of $W^s_{\bar{y}}$ and $W^u_{\bar{x}}$ at every point of the integral curve through $\bar{z}$. □

<u>Definition 8.3.1</u>. Let M, f, $\mathcal{R}$ satisfy the main assumptions of this section. Then, $\mathrm{grad}_{\mathcal{R}} f_{|M}$ is said to be in <u>general position</u> if <u>all stable and unstable manifolds intersect transversally</u>.

<u>Remark 8.3.2</u>. Suppose that $\mathrm{grad}_{\mathcal{R}} f_{|M}$ is in general position. Let $\bar{x},\bar{y} \in M$ be different critical points for $f_{|M}$ with (quadratic) index $k(\bar{x})$, $k(\bar{y})$.
If $W^u_{\bar{x}} \cap W^s_{\bar{y}} \neq \emptyset$, then the intersection contains at least one integral curve of $\mathrm{grad}_{\mathcal{R}} f_{|M}$, so $\dim(W^u_{\bar{x}} \cap W^s_{\bar{y}}) \geq 1$.
Choose $\bar{z} \in W^u_{\bar{x}} \cap W^s_{\bar{y}}$. From transversality we obtain with $m = \dim M$:

$$\begin{aligned} m &= \dim(T_{\bar{z}}W^u_{\bar{x}} + T_{\bar{z}}W^s_{\bar{y}}) = \\ &= \underbrace{\dim T_{\bar{z}}W^u_{\bar{x}}}_{m-k(\bar{x})} + \underbrace{\dim T_{\bar{z}}W^s_{\bar{y}}}_{k(\bar{y})} - \underbrace{\dim(T_{\bar{z}}W^u_{\bar{x}} \cap T_{\bar{z}}W^s_{\bar{y}})}_{\geq 1} . \end{aligned}$$

Hence, we obtain the inequality $m \le m - k(\bar{x}) + k(\bar{y}) - 1$, and consequently the implication:

$$\boxed{W^u_{\bar{x}} \cap W^s_{\bar{y}} \neq \emptyset \quad \text{and} \quad \bar{x} \neq \bar{y} \Rightarrow k(\bar{x}) < k(\bar{y})} \tag{8.3.19}$$

(Note that also $f(\bar{x}) < f(\bar{y})$ in the r.h.s. of (8.3.19)).
If we represent a critical point $\bar{x}$ for $f_{|M}$ by means of $\frac{S^\alpha}{S^\beta}$, where $S^\alpha(S^\beta)$ is a generating sphere for the unstable(stable) manifold to $\bar{x}$ corresponding to $\text{grad}_R f_{|M}$, then formula (8.3.19) suggests that we can make a table of the critical points $f_{|M}$ as for example depicted in Fig. 8.3.6, where $M = S^3$ and f some specific function (in accordance with the Morse relations; see Section 5.2).

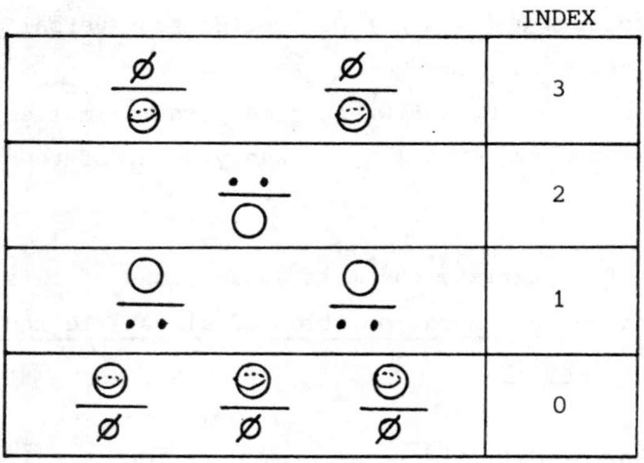

$M = S^3$

Fig. 8.3.6

Figure 8.3.6 has a deeper meaning. In fact, given a nondegenerate function $f_{|M}$, S. Smale showed (cf. [69]) that we can perturb $f_{|M}$ into a function $\tilde{f}_{|M}$ having the same critical points as $f_{|M}$ but with critical values coinciding with the corresponding (quadratic) indices. Such a function is also called "self-indexing". The construction of a self-indexing function, starting with a given function, gives a first insight to what extent critical values can be changed in relation to each other (without destroying the nondegeneracy of the critical points).
Let us consider such a construction in main steps, neglecting details.
So, let $f_{|M}$ be nondegenerate. Firstly, we note that the value of a local

minimum (index 0) can be lowered as we like. In fact, take local coordinates as in the Morse Lemma and then the 1-dimensional situation is already representative (cf. Fig. 8.3.7.a).

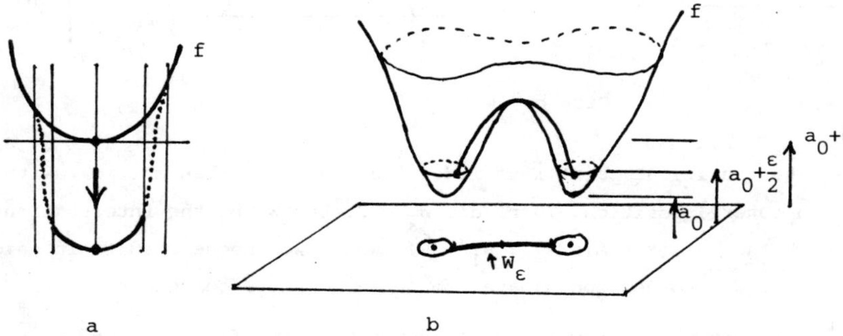

Fig. 8.3.7

So, suppose that all critical points for $f_{|M}$ of <u>index 0</u> have <u>the same value</u>, say a_0.
Next, we can lower the functional value of the critical points of <u>index 1</u> to the value $a_0+\varepsilon$, where $\varepsilon > 0$ arbitrarily small. For this we choose a Riemannian metric R such that $\mathrm{grad}_R f_{|M}$ is in general position. Let $\bar{x}$ be a critical point of index 1 and consider the stable manifold $W^s_{\bar{x}}$.
Let W_ε be the intersection of $W^s_{\bar{x}}$ with the set $\{x \in M | f(x) \geq a_0 + \frac{\varepsilon}{2}\}$ (cf. Fig. 8.3.7.b). Then W_ε is diffeomorphic with a compact interval, say the unit ball D^1 in $\mathbb{R}^1$. Next, we construct a diffeomorphism of a neighborhood $\mathcal{O}$ of W_ε in M to the set $D^1 \times D^{m-1}$ such that f in the new coordinates takes the form $f(\bar{x}) - x_1^2 + \sum_{j=2}^{m} x_j^2$ (cf. Fig. 8.3.8.a).
Then we replace $f(\bar{x}) - x_1^2$ by means of the function $\tilde{f}$, <u>dotted</u> in Fig. 8.3.8.b, with maximal value $a_0+\varepsilon$. Finally we taper off this perturbation in $(x_2,\ldots,x_m)$-coordinates (along the x_1-axis) as in Fig. 8.3.7.a.

So, repeating this process, suppose that all critical points for $f_{|M}$ of index 1 have value $a_1 := a_0+\varepsilon$.

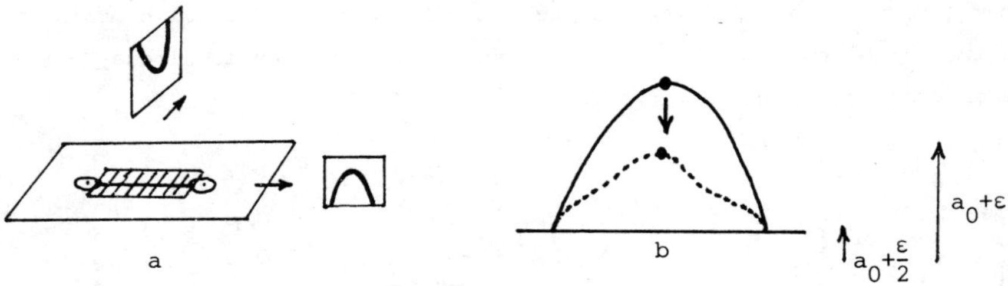

Fig. 8.3.8

Now we take a critical point $\bar{x}$ of index 2 and a Riemannian metric R with $\mathrm{grad}_R f$ in general position. Then, $\dim W^s_{\bar{x}} = 2$. Let W be the intersection of $W^s_{\bar{x}}$ with the set $\{x \in M \mid f(x) \geq a_1 + \frac{\varepsilon}{2}\}$. Next, we choose a suitable neighborhood O of W_ε and new coordinates for O such that O corresponds to $D^2 \times D^{m-2}$, whereas f is new coordinates has the form $f(\bar{x}) - \sum_{i=1}^{2} x_i^2 + \sum_{j=3}^{m} x_j^2$. Taking (x_1,x_2) resp. $(x_3,\ldots,x_m)$ together we can proceed as in the foregoing case in order to lower the functional value at $\bar{x}$ to $a_1+\varepsilon$, etc. etc.

Finally, suppose that the value of f at every critical point of index i equals a_i, $i = 0,\ldots,m$ (= dim M), where $a_0 < a_1 <\ldots< a_m$.
Then take a diffeomorphism $\psi\colon \mathbb{R} \to \mathbb{R}$ (monotone increasing) such that $\psi(a_i) = i$. Consequently, the function $\tilde{f}_{|M} := \psi \circ f_{|M}$ is self-indexing.

A next point to note is that sometimes we can cancel a pair of critical points of (quadratic) index (k, k+1). The basic idea in dimension 1 is depicted in Fig. 8.3.9 (cf. also [62]).

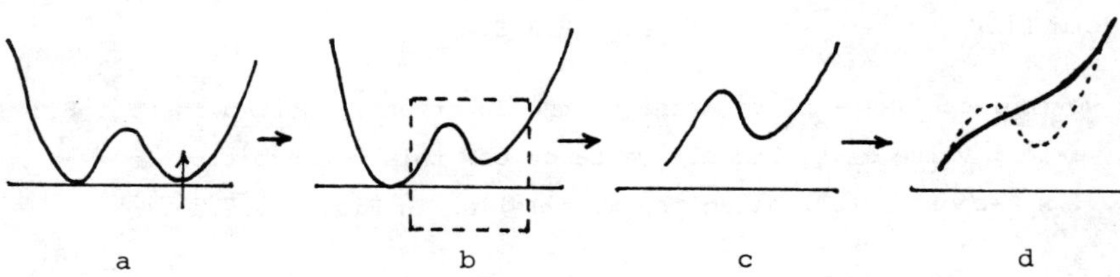

Fig. 8.3.9

Let ϕ denote the function in Fig. 8.3.9.a. Then, we can achieve a similar cancellation for the function $\phi(x_1) + \sum_{i=2}^{m} x_i^2$. This shows, at least in principle, how to cancel critical points of index 0 (local minima) against certain critical points of index 1. In fact, if M is <u>connected</u>, then we can cancel all local minima, <u>except for one</u> (which should exist in view of the Weierstrass Theorem).

Finally, it is <u>sometimes</u> possible to lower the functional value of a critical point of (quadratic) index k thereby passing the level of a critical point of index ℓ with $\ell < k$. For example, let M be 3-dimensional and let $\bar{x}, \bar{y}$ be critical points for f of index 1,2 respectively. Suppose that $\mathcal{R}$ is a Riemannian metric such that $\text{grad}_{\mathcal{R}} f_{|M}$ is in general position, and that $W^u_{\bar{x}}$ and $W^s_{\bar{y}}$ intersect according to Fig. 8.3.10. Then it is not difficult to understand that the metric $\mathcal{R}$ can be changed into a new metric $\tilde{\mathcal{R}}$ such that for $\text{grad}_{\tilde{\mathcal{R}}} f_{|M}$ the manifolds $W^u_{\bar{x}}$ and $W^s_{\bar{y}}$ have an empty intersection. But then, just as in the discussion above, the value at $\bar{y}$ can be lowered and the critical point $\bar{x}$ is <u>not an obstruction</u> for this process. □

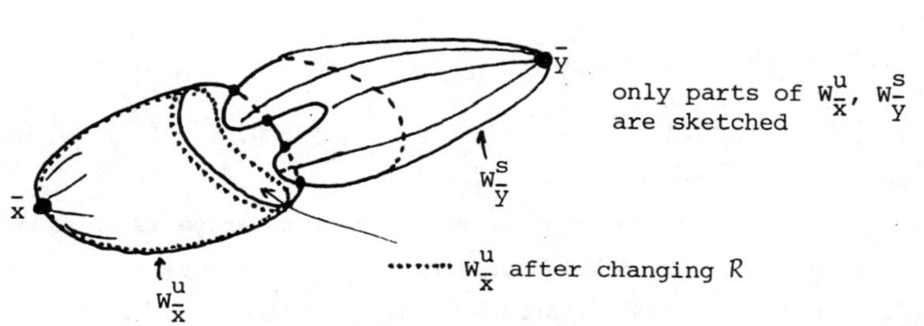

Fig. 8.3.10

<u>Remark 8.3.3</u>. Theorem 8.3.1 can easily be extended to the case that M is a compact manifold with <u>smooth</u> boundary ∂M, provided that $\text{grad}_{\mathcal{R}} f_{|M}$ is <u>nowhere tangent</u> ("transversal") to ∂M; this transversality with respect to the boundary is also preserved under small C^1-perturbations of the vector field. In the latter transversal case, the connected components of ∂M can be divided into two subsets according to the fact whether $\text{grad}_{\mathcal{R}} f_{|M}$ points outward or inward. The main difference with the case of an empty boundary is the fact that a boundary point can be reached from the interior in a finite integration time.

Remark 8.3.4. In case that M is a regular constraint set (or an MGB), the idea of a (negative) gradient vector field for the minimization of some smooth function f on M has to be adjusted in an appropriate way along the boundary of M. In fact, at a boundary point one might like to decide which active (= binding) inequality constraint has to be regarded as an equality constraint and which one can be omitted in order that the new vector field lies at each point in the tangent cone. Note that such a vector field becomes discontinous along the boundary. We will not go into details here, but merely point out certain phenomena w.r.t. standard negative gradients $-D^T f(x)$ (no equality constraints are present):

a). Suppose that $M = \{(x_1,\dots,x_n) \mid x_1 \geq 0,\dots,x_p \geq 0\}$, so ∂M is not smooth if $p > 1$. Moreover, let $f \in C^\infty(\mathbb{R}^n,\mathbb{R})$ and suppose that for some $r \leq p$:

$$\frac{\partial f}{\partial x_i}(0) \geq 0,\ i = 1,\dots,r\ ,\ \frac{\partial f}{\partial x_j}(0) < 0,\ j = r+1,\dots,p\ . \qquad (8.3.20)$$

Then, at the origin we will regard the inequalities $x_i \geq 0$, $i = 1,\dots,r$ as equality constraints and we omit $x_j \geq 0$, $j = r+1,\dots,p$, as constraints; a candidate for the resulting "negative gradient" from $D^T f(x)$ at $x = 0$ then becomes the following vector (cf. Fig. 8.3.11.a).

$$(0,\dots,0, -\frac{\partial f}{\partial x_{r+1}},\dots,-\frac{\partial f}{\partial x_n})_{x=0}\ .$$

Let $f, g_1,\dots,g_p \in C^\infty(\mathbb{R}^n,\mathbb{R})$, let $M := \{x \in \mathbb{R}^n \mid g_i(x) \geq 0,\ i = 1,\dots,p\}$ be a regular constraint set and $g_i(\bar{x}) = 0$, $i = 1,\dots,p$.
The question on which constraint g_i at $\bar{x}$ has to be regarded as an equality constraint, resp. has to be omitted as a constraint in order to construct an appropriate negative gradient from $-D^T f(\bar{x})$ can be decided as follows:

Consider the Euclidean distance function

$$(\mu_1,\dots,\mu_p) \mapsto \|D^T f(\bar{x}) - \sum_{j=1}^{p} \mu_j D^T g_j(\bar{x})\| \qquad (8.3.21)$$

Let $(\bar{\mu}_1,\dots,\bar{\mu}_p)$ be the unique minimum for the above function then, at $\bar{x}$, the constraint g_j for which $\bar{\mu}_j \geq 0$ (resp. < 0) will be regarded as an equality constraint (resp. will be omitted as a constraint).
To see this, let $\{\xi_{p+1},\dots,\xi_n\} \subset \mathbb{R}^n$ be a basis for the orthogonal complement of $\{D^T g_1(\bar{x}),\dots,D^T g_p(\bar{x})\}$ and consider the local coordinate transformation

$y = \psi(x)$, where $y_i = g_i(x)$, $i = 1,\ldots,p$ and $y_j = \xi_j^T(x-\bar{x})$, $j = p+1,\ldots,n$. Then, ψ sends $\bar{x}$ to the origin, the set M is locally described (under ψ) by means of the inequalities $y_i \geq 0$, $i = 1,\ldots,p$, and, moreover, for the composed function $\phi(y) := f\circ\psi^{-1}(y)$ we obtain:

$$\frac{\partial\phi}{\partial y_i}(0) = \bar{\mu}_i, \quad i = 1,\ldots,p . \tag{8.3.22}$$

For the verification of (8.3.22), note that there exist unique numbers $\alpha_{p+1},\ldots,\alpha_n$ such that, with $L(x) = f(x) - \sum_{i=1}^{p} \bar{\mu}_i g_i(x) - \sum_{j=p+1}^{n} \alpha_j \xi_j^T(x-\bar{x})$, the point $\bar{x}$ is a critical point for L (i.e. $DL(\bar{x}) = 0$). In new coordinates, under ψ, the function L takes the form

$$\tilde{L}(y) = \phi(y) - \sum_{i=1}^{p} \bar{\mu}_i y_i - \sum_{j=p+1}^{n} \alpha_j y_j . \tag{8.3.23}$$

Moreover, since $\bar{x}$ is mapped onto the origin, the origin is a critical point for $\tilde{L}$. But then, (8.3.22) follows immediately from (8.3.23) by differentiation; see also Example 3.2.6.

By means of an "active constraint strategy" as described above, the (+) Kuhn-Tucker points become precisely the zeros of the new vector field.

b). (+) Kuhn-Tucker points at the boundary may be reached in a <u>finite</u> integration time. See Fig. 8.3.11.b, where $M = \{(x_1,x_2) \mid x_1 \geq 0\}$, $f(x_1,x_2) = x_1+x_2^2$ and consider the dotted integral curve.

c). If M has a smooth boundary, then the boundary can pass from "binding" to "non-binding" at $x \in \partial M$ if $Df(x)$ is tangent to ∂M; this has the consequence that several integral curves may pass through the same (boundary) point; see Fig. 8.3.11.c.

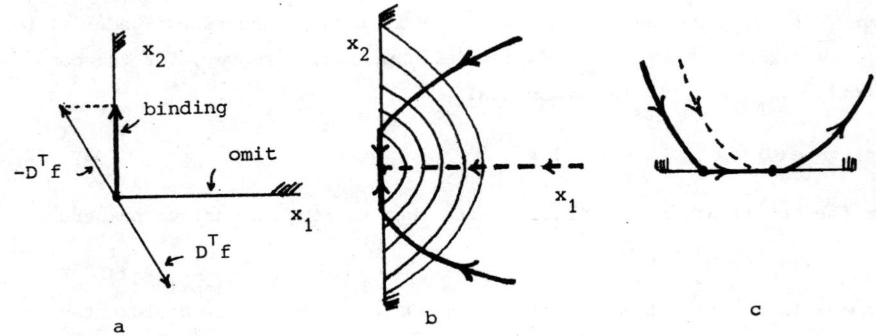

Fig. 8.3.11

Remark 8.3.5. A smooth vector field χ on $M \subset \mathbb{R}^n$ (without boundary) is called a gradient field if there exist $f \in C^\infty(\mathbb{R}^n, \mathbb{R})$ and some smooth Riemannian metric $\mathcal{R}$ on some neighborhood of M such that $\chi = \text{grad}_{\mathcal{R}} f_{|M}$ (f not necessarily nondegenerate within this remark).
Being a subset of all smooth vector fields, the C^k-topology for the set of gradient vector fields is clear (by relativation).
Two gradient fields χ_1, χ_2 are called topologically equivalent if there exists a homeomorphism h: M → M mapping trajectories of χ_1 onto trajectories of χ_2, preserving the orientation of integration. A gradient field χ is called structurally stable if a C^1-neighborhood $\mathcal{O}$ of χ exists in the space of gradient fields, such that χ is topologically equivalent with every $\tilde{\chi} \in \mathcal{O}$.

Let $\chi = \text{grad}_{\mathcal{R}} f_{|M}$ and M compact. We mention the following basic result (cf. [65], but also [66]):
Necessary and sufficient for structural stability of χ is:

a. $f_{|M}$ is nondegenerate

b. all stable and unstable manifolds intersect transversally. □

8.4. The graphs 0-1-0 and 0-n-0.

Throughout this section we assume $M \subset \mathbb{R}^m$ to be a compact connected C^∞-manifold of dimension n (with $\partial M = \emptyset$); $n \geq 2$. Moreover, the word "index" will refer to "quadratic index".

Let $f \in C^\infty(\mathbb{R}^m,\mathbb{R})$ be such that $f_{|M}$ is nondegenerate and separating (cf. also Theorem 7.1.4).

For a gradient vector field $\mathrm{grad}_{\tilde{R}} f_{|M} = \tilde{\chi}$ we are interested in the following problem: is it possible to "walk" from one local minimum to an arbitrary other one, thereby "following" trajectories of $\tilde{\chi}$ and passing successively points of index 1 and 0 ?

As a consequence of Theorem 8.3.1 we may approximate $\tilde{\chi}$ by means of a gradient vector field $\chi = \mathrm{grad}_R f_{|M}$ in general position (cf. Definition 8.3.1); we denote by $\Phi(t,\cdot)$ or $\Phi_t(\cdot)$ the flow on M associated to χ.

Now, let $\bar{x}$ be a critical point for $f_{|M}$ with index k and $S^{k-1} \subset W^s_{\bar{x}}$, resp. $S^{n-k-1} \subset W^u_{\bar{x}}$ generating spheres as introduced in Section 8.3. A point $x \in S^{n-k-1}$ is called a continuity point if there exists a critical point $\bar{y}$ of index n (local maximum) and an $y \in S^{n-1} \subset W^s_{\bar{y}}$ such that x and y lie on the same trajectory of χ, i.e. $y = \Phi(t,x)$ for some t (another way of saying might be: there exists a trajectory of χ through x "born" in $\bar{x}$ and "dying" in $\bar{y}$). As to these continuity points in S^{n-k-1} we have the following result:

Lemma 8.4.1. Let $0 \leq k \leq n-1$; then the set $K_{\bar{x}}$ of continuity points in S^{n-k-1} is an open and dense subset of $S^\alpha := S^{n-k-1}$.

Proof. Let $k < n-1$, and let $\bar{z}$ ($\neq \bar{x}$) be a critical point of index $k' \leq n-1$. We consider $N_{\bar{z}} = S^\alpha \cap W^s_{\bar{z}}$. In case $N_{\bar{z}} \neq \emptyset$ (and thus $k' > k$ by (8.3.19)), $W^u_{\bar{x}} \pitchfork W^s_{\bar{z}}$ implies $S^\alpha \pitchfork W^s_{\bar{z}}$ and hence, $N_{\bar{z}}$ is a submanifold of S^α of codimension $n-k' \geq 1$ (see Remark 7.2.1). Now, from Remark 7.15 it follows that $N_{\bar{z}}$ is a thin subset of S^α and hence, also $N = \bigcup_{\bar{z} \neq \bar{x}} N_{\bar{z}}$ is thin in S^α (the union being taken over all critical points $\neq \bar{x}$ with index $< n$). In case $k = n-1$ we have $S^\alpha = S^0$ and $N = \emptyset$.

We claim that $S^\alpha \setminus N = K_{\bar{x}}$, a dense subset of S^α. In fact, let $\tilde{x} \in S^\alpha \setminus N$ and $x(t) := \Phi(t,\tilde{x})$; since M is compact, a $\bar{y} \in M$ and a sequence $t_n \to \infty$ exist such that $\lim_{n\to\infty} x(t_n) = \bar{y}$. Since $f(x(t))$ is strictly increasing (cf. also (8.2.14)), it follows from this and the continuity of f that $f(\bar{y}) = \sup_t f(x(t))$.
Now, assume $\chi(\bar{y}) \neq 0$ and take $s > 0$. We have

$$\Phi_s(\bar{y}) = \lim_{n\to\infty} \Phi_s(x(t_n)) = \lim_{n\to\infty} x(t_n+s),$$

hence $f(\Phi_s(\bar{y})) \leq f(\bar{y})$. However this is a contradiction: consequently, $\chi(\bar{y}) = 0$ and thus, $\bar{y}$ is a critical point of f. With f separating it now follows that any sequence $t_n \to \infty$ with $x(t_n)$ converging has $\bar{y}$ as its limit; this implies $\lim_{t\to\infty} x(t) = \bar{y}$ and thus $\tilde{x} \in W^s_{\bar{y}}$ (see also (8.1.17)).
Recalling the definition of N we conclude: $\bar{y}$ is a (local) maximum. Finally, the openess of $K_{\bar{x}}$ follows from the continuity of $\Phi(\cdot,\cdot)$ and the fact that $\bar{y}$ is an attractor for χ. □

<u>Remark 8.4.1.</u> In a dual way we can define the set of continuity points $\tilde{K}_{\bar{x}} \subset S^\beta := S^{k-1}$ with respect to local minima, and again, for $k \geq 1$, we have: $\tilde{K}_{\bar{x}}$ is an open and dense subset of S^β.

<u>Remark 8.4.2.</u> In the proof of Lemma 8.4.1 we assumed that f is separating in order to prove $\lim_{t\to\infty} x(t) = \bar{y}$. This assumption is not necessary. In fact we have (<u>exercise</u>):
Let M be compact, $f_{|M}$ a function with isolated critical points and $\tilde{\chi} = \mathrm{grad}_{\tilde{R}} f_{|M}$ a gradient vector field on M; then, for each trajectory $x(t)$ of $\tilde{\chi}$ the limit $\lim_{t\to\infty} x(t)$, resp. $\lim_{t\to-\infty} x(t)$, exists and is a critical point of f (see also Corollary 3.3.1). □

Following [35] we now define the so called <u>0-1-0 graph</u> G (on M):
- the vertices of G are the local minima $p^0_1,\dots,p^0_r$ and the critical points of index 1, say $p^1_1,\dots,p^1_s$, of $f_{|M}$.
- an edge of G can only connect vertices of different index; p^0_i is connected with p^1_j iff $W^s_{p^1_j} \cap S^{n-1} \neq \emptyset$, where S^{n-1} is the generating sphere of $W^u_{p^0_i}$.

Note that, as a consequence of Remark 8.4.1 each vertex p^1_j is incident with one or two edges. Moreover, it follows from the Morse-relations (see (5.2.6)) that $r \geq 1$ and $s \geq r-1$; hence, G has at least one vertex, and in case $r > 1$ it has at least one edge.

Theorem 8.4.1. The 0-1-0 graph G is connected.

Example 8.4.1. Consider the manifold $M = S^2$, embedded in $\mathbb{R}^3$ as is shown in Fig. 8.4.1. We take as function f the distance function w.r.t. the plane V: $f_{|M}$ is a separating, nondegenerate function. With $\mathcal{R}$ the standard inner product on $\mathbb{R}^3$, the (0-1-0) graph G with respect to $\text{grad}_{\mathcal{R}} f_{|M}$ is depicted in Fig. 8.4.1.a, and it is indeed connected.

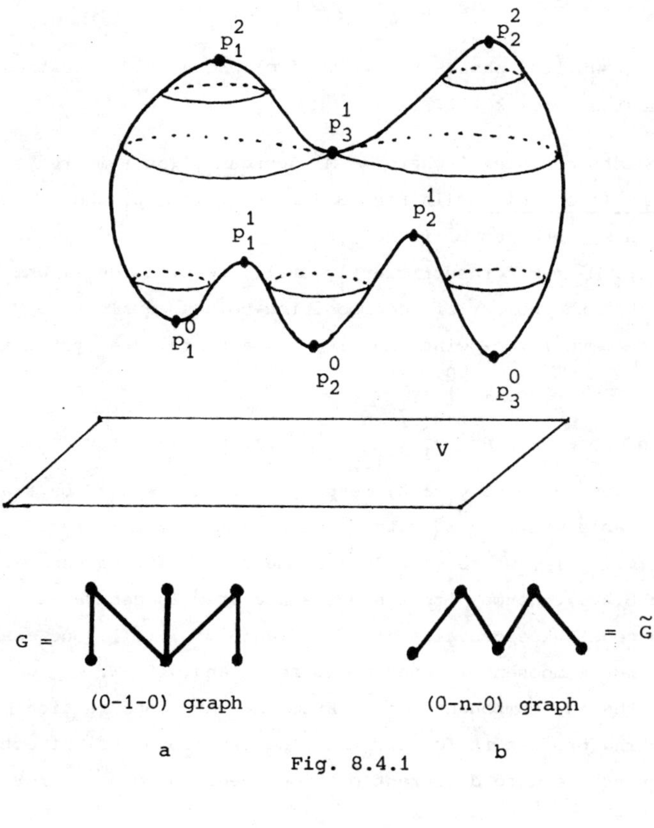

Fig. 8.4.1

Proof of Theorem 8.4.1. In case $r = 1$ it is easily seen that G is in fact $\{p_1^0\}$ or a tree (= connected graph without cycles). From now on we assume $r > 1$. We recall the definition of lower level sets M^a,

$$M^a = f_{|M}^{-1}((-\infty,a]) = \{x \in M | f(x) \leq a\},$$

and we denote by $\# M^a$ the number of pathcomponents of M^a. It follows from Theorem 3.3.7 that, with a increasing, a possible change in $\# M^a$ can only take place by passing a critical level of f. From Theorem 3.3.8 (see also Lemma 2.9.3 and relations (5.2.3), resp. (5.2.4)) we conclude:

$\# M^a$ increases by one iff $a \in \{f(p_1^0),\ldots,f(p_r^0)\}$;

$\# M^a$ decreases by one iff $a \in \{f(p_1^1),\ldots,f(p_s^1)\}$ and also relation (5.2.4) holds (with $a_{i-1} = a-\varepsilon$, $a_i = a+\varepsilon$, $k = 1$, $q = 0$).

$\# M^a$ remains constant if a passes the value f(p) where p is a critical point for $f_{|M}$ having index $k > 1$.

Those critical points of index 1 where $\# M^a$ decreases (as a increases) are called decomposition-points (in Fig. 8.4.1 the points p_1^1 and p_2^1 are decomposition-points, whereas p_3^1 is not).
Since $M^a = \emptyset$, resp. $M^a = M$ for sufficiently small, resp. large a, one easily derives that the number of decomposition-points equals r-1. Now, let $\bar{p} = p_i^1$ be a decomposition-point and $\psi\colon U \to V$ a coordinate system around $\bar{p}$ such that $f\circ\psi^{-1}(0,y) = -y_1^2 + \sum_{i=2}^{n} y_i^2 + f(\bar{p})$.
We denote by K the cone $\{y \in \mathbb{R}^n | -y_1^2 + \sum_{i=2}^{n} y_i^2 \leq 0\}$; its interior $\mathring{K}$ consists of two components according to $y_1 > 0$, resp. $y_1 < 0$. Let $W_{\bar{p}}^s \cap U$ be transformed into the stable manifold W_0^s with respect to the vector field $\bar{X} = \bar{R}(y)\cdot(-y_1,y_2,\ldots,y_n)^T$, where $\bar{R}(y)$ is a symmetric, positive $n\times n$ matrix (see also Remark 8.2.3). Hence, its tangent space at 0 is generated by an eigenvector with negative eigenvalue of $\bar{R}(0)\cdot\mathrm{diag}(-1,1,\ldots,1)$; one easily shows that its first component is unequal to zero, and with $W_0^s \subset K$ we see that W_0^s connects the two components of $\mathring{K}$. From the very construction of the (one-)cell attaching process it follows that the two (opposite) directions of quadratic decrease lead to different pathcomponents of $M^{f(\bar{p})-\varepsilon}$ for $\varepsilon > 0$

sufficiently small; hence, the same holds with respect to $W^s_{\bar{p}}$. These facts imply the following consequences w.r.t. the subgraph Γ with vertices $p^0_1,\ldots,p^0_r$ and the decomposition-points:

1) Each decomposition-point is incident with two edges; the number of edges of Γ equals $2(r-1)$ and the number of vertices of Γ equals $2r-1$.

2) There are no cycles in Γ. In fact, assume that $\tilde{\Gamma}$ is a cycle in Γ and $a_0 := f(p^1_{j_0})$ is maximal for $f(p^1_j)$ with $p^1_j \in \tilde{\Gamma}$.
All vertices different from $p^1_{j_0}$ then belong to $M^{a_0-\varepsilon}$ with $\varepsilon > 0$ sufficiently small, thus implying that both parts of $W^s_{p^1_{j_0}}$ belong to one pathcomponent: a contradiction!

3) It is well-known that a graph without cycles has cyclomatic number $\nu(\Gamma) = 0$, where

$$\nu(\Gamma) = \#\text{ edges} - \#\text{ vertices} + \#\text{ components.}$$ (cf. [24])

Hence, Γ has one component.

So, it follows that Γ is a tree and G is connected. □

Remark 8.4.3. With Γ as in the proof of Theorem 8.4.1 we have: there exists a unique minimal walk in Γ connecting the vertices p^0_i and p^0_j. □

We now turn to another graph, the so called 0-n-0 graph $\tilde{G}$ ([35]):
- the vertices of $\tilde{G}$ are the local minima $p^0_1,\ldots,p^0_r$ and the local maxima $p^n_1,\ldots,p^n_t$ of $f_{|M}$.
- an edge of $\tilde{G}$ can only connect vertices of different index; p^0_i is connected with p^n_j iff a trajectory of χ, is born in p^0_i and dies in p^n_j.

Example 8.4.2. In the situation of Figure 8.4.1 it turns out that the 0-n-0 graph is connected; see Fig. 8.4.1.b.

Theorem 8.4.2. The 0-n-0 graph $\tilde{G}$ is connected.

Proof. We have $r \geq 1$ and $t \geq 1$; from Lemma 8.4.1 and Remark 8.4.1 it follows that each vertex of $\tilde{G}$ is incident with at least one edge of $\tilde{G}$.

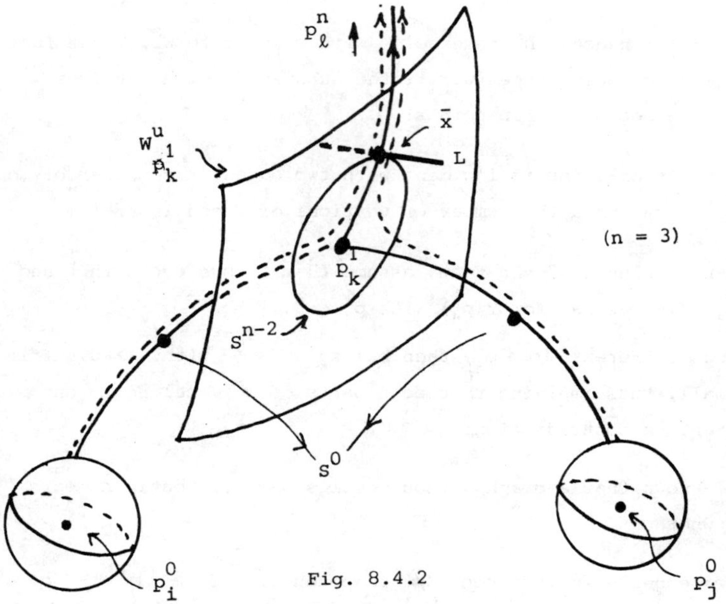

Fig. 8.4.2

In case r = 1, it is easily seen that $\tilde{G}$ is a tree with t edges. So, we assume r > 1. Let Γ be the graph as introduced in the proof of Theorem 8.4.1 and let $p_k^1 \in \Gamma$ be connected with p_i^0 and p_j^0. Let ψ: U → V be a coordinate system around p_k^1 as in the proof of Theorem 8.4.1. Let S^{n-2} be chosen such that $S^{n-2} \subset U$. Then there exists $\bar{x} \in K_{p_k^1} \subset S^{n-2}$ according to Lemma 8.4.1 such that the trajectory of χ through $\bar{x}$ tends to some local maximum p_ℓ^n. Moreover, it follows from the continuity of $\Phi(\cdot,\cdot)$ and the fact that p_ℓ^n is an attractor, resp. p_i^0 and p_j^0 are repellors: there exist open $\mathbb{R}^m$-neighborhoods $\mathcal{O}$ of $\bar{x}$ in U resp. $\mathcal{O}_1$, $\mathcal{O}_2$ of the (two) points of S^0 in U such that the trajectories of χ through points of M in these neighborhoods tend to p_ℓ^n, resp. come from p_i^0, p_j^0.
Now, let $L \subset \mathcal{O} \cap M$ be a curve-segment through $\bar{x}$ and transversal to $W^u_{p_k^1}$. Let $\tilde{x} \to \bar{x}$ on L; then $\Phi(\cdot,\tilde{x})$ remains on one side of the hyperplane $W^u_{p_k^1}$ in $U \cap M$; moreover, for $\tilde{x} \in L$ sufficiently close to $\bar{x}$ we have $\underline{\Phi(t,\tilde{x}) \in \mathcal{O}_1 \text{ or } \mathcal{O}_2}$ for some t (this can be seen by considering $\bar{R}(y)(-y_1,y_2,\ldots,y_n)^T$ as the vector field on $(\{0_{m-n}\} \times \mathbb{R}^n) \cap V$; cf. proof of Theorem 8.4.1). Hence, p_ℓ^n is connected to p_i^0, p_j^0; since Γ is a tree, $\tilde{G}$ is connected. □

Remark 8.4.4. If χ is not in general position, then the graph $\tilde{G}$ need not be connected. We illustrate this on $M = S^3$; see Fig. 8.4.3 where, of course, only a part of S^3 is depicted. Here, the function $f_{|M}$ is nondegenerate and has 6 critical points: two local minima (p_1^0, p_2^0), one point (p_1^1) of index 1, one point (p_1^2) of index 2 and two local maxima (p_1^3, p_2^3), where p_1^3 is not depicted. The stable manifold to p_1^2 and the unstable manifold to p_1^1 together form a sphere S^2 which decomposes S^3 into two components and hence, $\tilde{G}$ cannot be connected.

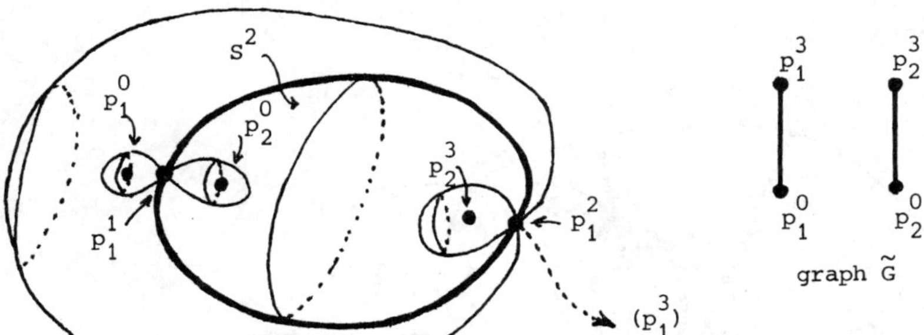

Fig. 8.4.3

Remark 8.4.5. Finally we discuss a somewhat different situation. Let M be a compact, connected n-dimensional manifold in $\mathbb{R}^n$ with smooth boundary ∂M; the interior of M will be denoted by $\mathring{M}$. Let $f \in C^\infty(\mathbb{R}^n, \mathbb{R})$ be nondegenerate and separating; let $\tilde{R}$ be a smooth Riemannian metric and suppose that $\tilde{\chi} := \mathrm{grad}_{\tilde{R}} f$ is transversal to ∂M and points outwards on ∂M.
Again (see also Remark 8.3.3) we can approximate $\tilde{\chi}$ by a gradient vector field $\chi := \mathrm{grad}_R f$ in general position and preserving $\chi \pitchfork \partial M$.
In a similar way we can define the 0-1-0 graph with respect to $\mathring{M}$ and Theorem 8.4.1 remains valid:

1. Let $x(t) \subset W^s_{\bar{x}}$ be a trajectory of χ where $\bar{x}$ is a critical point of index 1; then, for t decreasing, x(t) remains in $\mathring{M}$ and hence, x(t) is defined for $t \to -\infty$ (see also Lemma 2.3.1) with $\lim_{t\to-\infty} x(t)$ being a local minimum.

2. There are no (+) Kuhn-Tucker points on ∂M for $f_{|M}$: in fact, if $\bar{y} \in \partial M$, then $-\chi(\bar{y}) \in C_{\bar{y}}M$ and

$$-Df(\bar{y})\chi(\bar{y}) = -\chi(\bar{y})^T R(\bar{y})\chi(\bar{y}) < 0;$$

hence, $Df(\bar{y})[C_{\bar{y}}M] \not\subset \mathbb{H}$. Conclusion: Theorem 3.3.8 and Remark 3.3.12 can be used again, and there are no local minima on ∂M.

It is also possible to define a 0-n-0 graph w.r.t. M, simply by treating each component of ∂M, as an entity, as an additional local maximum. For an illustration, see Fig. 8.4.4 (where p_i^k stands for a critical point of index k).

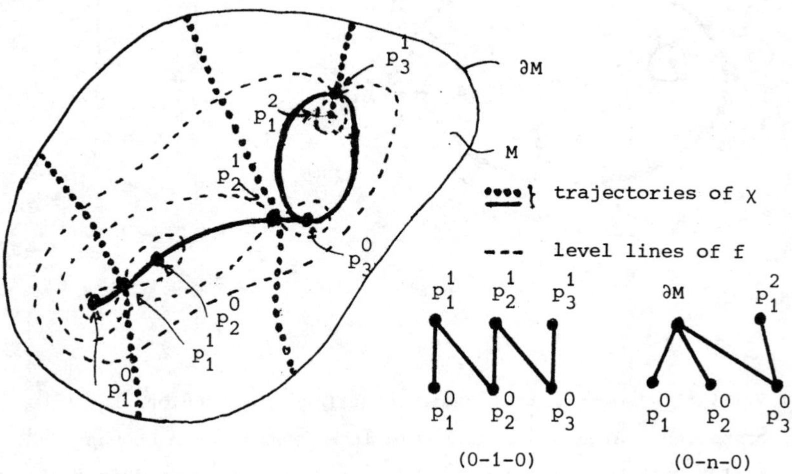

Fig.8.4.4

8.5. Reflected gradients.

In this section the word "index" will refer to "quadratic index".
As we have seen in Section 8.4, the search for various local minima of a differentiable function f in n variables can be done with the aid of critical points of index 1 (see also Remark 8.4.5), by means of a descending/ascending method; hence, it makes sense to determine such critical points. One way to this end is a desingularized version of Newton's method, viewed at infinitesimally as an autonomous differential equation

$$\frac{dx}{dt} = -\widetilde{D^2 f}.D^T f \ , \tag{8.5.1}$$

where $\widetilde{D^2 f}$ is the adjoint matrix of the matrix $D^2 f$. It turns out that the critical points of even index become attractors, and those of odd index become repellors. Necessarily a new set of singular points appears, generically being a stratifiable set of dimension $\leq n-2$; see Chapter 9.

In this section we consider an approach as introduced in [39]. For each k $(0 \leq k \leq n)$ we alter the gradient vector field $D^T f$ - by means of a partial reflection - into a new vector field F_k. Now, restricted to the critical point set of f, <u>only the critical points of index k are attractors</u> of F_k.

Let $f \in C^\infty(\mathbb{R}^n, \mathbb{R})$ and let $\lambda_1(x) \leq \ldots \leq \lambda_n(x)$ be the eigenvalues of the Hessian $D^2 f(x)$. By $\mathcal{O}_f$ we denote the maximal subset (by inclusion) of $\mathbb{R}^n$ such that $\lambda_i(x) < \lambda_j(x)$ for $i < j$ and $x \in \mathcal{O}_f$. Since eigenvalues depend continuously on the matrix elements, the set $\mathcal{O}_f$ is open.
Now, let $\bar{x} \in \mathcal{O}_f$ and $\bar{\lambda} := \lambda_i(\bar{x})$ with $\bar{v}$ an eigenvector of $D^2 f(\bar{x})$ for $\bar{\lambda}$ such that $\bar{v}^T \bar{v} = 1$.
Consider the C^∞-mapping $\mathcal{T}$: $\mathbb{R}^{2n+1} \to \mathbb{R}^{n+1}$ defined by

$$\mathcal{T}: \begin{pmatrix} v \\ \lambda \\ x \end{pmatrix} \mapsto \begin{pmatrix} (\lambda I_n - D^2 f(x))v \\ \bar{v}^T v - 1 \end{pmatrix} .$$

We have $\mathcal{T}(\bar{v}, \bar{\lambda}, \bar{x}) = 0$ and, with $A = D_{\binom{v}{\lambda}} \mathcal{T}(\bar{v}, \bar{\lambda}, \bar{x})$,

$$A = \left(\begin{array}{c|c} H & \bar{v} \\ \hline \bar{v}^T & 0 \end{array} \right) ,$$

where H is the symmetric matrix $(\bar{\lambda}I_n - D^2f(\bar{x}))$ with eigenvalues 0 and $\mu_1,\ldots,\mu_{n-1}$ ($\neq 0$); note that $H\bar{v} = 0$. Denote an eigenvector of H associated with μ_j by ξ_j. Then, one easily verifies that $\begin{pmatrix}\xi_1\\0\end{pmatrix},\ldots,\begin{pmatrix}\xi_{n-1}\\0\end{pmatrix},\begin{pmatrix}\bar{v}\\1\end{pmatrix},\begin{pmatrix}\bar{v}\\-1\end{pmatrix}$ constitute an (orthogonal) basis of eigenvectors of A with respective eigenvalues $\mu_1,\ldots,\mu_{n-1},1,-1$; hence, $\det A = -\prod_{j=1}^{n-1}\mu_j \neq 0$, i.e. A is non-singular.

Exercise 8.5.1. Prove the nonsingularity of A by using Exercise 3.2.3. □

In virtue of the implicit function theorem, C^∞-mappings $\lambda_i(x)$ and $v(x)$ exist on a neighborhood of $\bar{x}$ such that $D^2f(x)\cdot v(x) = \lambda_i(x)v(x)$. We denote by $P_i(x)$ the matrix $(v^T(x)v(x))^{-1}v(x)v^T(x)$; then, $P_i(x)$ is the matrix of the orthogonal projection on the eigenspace associated with $\lambda_i(x)$, and $P_i(x)$ is as such uniquely determined. It follows that on $\mathcal{O}_f$ the functions $\lambda_i(x)$ and $P_i(x)$ are defined as smooth functions (matrices).
As it is well-known, the matrix $P_i(x)$ is idempotent and $P_i(x)\cdot P_j(x) = 0$ for $i \neq j$, while the socalled spectral representation for $D^2f(x)$ gives:

$$D^2f(x) = \sum_{j=1}^{n} \lambda_j(x)P_j(x). \tag{8.5.2}$$

Now, let $k = 0,1,\ldots,n$ and $x \in \mathcal{O}_f$; we define $F_k(x)$ by

$$F_k(x) = [P_1(x) +\ldots+ P_k(x) - P_{k+1}(x) -\ldots- P_n(x)]D^Tf(x) . \tag{8.5.3}$$

Obviously, F_k is a smooth vector field defined on $\mathcal{O}_f$, where $F_0 = -D^Tf$ and $F_n = D^Tf$.

Let $\mathcal{A}$ denote the set of symmetric n×n matrices and $\tilde{\mathcal{A}}$ the subset of $\mathcal{A}$ consisting of those matrices which have at least one eigenvalue of (algebraic) multiplicity greater than one. Now, it is easily seen that $\mathcal{A}\backslash\tilde{\mathcal{A}}$ is an open and dense subset of $\mathcal{A}$; moreover $\tilde{\mathcal{A}}$ is an algebraic set since it can be described by the socalled discriminant of the characteristic polynomial of the matrices (see [54]). Hence, $\tilde{\mathcal{A}}$ is a closed set and, moreover, $\tilde{\mathcal{A}}$ admits a Whitney-regular stratification (cf. [16]) of codimension greater than zero ($\mathcal{A}\backslash\tilde{\mathcal{A}}$ being dense). We define the set $N := \mathbb{R}^n \times \mathbb{R} \times \{0_n\} \times \tilde{\mathcal{A}}$ in the jet-space $J(n,1,2)$; a Whitney-regular stratification of N is obtained by taking the sets $\mathbb{R}^n \times \mathbb{R} \times \{0_n\} \times \Sigma$, Σ a stratum of $\tilde{\mathcal{A}}$, as strata ("product-stratification"). Note that the codimension of N in J(n,1,2) is greater than n. We put:

$$\mathcal{M} = \bar{\pitchfork}^2 N = \{f \in C^\infty(\mathbb{R}^n,\mathbb{R}) \mid j^2 f \,\bar{\pitchfork}\, N\}.$$

It follows from Theorem 7.5.1 that $\mathcal{M}$ is a C^3-open and dense subset of $C^\infty(\mathbb{R}^n,\mathbb{R})$. Since codim(N) > n, for $f \in \mathcal{M}$ we have $j^2 f[\mathbb{R}^n] \cap N = \emptyset$; from this and the fact that N in closed we see that $\mathcal{M}$ is also C^2-open.

Note: if $f \in \mathcal{M}$, then every critical point of f belongs to $\mathcal{O}_f$.

Finally, we define

$$\mathcal{M}^* = \{f \in \mathcal{M} \mid f \text{ is nondegenerate}\}.$$

Together with Theorem 7.1.3 we obtain that $\mathcal{M}^*$ is C^2-open and dense in $C^\infty(\mathbb{R}^n,\mathbb{R})$.

<u>Remark 8.5.1</u>. The density of $\mathcal{M}^*$ can also be obtained as follows. First, approximate $f \in C^\infty(\mathbb{R}^n,\mathbb{R})$ with a nondegenerate $\tilde{f}$. Then, at each critical point $\bar{x}$ of $\tilde{f}$ we add a suitable quadratic function which has to be tapered off outside a neighborhood of $\bar{x}$ (exercise).

<u>Lemma 8.5.1</u>. Let $f \in \mathcal{M}^*$, let $\bar{x}$ be a critical point of f and $k \in \{0,1,\ldots,n\}$. Then, it holds:

$$DF_k(\bar{x}) = \sum_{i=1}^{k} \lambda_i(\bar{x}) P_i(\bar{x}) - \sum_{j=k+1}^{n} \lambda_j(\bar{x}) P_j(\bar{x}) \tag{8.5.4}$$

<u>Proof</u>. From (8.5.3) and $Df(\bar{x}) = 0$ it follows that

$$DF_k(\bar{x}) = [P_1(\bar{x}) + \ldots + P_k(\bar{x}) - P_{k+1}(\bar{x}) - \ldots - P_n(\bar{x})] D^2 f(\bar{x}).$$

Then, use (8.5.2) and the properties of the projection matrices $P_i(\bar{x})$. □

Let $f \in \mathcal{M}^*$, $\bar{x}$ a critical point for f and $\mathcal{O} \subset \mathcal{O}_f$ an open neighborhood of $\bar{x}$ such that on $\mathcal{O}$ the local stable manifold $W^s_{\bar{x}}$ with respect to the vector field F_k is defined (cf. (8.1.17)).

Theorem 8.5.1. Let $f \in M^*$, $\bar{x}$ a critical point for f of index i; $\bar{x} \in \mathcal{O} \subset \mathcal{O}_f$ and $W^s_{\bar{x}}$ as above. Then,

$W^s_{\bar{x}}$ is a smooth manifold of dimension $n - |k-i|$.

Proof. Using the properties of the matrices $P_i(\bar{x})$, Formula (8.5.4) can be interpreted as the spectral representation of the (symmetric) matrix $DF_k(\bar{x})$; hence the eigenvalues of $DF_k(\bar{x})$ are $\lambda_1(\bar{x}),\ldots,\lambda_k(\bar{x})$, $-\lambda_{k+1}(\bar{x}),\ldots,-\lambda_n(\bar{x})$ respectively. Now, dim $W^s_{\bar{x}}$ equals index $DF_k(\bar{x})$ (cf. Section 8.1); since the index of $\bar{x}$ (as a critical point) equals i, we have $\lambda_1(\bar{x})<\ldots<\lambda_i(\bar{x}) < 0 < \lambda_{i+1}(\bar{x}) <\ldots<\lambda_n(\bar{x})$ and hence, the number of negative eigenvalues for $DF_k(\bar{x})$ equals $n - |k-i|$. □

Corollary 8.5.1. Let $f \in M^*$ and let $\bar{x}$ be a critical point for f of index i. Then, with respect to the vector field F_k we have:

a. $\bar{x}$ is an attractor iff $k = i$,

b. $\bar{x}$ is a repellor iff $|k-i| = n$ (i.e. iff $i = n$, $k = 0$, resp. $i = 0$, $k = n$). □

Remark 8.5.2. Let $n = 2$. Then $\bar{x} \notin \mathcal{O}_f$ iff $D^2f(\bar{x}) = \lambda_1(\bar{x})I_2$, or otherwise stated, $f_{11} = f_{22}$ and $f_{12} = 0$ in $\bar{x}$ ($f_{ij} := \frac{\partial^2}{\partial x_i \partial x_j} f$). The latter two equations define a closed (linear) manifold in the 2-jet space of codimension 2. Let $\tilde{M}$ denote the subset of $C^\infty(\mathbb{R}^2,\mathbb{R})$ consisting of those functions whose 2-jet extension is transversal to this manifold. Then, $\tilde{M}$ (and hence, also $\tilde{M} \cap M^*$) is a C^3-open and dense subset of $C^\infty(\mathbb{R}^2,\mathbb{R})$. Now, let $f \in \tilde{M} \cap M^*$. Then, $E_f := \mathbb{R}^2\backslash\mathcal{O}_f$ is a <u>discrete</u> set, disjoint from the critical point set of f. One can ask about the behaviour of $F_1(x) = (P_1(x) - P_2(x))D^Tf(x)$ in a neighborhood of points in E_f (see [39]). Similar questions are (partially) the subject of Chapter 9, but then for the Newton-system (8.5.1) which has the following form in the present notation:

$$\frac{dx}{dt} = (\lambda_2(x)\cdot P_1(x) + \lambda_1(x)\cdot P_2(x))D^Tf(x). \tag{8.5.5}$$

9. NEWTON FLOWS

Let F be a smooth mapping from $\mathbb{R}^n$ to $\mathbb{R}^n$; $n \geq 2$.
In this chapter we study a special autonomous differential equation associated with F. The equation enjoys the property that - on the subset of $\mathbb{R}^n$ where the Jacobian matrix of F is regular - its Euler discretization just constitutes a (relaxed) Newton-Raphson iteration formula for finding the zeros of F. That is why we call this differential equation (or more appropriate: the flow of the associated smooth vector field) a Newton-flow. Moreover, it is defined on the whole $\mathbb{R}^n$.

Local and global features of Newton-flows will be discussed (Sections 9.1, 9.2). Apparently, the case where the above mapping F is the derivative of a smooth function on $\mathbb{R}^n$, is of particular interest in optimization. In this special case the Newton-flow will be referred to as to a Gradient Newton-flow; to this type of Newton-flows the main attention will be paid (Section 9.3).

We conclude this chapter by giving a review of some of the results we have obtained on another special case, namely the case where the underlying mapping F is a meromorphic function from $\mathbb{C}$ to $\mathbb{C}$. In this case, the associated Newton-flows are called Meromorphic Newton-flows and in the subcase where F is rational we call them Rational Newton-flows (Section 9.4).

9.1. Introduction; essential and extraneous singularities.

Throughout this section, let F be a smooth mapping from $\mathbb{R}^n$ to $\mathbb{R}^n$, i.e. $F \in C^\infty(\mathbb{R}^n, \mathbb{R}^n)$ and $n \geq 2$. As usual, $DF(x)$ stands for the matrix of first order derivatives at x (Jacobian matrix).
We define the critical set of F as:

$$\mathrm{Crit}(F) = \{x \in \mathbb{R}^n \mid \det DF(x) = 0\}.$$

On $\mathbb{R}^n \backslash \mathrm{Crit}(F)$, we consider the autonomous differential equation:

$$\frac{dx}{dt} = -DF^{-1}(x) \cdot F(x). \tag{9.1.1}$$

Along a trajectory $x(t)$ of (9.1.1) we have:

$$\frac{d}{dt} F(x(t)) = DF(x(t)) \cdot \frac{dx}{dt} = -F(x(t)).$$

So,

$$F(x(t)) = e^{-t} F(\bar{x}) \quad , \quad \bar{x} = x(0). \tag{9.1.2}$$

Hence, along $x(t)$, $\|F\|$ decreases exponentially, whereas the direction of F remains constant.

Note that Euler's approximation to equation (9.1.1) yields

$$x_{k+1} = x_k - h_k DF^{-1}(x_k)\cdot F(x_k), \quad \text{(Newton-Raphson iteration)}$$

where the steplenghts h_k may be suitably chosen. Compare also Example 8.2.1.

The fact that (9.1.1) is not defined on the whole $\mathbb{R}^n$ causes a lot of troubles, both from a theoretical as well as from a computational point of view: in fact, Crit(F) may be very irregular and near Crit(F) the r.h.s. of (9.1.1) may "blow up". We shall approach this problem by "desingularizing" the system (9.1.1) (see also [6], [13], [19], [32], [36], [37] and [71]). Firstly, let us introduce the concept of <u>adjoint matrix</u> (cf. [12]). Let A be an n×n-matrix. Then, the adjoint matrix - say $\tilde{A}$ - of A is defined as $((-1)^{i+j} m_{ij})^T$, where m_{ij} is the minor of A obtained by deleting from A the ith row and the jth column and taking the determinant. Then, consider the following autonomous differential equation which is globally defined on $\mathbb{R}^n$:

$$\frac{dx}{dt} = -\widetilde{DF}(x)\cdot F(x). \tag{9.1.3}$$

From the very definition of "adjoint matrix" we conclude that

$$\widetilde{DF}(x)\cdot DF(x) = \det DF(x)\cdot I_n \ , \tag{9.1.4}$$

where I_n = n×n-identity matrix. Hence, on $\mathbb{R}^n\backslash\text{Crit}(F)$, the phase-portraits of (9.1.1) and (9.1.3) are equal (up to orientation). In fact, at a point $x \in \mathbb{R}^n\backslash\text{Crit}(F)$ the vector $-\widetilde{DF}(x)\cdot F(x)$ points into the same (resp. in the opposite) direction as the vector $-DF^{-1}(x)\cdot F(x)$ in case $\det DF(x) > 0$ (resp. $\det DF(x) < 0$). See also Fig. 9.1.1. Consequently, the system (9.1.3) may be regarded as an extension (or desingularization) of (9.1.1) to the whole $\mathbb{R}^n$. The smooth vectorfield which is associated with (9.1.3), i.e. the vectorfield given by the r.h.s. of this equation will be denoted by $N(F)$, the Newton-system for F. The flow (cf. Section 8.1) associated with $N(F)$ is called Newton-flow for F and will sometimes also be denoted by $N(F)$.

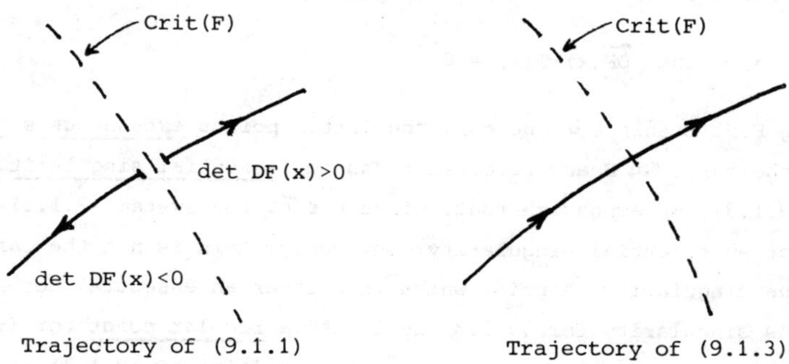

Fig. 9.1.1

As it was already mentioned, the set Crit(F) may have a bizarre structure. Therefore, our first target is to reject the pathologies (as far as they are not generic). To this aim, we shall assume, from now on, that the following Condition (*) is fulfilled.

Condition (*): $\|F(x)\| + |\det DF(x)| \neq 0$, all $x \in \mathbb{R}^n$.

From the definition of transversality (cf. Chapter 7) we see that Condition (*) is equivalent with

$$F \pitchfork \{0\},$$

where 0 is the origin in $\mathbb{R}^n$.

The set of all mappings F which fulfil Condition (*) is denoted by $\mathcal{F}$.

In view of Theorems 7.3.3, 7.3.1 it follows:

Lemma 9.1.1. The set $\mathcal{F}$ is C^k-dense (for all k) and C^k-open (for all $k \geq 1$) in $C^\infty(\mathbb{R}^n,\mathbb{R}^n)$.
Moreover, the zeros for a mapping F in $\mathcal{F}$ are isolated.

In comparison with (9.1.1) the system (9.1.3) has the advantage that it is well defined everywhere on $\mathbb{R}^n$. But there is also an extra complication:

not only the zeros are equilibrium states for (9.1.3) - as it is the case for (9.1.1) - but also those points x for which

$$F(x) \neq 0 \quad \text{and} \quad \widetilde{DF}(x)\cdot F(x) = 0.$$

Following F.H. Branin [6] we call the latter points *extraneous singularities*, whereas the zeros for F are referred to as to *essential singularities* for system (9.1.3). We emphasize that, since $F \in \mathcal{F}$, the system (9.1.1) is well-defined at an essential singularity; obviously, this is not the case at any extraneous singularity. A point which is neither an essential nor an extraneous singularity for (9.1.3) is called a *regular point* for (9.1.3). The sets of all extraneous (essential) singularities of (9.1.3) is denoted by resp, Ext(F) and Ess(F), whereas Reg(F) stands for the set of all regular points of (9.1.3). Note that the sets Reg(F), Ext(F) and Ess(F) constitute a partition of $\mathbb{R}^n$.

In view of Section 8.1, the local phase-portrait of system (9.1.3) around a regular point $\tilde{x}$ is of the form as depicted in Fig. 9.1.2.a. In a neighborhood of a point $\bar{x} \in \text{Ess}(F)$ we may linearize the r.h.s. of (9.1.3) as follows

$$-\widetilde{DF}(x)\cdot F(x) \approx -\widetilde{DF}(\bar{x})\cdot DF(\bar{x})\cdot(x-\bar{x}) = -\det DF(\bar{x})\cdot(x-\bar{x}).$$

Since $F \in \mathcal{F}$ (and thus $\det DF(\bar{x}) \neq 0$) we conclude (cf. Section 8.1) that the local phase-portrait of (9.1.3) around $\bar{x}$ is - upto topological equivalence - of the form as depicted in Fig. 9.1.2.b or 9.1.2.c: if $\det DF(\bar{x})$ is positive (negative), then $\bar{x}$ is an attractor (repellor) for $\mathcal{N}(F)$.

Local phase-portraits of (9.1.3)

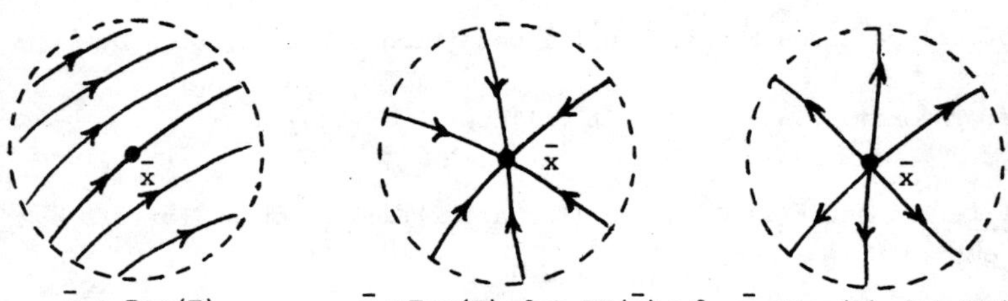

$\bar{x} \in \text{Reg}(F)$ — a; $\bar{x} \in \text{Ess}(F); \det DF(\bar{x}) > 0$ — b; $\bar{x} \in \text{Ess}(F); \det DF(\bar{x}) < 0$ — c

Fig. 9.1.2

A next and natural step should be the investigation of the possible local phase-portraits of (9.1.3) around an extraneous singularity. However, as it will be clear in the sequel, the answer to this question is - by far - not as easy as in the case of essential singularities. In Section 9.3 (the case of Gradient Newton-flows) we shall deal with this problem under additional transversality conditions. In the case of Meromorphic Newton-flows, the possible forms of the local phase-portraits around an extraneous singularity are easily to detect (cf. Section 9.4).

For the moment we content ourselves with deriving two criteria (of algebraic resp. geometrical nature) for a point to be an extraneous singularity. Consider the $n\times(n+1)$ matrix $[DF(x) \vdots F(x)]$, where $F(x)$ is the last column. Then, we have the following algebraic characterization of points in Ext(F).

Lemma 9.1.2. Let $F \in \mathcal{F}$ and let $\bar{x} \in \mathbb{R}^n$ be given. Then,

$$\bar{x} \in \text{Ext}(F) \quad \text{iff} \quad \text{rank}[DF(\bar{x}) \vdots F(\bar{x})] \leq n-1.$$

Proof. In the case where rank $DF(\bar{x}) < n-1$, all coefficients of the matrix $\widetilde{DF}(\bar{x})$ are equal to zero and we have $F(\bar{x}) \neq 0$ (use $F \in \mathcal{F}$). Consequently, both sides of the equivalence are satisfied by $\bar{x}$. However, if rank $DF(\bar{x}) = n$, then one easily verifies that no side of this equivalence is satisfied by $\bar{x}$. So it remains to treat the case where rank $DF(\bar{x}) = n-1$ and $F(\bar{x}) \neq 0$. Let V denote the linear space spanned by the columns of $DF(\bar{x})$. Since, in this case, $\widetilde{DF}(\bar{x}) \neq 0$ it follows from Relation (9.1.4) that $V = \ker \widetilde{DF}(\bar{x})$. Now, the assertion is a consequence of the following chain of equivalences: $\bar{x} \in \text{Ext}(F)$ iff $F(\bar{x}) \in \ker \widetilde{DF}(\bar{x})$ iff $F(\bar{x}) \in V$ iff $\text{rank}[DF(\bar{x}) \vdots F(\bar{x})] = n-1$. □

As a corollary of the above lemma we have: for $F \in \mathcal{F}$, the set Ext(F) is closed in $\mathbb{R}^n$. From the continuity of F resp. $\widetilde{DF}\cdot F$, it follows: Ess(F) is closed and Reg(F) is open.

A geometrical characterization of points in Ext(F) can be obtained as follows.

Recall that, along a trajectory of (9.1.1), the direction $\frac{F(x)}{\|F(x)\|}$ remains constant. Thus, it is plausible to consider the so called trajectory map TF (see also Smale [71]) which is defined as follows:

$$TF: \mathbb{R}^n \backslash Ess(F) \rightarrow S^{n-1} \text{ with } TF(x) = \frac{F(x)}{\|F(x)\|}, \tag{9.1.5}$$

where $S^{n-1} = \{(x_1,\dots,x_n) \in \mathbb{R}^n | \Sigma x_i^2 = 1\}$ stands for the usual $(n-1)$-sphere in $\mathbb{R}^n$.

We emphasize that <u>$TF(x(t))$ = constant</u>; $x(t)$ = trajectory of (9.1.1). We shall characterize the set $Ext(F)$ by means of the map TF. To this aim, we note that the concept of <u>critical point</u> for a smooth mapping between <u>Euclidean spaces</u> (cf. Chapter 7) can be easily extended - by using local coordinates - to the case of a <u>smooth map between C^∞-manifolds.</u>

Now, we have the following geometrical characterization of $Ext(F)$:

<u>Lemma 9.1.3.</u> Let $F \in \mathcal{F}$ and let $\bar{x} \in \mathbb{R}^n \backslash Ess(F)$ be given. Then,

$$\bar{x} \in Ext(F) \quad \text{iff} \quad \bar{x} \text{ is critical point for TF.}$$

Before proving Lemma 9.1.3 we use a preliminary step.

Firstly, put $F(x) = (F_1(x),\dots,F_n(x))^T$.

With this notation, the augmented matrix $[DF(x) \vdots F(x)]$ takes the form

$$i^{th}\text{ row} \qquad \overset{\leftarrow n \rightarrow \qquad \leftarrow 1 \rightarrow}{\left(\begin{array}{c|c} DF_1(x) & F_1(x) \\ \cdots & \cdots \\ DF_i(x) & F_i(x) \\ \cdots & \cdots \\ \hline DF_n(x) & F_n(x) \end{array}\right)} \; \updownarrow n$$

For $x \notin Ess(F)$, we may assume (no loss of generality) that $F_n(x) \neq 0$.

Put $G_i(x) = \dfrac{F_i(x)}{F_n(x)}$, $i = 1,\dots,n-1$ and $G(x) = (G_1(x),\dots,G_{n-1}(x))^T$.

Subtracting the row vector $G_i(x) \cdot [DF_n(x) \vdots F_n(x)]$ from the ith row $[DF_i(x) \vdots F_i(x)]$ in the above matrix, $i = 1,\dots,n-1$, and deviding each entry of the resulting matrix by $F_n(x)$, we obtain:

$$\text{rank}[DF(x) \vdots F(x)] = \text{rank} \overset{\longleftarrow n \longrightarrow \quad \leftarrow 1 \rightarrow}{\left(\begin{array}{c|c} & 0 \\ DG(x) & \vdots \\ & 0 \\ \hline F_n^{-1}(x) \cdot DF_n(x) & 1 \end{array}\right)} \; \updownarrow n-1$$

From this, it follows that for $x \notin \mathrm{Ess}(F)$:

$$\mathrm{rank}[DF(x) \,\vdots\, F(x)] = \mathrm{rank}[DG(x)] + 1 \,. \qquad (9.1.6)$$

Proof (of Lemma 9.1.3).
For $y = (y_1, \ldots, y_{n-1})$, $y_i \in \mathbb{R}$ and $z = [y \,\vdots\, 1]^T \in \mathbb{R}^n$, we define the mapping $\psi: \mathbb{R}^{n-1} \to S^{n-1}$ by $\psi(y) = \frac{z}{\|z\|}$.
Then, ψ induces a local parametrization around each point $x \in S^{n-1}$ with $x_n > 0$, cf. Fig. 9.1.3.

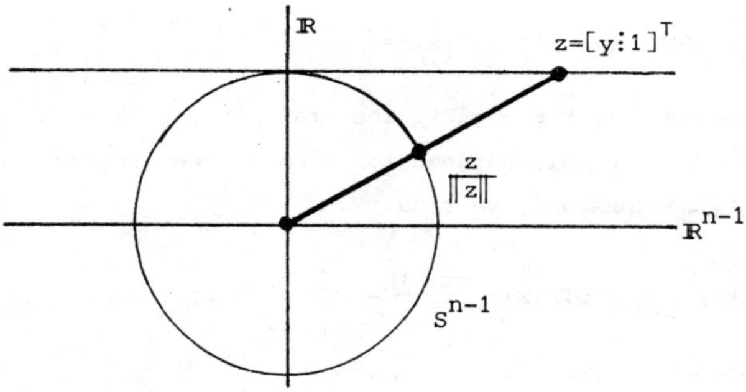

Fig. 9.1.3

For $\bar{x} \in \mathbb{R}^n \setminus \mathrm{Ess}(F)$ we assume (no loss of generality) that $F_n(\bar{x}) > 0$.
In a neighborhood of such a point $\bar{x}$ we have:

$$TF = \psi \circ G \qquad (9.1.7)$$

and hence $\bar{x}$ is critical point for TF iff rank $DG(\bar{x}) \leq n-2$.
Now, application of Lemma 9.1.2, using (9.1.6) yields the desired result. □

We proceed with a characterization of points in Reg(F) and of the trajectories of $N(F)$ through such points.

Lemma 9.1.4. Let $F \in \mathcal{F}$ and let $\tilde{x} \in \mathbb{R}^n \setminus \mathrm{Ess}(F)$ be given. Then,

a. A point $\tilde{x}$ belongs to Reg(F) iff $\dim \mathrm{Ker}\, DTF(\tilde{x}) = 1$. Moreover, if $\tilde{x} \in \mathrm{Reg}(F)$, then $\ker DTF(\tilde{x})$ is spanned by $N(F)_{|\tilde{x}}$ $(= -\widetilde{DF}(\tilde{x}) \cdot F(\tilde{x}))$.

b. Let $\tilde{x} \in \mathrm{Reg}(F)$, then the trajectory of $N(F)$ through $\tilde{x}$ is (locally around $\tilde{x}$) given as the solution set of the following system of equations:

$$G_i(x) = G_i(\tilde{x}) \quad , \quad i = 1,\dots,n-1 \; . \qquad (9.1.8)$$

Proof.

a. From Lemma 9.1.3 it follows that

$$\tilde{x} \in \mathrm{Ext}(F) \quad \text{iff} \quad \dim \mathrm{Ker}\, DTF(\tilde{x}) > 1$$

and hence

$$\tilde{x} \in \mathrm{Reg}(F) \quad \text{iff} \quad \dim \mathrm{Ker}\, DTF(\tilde{x}) = 1.$$

Now, let us suppose that $\tilde{x} \in \mathrm{Reg}(F)$. The trajectory of $N(F)$ through $\tilde{x}$ - say $x(t)$ with $x(0) = \tilde{x}$ - is well-defined for t on an open interval containing 0. Since $TF(x(t))$ = constant, we find

$$\frac{d}{dt}[TF(x(t))]_{t=0} = DTF(\tilde{x})\cdot \frac{dx(0)}{dt} = DTF(\tilde{x})\cdot N(F)_{|\tilde{x}} = 0.$$

Thus, $N(F)_{|\tilde{x}}$ spans $\mathrm{Ker}\, DTF(\tilde{x})$.

If $\tilde{x} \in \mathrm{Reg}(F)$, then we have $\mathrm{rank}[DF(\tilde{x}) \vdots F(\tilde{x})] = n$, cf. Lemma 9.1.2. So, in view of (9.1.6) we see that $\mathrm{rank}[DG(\tilde{x})] = n-1$. Since $DG(\tilde{x})$ is the Jacobi-matrix of system (9.1.8) at $\tilde{x}$, the Implicit Function Theorem yields that - locally around $\tilde{x}$ - the solution of this system is a uniquely determined 1-dimensional C^∞-manifold, say γ. On the other hand, on the $N(F)$-trajectory through $\tilde{x}$ we have $TF(\cdot)$ = constant. Hence, in view of (9.1.7), this trajectory coincides (locally around $\tilde{x}$) with γ. □

9.2. Some global results.

In this section, we adopt the notations and terminology as introduced in Section 9.1. Especially, F stands (again) for a smooth mapping from $\mathbb{R}^n$ to $\mathbb{R}^n$; $n \geq 2$.

We present some results - due to Smale [71] and to Jongen, Jonker and Twilt [36] - on the global behaviour of Newton-flows.

Let M be a compact, connected C^{∞}-manifold in $\mathbb{R}^n$ of codimension one. As it is well-known (see e.g. [22]) the complement $\mathbb{R}^n \backslash M$ consists of two open components, one of which being bounded. The bounded component is denotes by M_{int} (the interior of M). Note that $M \cup M_{int}$ is an n-dimensional C^{∞}-manifold with boundary (= M).

We assume that M fulfils the so called Boundary Condition w.r.t. the Newton-flow $N(F)$.

Boundary Condition. $N(F) \pitchfork M$, i.e. at any point $x \in M$, the vectors of the tangent space $T_x M$ together with $N(F)_{|x}$ span the whole $\mathbb{R}^n$.

If M fulfils this Boundary Condition w.r.t. $N(F)$, then M is called a Global Boundary for $N(F)$. In the sequel we always shall assume (no loss of generality) that along a global boundary M the vector field $N(F)$ points inward to M_{int} (i.e. for each $x \in M$, the component of the vector $N(F)_{|x}$ which is perpendicular to $T_x M$, is directed according to the inward normal vector (cf. [22]) to M at x). This assumption will be referred to as to the Orientation Assumption.

From now on - throughout this whole chapter - by $\phi_x(t)$ we shall denote the maximal trajectory (cf. Section 2.3) of $N(F)$ through a point x; in particular $\phi_x(0) = x$. Let $\gamma_x = \{\phi_x(t) \mid t \geq 0\}$. By the Orientation Assumption, the closure $\bar{\gamma}_x$ of γ_x is contained in the compact set $M \cup M_{int}$ and thus $\phi_x(t)$ is defined for every $t \geq 0$ (cf. Lemma 2.3.1).

Lemma 9.2.1. Let $F \in \mathcal{F}$ and suppose that $M \subset \mathbb{R}^n$ is a Global Boundary for $N(F)$. Then, the trajectory map TF restricted to M is locally a C^{∞}-diffeomorphism and maps M onto S^{n-1}.

Proof. Since M is a Global Boundary for $N(F)$, we have $M \subset Reg(F)$. From Lemma 9.1.4.a it follows that the restriction $TF_{|M}$ is an immersion. Since $\dim M = \dim S^{n-1} = n-1$, the Inverse Function Theorem yields that $TF_{|M}$ is locally a C^{∞}-diffeomorphism and consequently $TF_{|M}$ is an open map. From this, we conclude that TF(M) is open in S^{n-1}. But since M is compact, TF(M) is a closed subset of S^{n-1} as well. In view of the connectedness of S^{n-1} (note that $n > 1$!) it follows that $TF(M) = S^{n-1}$. □

Lemma 9.2.2. Let $F \in \mathcal{F}$ and suppose that M is a global boundary for $\mathcal{N}(F)$. For a point $x \in \text{Reg}(F)$ which is contained in $M \cup M_{int}$ there are two possibilities, namely

1. The trajectory $\phi_x(t)$ is periodic , or
2. The topological closure of γ_x contains at least one equilibrium-state for $\mathcal{N}(F)$ in M_{int}.

Proof. Suppose that the trajectory through x is not periodic, i.e. $\phi_x: [0,\infty) \to \gamma_x$ is injective. In that case, the (infinite) subset $\{\phi_x(n) \mid n = 1,2,\dots\}$ of γ_x has at least one accumulation point, say α, in the (compact) set $M \cup M_{int}$. We shall prove that this point α is an equilibrium state for $\mathcal{N}(F)$, and hence $\alpha \in M_{int}$. To this aim we suppose that $\alpha \in \text{Reg}(F)$, i.e. $\mathcal{N}(F)_{|\alpha} \neq 0$. In view of Section 8.1, there exists a local C^∞-coordinate neighborhood $\mathcal{V}$ around α (a so-called flow box) with coordinates $y = (y_1,\dots,y_n)$ such that, w.r.t. these coordinates, $\alpha = 0$ and the system (9.1.3) is given by

$$\frac{dy}{dt} = (1,0,\dots,0)^T.$$

In $\mathcal{V}$ we choose an open ball B centered at α and with radius $\delta > 0$. Let $U \subset B$ be the local (n-1)-dim C^∞-manifold defined by $y_1 = 0$; see Fig. 9.2.1. U is transversal to the trajectories of $\mathcal{N}(F)$. Lemma 9.1.4 yields (for δ small): TF is injective on U. Thus there exists a constant n-tuple $(c_1,\dots,c_n)^T$ with the property that - w.r.t. the y-coordinates - the intersection $\gamma_x \cap B$ is a line segment (ℓ) given by $y(t) = (t+c_1,c_2,\dots,c_n)^T$, where t is contained in an interval of length $< \delta$. If the trajectory γ_x is traversed (for increasing t), then we will reach ℓ, but later we will leave ℓ (and thus B) as well. However, once we have left B, it is impossible to return to B by following γ_x for increasing t. In fact, "returning to B" means "returning to ℓ" and that is impossible since ϕ_x is injective. We conclude that a real $t_0 > 0$ exists, such that $\phi_x(t) \notin B$ if $t \geq t_0$. This however is in contradiction with the assumption that the points $\phi_x(n)$, n = 1,2,... do accumulate at α. □

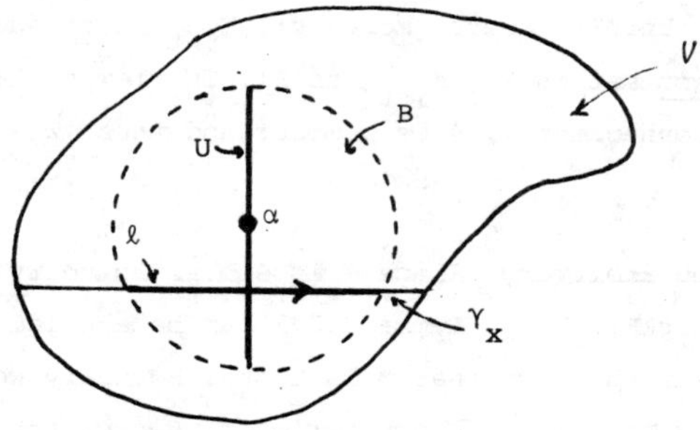

Fig. 9.2.1

Corollary 9.2.1. If - in the situation of Lemma 9.2.2 (proof) - we have $\alpha \in \mathrm{Ess}(F)$, then α must be an attractor (i.e. there exists an ε-neighborhood $B_\varepsilon(\alpha)$ around α such that for each $z \in B_\varepsilon(\alpha)$: $\lim_{t\to+\infty} \phi_z(t) = \alpha$). Note that in this case we have $\det DF(\alpha) > 0$.

In the next theorem, we use the concept of a "thin subset of a C^∞-manifold". In the case where the manifold is just the Euclidean space $\mathbb{R}^k$, this concept is introduced in Chapter 7, especially in Definition 7.1.1. Now, let X be an arbitrary C^∞-manifold of dimension k. Then, we call $V \subset X$ thin if the set $\phi(U \cap V)$ is thin in $\mathbb{R}^k$ for all local X-coordinate systems $\phi: U \to \mathbb{R}^k$. From this definition it follows that the image of a thin subset of X under a (local) diffeomorphism from X to the manifold Y is thin in Y.
For smooth mappings between C^∞-manifolds, Sard's theorem holds: the set of critical values is thin in the target-manifold. For a proof of this statement, we refer to [31].

Theorem 9.2.1. Let $M \subset \mathbb{R}^n$ be a Global Boundary for $N(F)$ and let $F \in \mathcal{F}$. Then, there exists a closed, thin subset Σ of M, such that the trajectory through any point in $M\backslash\Sigma$ tends to a zero for F in M_{int}.

Proof. Let Λ be the set of critical values of $TF_{|M\cup M_{int}}$. Then, by Sard's theorem it follows that Λ is thin in S^{n-1}. The restriction of TF to M is denoted by g_1 and we define $\Sigma := g_1^{-1}(\Lambda)$. Since, g_1 is locally a diffeomorphism (cf. Lemma 9.2.1) it follows that Σ is thin in M. The set

$\mathrm{Ext}_M(F) := M_{int} \cap \mathrm{Ext}(F)$ is closed in $\mathbb{R}^n$ and thus it is closed in the ($\mathbb{R}^n$-open) set $M_{int}\backslash \mathrm{Ess}(F)$. Since, moreover, M_{int} is bounded we conclude that $\mathrm{Ext}_M(F)$ is a <u>compact</u> set in $M \cup M_{int}\backslash \mathrm{Ess}(F)$. In view of Lemma 9.1.3, we have $\Lambda = TF(\mathrm{Ext}_M(F))$. Consequently, Λ is compact and thus $\Sigma (= g_1^{-1}(\Lambda))$ is closed in M.

Now, let $x \in M\backslash\Sigma$ be arbitrary. Since M is Global Boundary for $N(F)$, it follows that $x \in \mathrm{Reg}(F)$. Thus, Lemma 9.2.2 can be applied w.r.t. to the point x. Using again the fact that M is Global Boundary we conclude that the $N(F)$-trajectory through x is not periodic. Hence, the topological closure of this trajectory must contain an equilibrium-state - say α - for $N(f)$, which is contained in M_{int}. Since $TF(x)$ is a regular value for TF, application of Lemma 9.1.3, yields that $\alpha \notin \mathrm{Ext}(F)$. Consequently, α is a zero for F (i.e. $\alpha \in \mathrm{Ess}(F)$). Moreover, from the local structure of the phase-portrait of $N(F)$ around an essential singularity, it follows that α is the only equilibrium-state for $N(F)$ which is contained in the intersection of M_{int} and the closure of trajectory through x. Finally, the Orientation Assumption yields that α is an attractor for $N(F)$ (see also Corollary 9.2.1). □

If - apart from the conditions of Theorem 9.2.1 - we demand that there are no extraneous singularities in M_{int}, then we have the following result:

<u>Theorem 9.2.2</u>. Let M be a Global Boundary for $N(F)$ with $F \in F$. If moreover, M_{int} does not contain extraneous singularities, then

1. M_{int} contains no periodic trajectories;
2. M is diffeomorphic with S^{n-1};
3. M_{int} contains exactly one zero for F (which is an attractor for $N(F)$).

<u>Proof</u>. In view of Theorem 9.2.1 there is at least one attractor for $N(F)$ which is contained in M_{int}. Denote the attractors of $N(F)$ in M_{int} by α_i, $i = 1,2,\ldots,m$. By A_i we mean the basin in M_{int} of α_i, i.e. $x \in A_i$ iff $x \in M_{int}$ and $\lim_{t\to+\infty} \phi_x(t) = \alpha_i$. By M_1 we denote the set $\{x \in M_{int}\backslash \mathrm{Ess}(F) \mid \phi_x(t)$ is periodic$\}$. Note that the solution of system $N(F)$ depends continuously on the initial conditions. From this fact it follows that M_1 as well as A_i are open in $M_{int}\backslash \mathrm{Ess}(F)$ (in the case of M_1 one needs also a flowbox argument similar to the one used in the proof of Lemma 9.2.2; in the case

of A_i one needs the property of attractors which is referred to in Corollary 9.2.1). In view of the absence of extraneous singularities and Lemma 9.2.2, $M_{int}\backslash Ess(F)$ is the union of the (disjoint) sets M_1, A_i, $i = 1,\dots,m$. Since $M_{int}\backslash Ess(F)$ is connected, only one of these sets is non-empty. If follows that $M_1 = \emptyset$ (i.e. there are no periodic trajectories) and that there is only one attractor - say α - in M_{int}. From the last conclusion we find that for every $x \in M$ we have $\lim_{t\to\infty} \phi_x(t) = \alpha$.
Let $B_\varepsilon(\alpha)$ be an ε-neighborhood as in Corollary 9.2.1 and let $\partial B_\varepsilon(\alpha)$ be its boundary. If ε sufficiently small, then $\partial B_\varepsilon(\alpha)$ is a Global Boundary for $N(F)$, cf. (9.1.4). The map $\psi: M \to \partial B_\varepsilon(\alpha)$, $\psi(x) = \gamma(x) \cap \partial B_\varepsilon(\alpha)$ and $\gamma(x) = N(F)$-trajectory through x is well defined, continuous and injective ([30]). Let $g_1 = TF_{|M}$ and $g_2 = TF_{|\partial B_\varepsilon}(\alpha)$, then $g_2 \circ \psi = g_1$. Using the fact that g_1, g_2 are local diffeomorphisms (cf. Lemma (9.2.1) and that ψ is continuous, one easily proves that ψ is a local diffeomorphism, and hence ψ is an open map. Because ψ is injective, $\partial B_\varepsilon(\alpha)$ is connected ($n > 1$!) and M is compact it follows that ψ is a diffeomorphism. If we are able to prove that there are no repellors in M_{int}, then we are done. Suppose, β is a repellor in M_{int}. A trajectory leaving from β must tend to α and thus it must intersect $\partial B_\varepsilon(\alpha)$ at a point, say x_0. On the other hand, the point x_0 must also lie on the trajectory through $\psi^{-1}(x_0)$. This however, is impossible in view of the Orientation Assumption. □

<u>Remark 9.2.1</u>. The result of the preceding theorem may be paraphrased in the following way: the local topological structure of the phase portrait of $N(F)$ around a zero for F is maintained globally within any Global Boundary M as long as no extraneous singularities are involved. (cf. Fig. 9.1.2.b).

<u>Remark 9.2.2</u>. (The case n = 1).
Let F be a smooth function from $\mathbb{R}$ to $\mathbb{R}$. Then, system (9.1.1) takes the form

$$\frac{dx}{dt} = - \frac{F(x)}{F'(x)}, \tag{9.2.1}$$

where $F'(x)$ stands for the usual derivative of F at x. Note that the definition of <u>adjoint matrix</u> - as given in Section 9.1 - does not make sense in the one-dimensional case. In order to maintain the - crucial - relation (9.1.4) we propose: $\widetilde{F'}(x) := 1$. With this convention, we find for the desingularized system (cf. (9.1.3)):

$$\frac{dx}{dt} = -F(x). \tag{9.2.2}$$

As in the case of smooth mappings from $\mathbb{R}^n$ to $\mathbb{R}^n$ with $n \geq 2$, we may introduce Condition (*) as well as the sets Crit(F), Ess(F), Ext(F) and Reg(F) for functions $F \in C^\infty(\mathbb{R},\mathbb{R})$. One easily verifies that Condition (*) now reads: <u>The graph of F intersects the x-axis transversally</u>. Lemma 9.1.1 remains valid. This is also true - in a trivial way - for Lemma 9.1.2 and 9.1.3; in particular we find that the set Ext(F) is empty. However, the results obtained in Section 9.2 do loose their relevance in the one dimensional case. (Note that: S^0 is not connected and the interior of a Global Boundary ($\subset \mathbb{R}$) of $N(F)$ is not defined). For an illustration, we refer to Figure 9.2.2.a and 9.2.2.b, where the arrows indicate the vector fields $-\frac{F(x)}{F'(x)}$ respectively $-F(x)$.

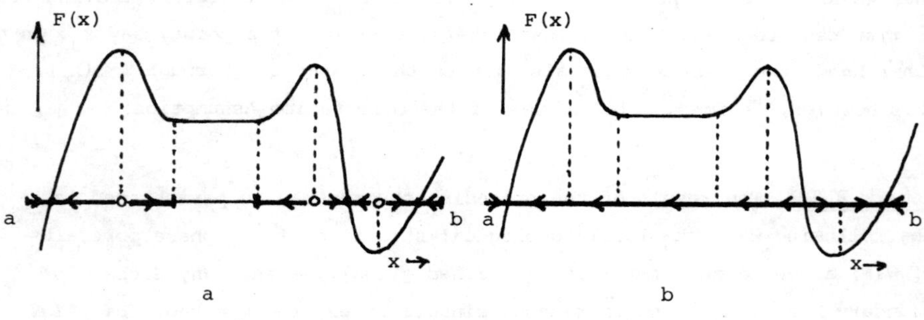

Fig. 9.2.2

In Fig. 9.2.2, the function F fulfils Condition (*). Note that the zeros for F are attractors for the vector field $-\frac{F}{F'}$, whereas these zeros are either attractors or repellors for the vector field -F. The non-connected boundary of the interval [a,b] fulfils the Orientation Assumption w.r.t. $N(F)$ and there are no extraneous singularities. Nevertheless, the number of essential singularities within the interior (a,b) is greater than one (cf. Theorem 9.2.2). □

3. Gradient Newton-flows.

Throughout this section, the mapping F will be of the form $F(x) = D^T f(x)$, where f is a smooth function on $\mathbb{R}^n$, $n \geq 2$. The associated Newton-flow will be called Gradient Newton-flow and will shortly be denoted by $N(f)$ (rather than $N(D^T f)$). The associated set of regular points, extraneous and essential singularities will be denoted by resp. Reg(f), Ext(f) and Ess(f). For the set Crit(F) we write in this case Crit(f); thus,
$\mathrm{Crit}(f) = \{x \in \mathbb{R}^n \mid \det F(x) \ (= \det D^2 f(x)) = 0\}$. Note that Condition (*) of Section 9.1 now reads:

$$\widetilde{j^1 f} \pitchfork \{0\} \ , \quad 0 \in \mathbb{R}^n,$$

where $\widetilde{j^1 f}$ stands for the map $\mathbb{R}^n \to \mathbb{R}^n$: $x \mapsto D^T f(x)$, i.e. $\widetilde{j^1 f}$ is a reduced 1-jet extension of f.
The set of all smooth functions f which fulfil this Condition (*) is referred to as to F_*. Obviously, F_* is a C^k-dense (for all k) and a C^k-open (for $k \geq 2$) subset of $C^\infty(\mathbb{R}^n, \mathbb{R})$, cf. Theorem 7.4.1.

Of course all the results obtained in the preceding sections do also hold in this section. Under transversality conditions, similar to Condition (*), the structure of the sets Crit(f) and Ext(f) will be discussed. In the case where n = 2, we describe the topological structure of the phase portrait of $N(f)$ around an extraneous singularity. Moreover, we give an indication how to generalize this result to the higher dimensional case. Finally we shall pay attention to structural stability aspects.

We begin with investigating the set Crit(f), i.e. the subset of $\mathbb{R}^n$ on which the desingularized Newton-system (9.1.3) is well-defined but not the system (9.1.1).
To this aim, let A $(= \mathbb{R}^{\frac{1}{2}n(n+1)})$ stand for the set of all symmetric n×n-matrices and let A_i be the submanifold of A consisting of all matrices with rank equal to i, i = 0,1,...,n; compare also Example 7.3.4. The reduced 2-jet $\widetilde{j^2 f}$ of f will be the mapping

$$\widetilde{j^2 f}: \mathbb{R}^n \to \mathbb{R}^n \times A (= \mathbb{R}^n \times \mathbb{R}^{\frac{1}{2}n(n+1)}): x \mapsto (D^T f(x), D^2 f(x)).$$

We propose the following transversality condition for smooth functions f

Condition (**) $\quad \widetilde{j^2 f} \pitchfork \mathbb{R}^n \times A_i \ , \quad i = 0,1,\ldots,n-1.$

The subset of F_* of all functions f which fulfil Condition (**) is denoted by F_{**}.

Let x be a critical point for $D^T f$ (i.e. $x \in Crit(f)$). Then we define $e(f,x) :=$ co-rank $D^2 f(x)$ and we put $E(f) = \max_{x \in Crit(f)} e(f,x)$.

Now, we have the following result:

Theorem 9.3.1.

a. The set F_{**} is C^k-dense (for all k) and C^k-open (for all $k \geq 3$) in $C^\infty(\mathbb{R}^n,\mathbb{R})$.

Let $f \in F_{**}$; then,

b. The set Crit(f) is a closed, Whitney-regular stratified subset of $\mathbb{R}^n$ of codimension one with strata: $(\widetilde{j^2 f})^{-1}(\mathbb{R}^n \times A_i)$, $i = 0,\ldots,n-1$.

c. The following inequality holds: $E(f) \leq -\frac{1}{2} + \frac{1}{2}\sqrt{1+8n}$.

Proof. a,b. The set F_* is C^k-dense (all k) and C^k-open ($k \geq 2$) in $C^\infty(\mathbb{R}^n,\mathbb{R})$. Note that $Crit(f) = (\widetilde{j^2 f})^{-1}[\bigcup_{i=0}^{n-1} \mathbb{R}^n \times A_i]$, where $\{\mathbb{R}^n \times A_i\}_{i=0,1,\ldots,n-1}$ constitutes a Whitney-regular stratification of the closed set $\bigcup_{i=0}^{n-1} \mathbb{R}^n \times A_i$ of codimension one, cf. Section 7.5 and Example 7.3.4. Application of Theorem 7.5.1 yields the desired result.

c. Let $x \in Crit(f)$. Then, there exists an index i, $0 \leq i \leq n-1$, such that A_i contains $D^2 f(x)$, i.e. $e(f,x) = n-i$. By a simple dimension argument, it follows from the transversality Condition (**) that (cf. also Example 7.3.4)

$$\text{codim } A_i = \frac{1}{2} e(f,x)(e(f,x) + 1) \leq n.$$

This inequality holds for any $x \in Crit(f)$. So, by considering the maximum of $e(f,x)(e(f,x) + 1)$ on Crit(f), we find $\frac{1}{2}E(f)(E(f) + 1) \leq n$. From this, the assertion follows straightforward. □

Corollary 9.3.1. Let $f \in C^\infty(\mathbb{R}^2,\mathbb{R})$ belong to F_{**}. Then, from the above theorem it follows that $E(f) < 2$. Thus, for a point $x \in Crit(f)$, we have rank $D^2 f(x) = 1$. So, we may conclude that $Crit(f) = (\widetilde{j^2 f})^{-1}[\mathbb{R}^2 \times A_1]$ and hence the set Crit(f) is either a closed one-dimensional smooth submanifold (without boundary) of $\mathbb{R}^2$ or it is empty. In the first case, Crit(f) consists of the union of unbounded smooth curves, and smooth curves diffeomorphic to a circle. □

Now, we pass on to the investigation of the set Ext(f) of extraneous singularities. In order to exclude pathologies, we have (again) to restrict ourselves to a <u>generic</u> subset of functions f. This is done by means of a transversality condition, similar to Conditions (*) and (**). To this aim we need the following result on "augmented symmetric" matrices.

Consider the set of all $n\times(n+1)$-matrices of the form $[A\vdots b]$, where A is a <u>symmetric</u> $n\times n$-matrix and b is the last column. Since $[A\vdots b]$ may be identified with the ordered pair (b,A) the set of all those matrices can be identified with $\mathbb{R}^n \times \mathbb{R}^{\frac{1}{2}n(n+1)} = \mathbb{R}^m$, $m = \frac{1}{2}n(n+3)$. By V_i we denote the subset in $\mathbb{R}^m$ of all matrices $[A\vdots b]$ with rank equal to i, $i = 0,\dots,n$.

<u>Lemma 9.3.1.</u>

a. Let V be a nonsingular $n\times n$ matrix.
   Then, $\text{rank}[V^TAV\vdots V^Tb] = \text{rank}\,[A\vdots b]$.

b. The set V_i constitutes a smooth submanifold of $\mathbb{R}^m$ $(m = \frac{1}{2}n(n+3))$ of co-dimension $\frac{1}{2}(n-i)(n-i+3)$, $i = 0,\dots,n$.

<u>Proof</u>. <u>a</u>. Given the matrix V, by $\hat{V}$ we denote the $(n+1)\times(n+1)$-matrix obtained from V by adding an $(n+1)^{\text{th}}$ column and an $(n+1)^{\text{th}}$ row, each of their entries being equal to zero, except the last one which is equal to 1. Then, we have the following identity:

$$V^T\cdot[A\vdots b]\cdot\hat{V} = [V^T\cdot A\vdots V^T\cdot b]\cdot\hat{V} = [V^T\cdot A\cdot V\vdots V^T\cdot b]$$

Assertion <u>a</u>. follows directly from the latter identity since V and $\hat{V}$ are both nonsingular.

<u>b</u>. Let $[\hat{A}\vdots\hat{b}] \in V_i$. We will distinguish between the two possible cases: rank $\hat{A} = i$ and rank $\hat{A} = i-1$.

<u>The case rank $\hat{A} = i$.</u>

For any non-singular $n\times n$ matrix V, the map Φ_V: $[A\vdots b] \mapsto [V^TAV\vdots V^Tb]$ is a C^∞-diffeomorphism on $\mathbb{R}^m$, leaving V_i invariant (cf. Part a of the lemma). Therefore, we may assume, without loss of generality, that in an $\mathbb{R}^m$-neighborhood of $[\hat{A}\vdots\hat{b}]$, say $\mathcal{O}$, the upper-left $i\times i$-submatrix of $[A\vdots b]$ is non-singular. Now we consider a partition of $[A\vdots b] \in \mathcal{O}$ into submatrices as indicated in Scheme (9.3.1)

$$\begin{array}{c|c|c|c|} & i & n-i & 1 \\ \hline i & \alpha & \gamma & \beta_1 \\ \hline n-i & \gamma^T & \delta & \beta_2 \\ \hline \end{array}$$

Scheme 9.3.1

In particular, the matrix α is a nonsingular symmetric $i\times i$-matrix with coefficients $a_{k\ell}$, $1 \le k,\ell \le i$. The $i\times(n-i)$-matrix γ has the coefficients $a_{k\ell}$, $1 \le k \le i$, $i < \ell \le n$, whereas $\delta = (n-i)\times(n-i)$-matrix $(a_{k\ell})$ with $i < k,\ell \le n$. Finally, $\beta_1 = (b_1,\dots,b_i)^T$ and $\beta_2 = (b_{i+1},\dots,b_n)^T$.
Now, the matrix $[A\vdots b]$ $(\in \mathcal{O})$ is an element of V_i iff the following relation holds (cf. Example 7.3.3)

$$[\delta\vdots\beta_2] = \gamma^T\cdot\alpha^{-1}\cdot[\gamma\vdots\beta_1]. \tag{9.3.1}$$

Relation (9.3.1) is equivalent with

$$\delta - \gamma^T\cdot\alpha^{-1}\cdot\gamma = 0 \tag{9.3.1.a}$$

$$\beta_2 - \gamma^T\cdot\alpha^{-1}\cdot\beta_1 = 0 \tag{9.3.1.b}$$

From these relations it follows: if $[A\vdots b] \in \mathcal{O} \cap V_i$, then the coefficients of the matrices δ and β_2 can be smoothly expressed in the coefficients of γ, α and β_1. Due to the symmetry in both sides of Relation (9.3.1.a), this relation can be considered as a system of $\frac{1}{2}(n-i)(n-i+1)$ equations in the "variables" $a_{k\ell}$, $k,\ell = 1,\dots,n$, $k \le \ell$, (and b_k, $k = 1,\dots,n$). Relation (9.1.3.b) gives rize to n-i equations in these "variables". Together, the left-hand sides of Relations 9.1.3.a,b form a defining system of functions (cf. Section 7.2) for $\mathcal{O} \cap V_i$, the actual number being $\frac{1}{2}(n-i)(n-i+3)$. (Note that these functions constitute an independent set since <u>each</u> of the coefficients $a_{k\ell}$, $k \le \ell$, $k,\ell = i+1,\dots,n$, and b_k, $k = i+1,\dots,n$ appears (as a linear variable) in exactly one of the functions).

The case rank $\hat{A} = i-1$.

The diffeomorphism Φ_V (cf. previous case) not only leaves V_i invariant, but also respects the rank of the left n×n-submatrix (since rank $V^T AV$ = rank A). Therefore, we may assume that the upper-left (i-1)×(i-1)-submatrix of $\hat{A}$ is nonsingular. We denote this matrix by $\hat{\alpha}_0$.
In order to simplify our notations, we shall represent - for the moment - the augmented matrices [A⋮b] as [b⋮A].
We denote the i×i-left upper submatrix of $[\hat{b}\vdots\hat{A}]$ by $\hat{\alpha}_*$.

Contention: without loss of generality we have rank $\hat{\alpha}_* = i$.
The proof of this contention runs as follows:
Among the last (n-i+1) rows of $[\hat{b}\vdots\hat{A}]$ there is at least one row - say the jth row - which constitutes together with the first (i-1) rows of this matrix an independent set. This follows from: rank$[\hat{b}\vdots\hat{A}]$ = i. The (permutation) matrix R_{ij} is obtained from the n×n-unit matrix by interchanging the ith and the jth row. Now, we consider the diffeomorphism Φ_V: $\mathbb{R}^m \to \mathbb{R}^m$ with $V = R_{ij}$. Apparently, the first i rows of

$$\Phi_V([\hat{b}\vdots\hat{A}]) = [V^T\hat{b}\vdots V^T\cdot\hat{A}\cdot V]$$

form a linearly independent set of rows. From this and from the facts that

(i) the (i-1)×(i-1) left upper submatrix of $V^T\cdot\hat{A}\cdot V$ just equals $\hat{\alpha}_0$;

(ii) rank $V^T\cdot\hat{A}\cdot V$ = rank $\hat{A}$ = i-1,

it follows that the left upper i×i submatrix of $\phi_V([\hat{b}\vdots\hat{A}])$ is nonsingular. This proves the contention.

We consider a partition of [b⋮A] according to Scheme 9.3.2.

In this partition, the submatrices have the dimensions as indicated in the scheme; the matrix within the bold (resp. dotted) lines is denoted by α_* (resp. α). On an $\mathbb{R}^m$-neighborhood - say $\mathcal{O}$ - of $[\hat{b}\vdots\hat{A}]$ we may assume:

$$\text{rank } \alpha_0 = i-1 \quad \text{and} \quad \text{rank } \alpha_* = i.$$

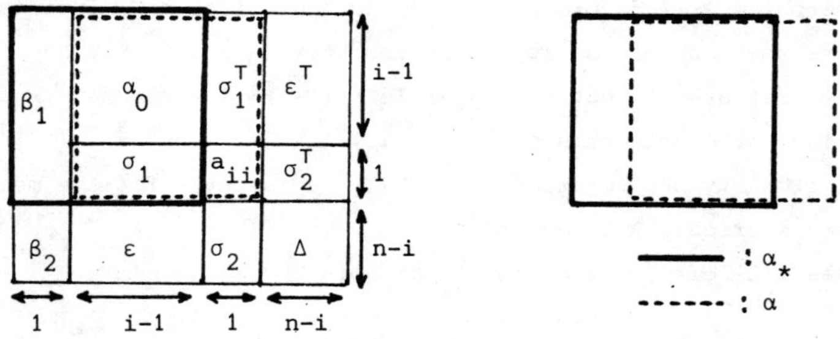

[b⁝A]

Scheme 9.3.2

As in the case rank $\hat{A} = i$, we have: the matrix [b⁝A] ($\epsilon\ \mathcal{O}$) has rank i if and only if

$$\Delta - [\beta_2 \vdots \varepsilon]\cdot\alpha_*^{-1}\cdot[\varepsilon \vdots \sigma_2]^T = 0 \tag{9.3.2.a}$$

$$\sigma_2 - [\beta_2 \vdots \varepsilon]\cdot\alpha_*^{-1}\cdot[\sigma_1 \vdots a_{ii}]^T = 0 \tag{9.3.2.b}$$

We may conclude: given the matrices $\alpha_*, \alpha, \beta_2, \varepsilon$ (in particular, rank $\alpha_* = i$) then there is only one $(n-i)\times 1$-matrix σ_2 and one $(n-i)\times(n-i)$ matrix Δ such that the corresponding (cf. Scheme 9.3.2) matrix [b⁝A] has rank i. However, it is not a priori clear that matrix Δ, determined in this way, is <u>symmetric</u> (cf. (9.3.2.a). In order to prove this, we fix the submatrices $\beta_1, \beta_2, \alpha, \varepsilon$ of [b⁝A] and determine σ_2 according to (9.3.2.b). Thus, the first (i+1) columns of matrix [b⁝A] are fixed and if we denote these columns by respectively $a_0 = \begin{pmatrix}\beta_1\\ \cdots \\ \beta_2\end{pmatrix}$, $a_1, \ldots, a_i$, then we have

$$a_i = \xi_0 a_0 + \xi_1 a_1 + \ldots + \xi_{i-1} a_{i-1}\ ,\ \xi_k \in \mathbb{R}. \tag{1}$$

Now, we distinguish between two cases: rank $\alpha = i$ and rank $\alpha = i-1$.
<u>If rank $\alpha = i$</u>, then there exists a <u>unique symmetric</u> $(n-i)\times(n-i)$-matrix, say $\tilde{\Delta}$, such that the matrix $\tilde{A}$ given by

$$\tilde{A} = \left[\begin{array}{cc|c} \multicolumn{2}{c|}{\alpha} & \varepsilon^T \\ & & \sigma_2^T \\ \hline \varepsilon & \sigma_2 & \tilde{\Delta} \end{array}\right]$$

has the rank i (in fact $\tilde{\Delta} := [\varepsilon\vdots\sigma_2]\cdot\alpha^{-1}\cdot[\varepsilon\vdots\sigma_2]^T$, cf. Example 7.3.4). The coefficient ξ_0 in the above relation (1) does not vanish, since in that case we should have: rank $\alpha < i$. So, the first column of $[b\vdots\tilde{A}]$, i.e. a_0 depends linearly on the first i columns of $\tilde{A}$. It follows that both $[b\vdots A]$ and $[b\vdots\tilde{A}]$ are of rank i. Since (eventually apart from Δ and $\tilde{\Delta}$) all corresponding submatrices, according to Scheme 9.3.2, in $[b\vdots A]$ and $[b\vdots\tilde{A}]$ are the same, we conclude that also $\Delta = \tilde{\Delta}$. Thus, Δ is symmetric.

<u>If rank α = i-1</u>, then we consider a matrix A_t with in the ith row and in the ith column the coefficient $a_{ii}+t$ and all other coefficients equal to the corresponding coefficients of the matrix A. In the partition (cf. Scheme 9.3.2) of $[b\vdots A_t]$ the submatrices are denoted with a subscript t. We obviously have $\det \alpha_t = t\cdot\det \alpha_0 \neq 0$ if $t \neq 0$. So for $t \neq 0$ we find - as in the case rank α = i - that Δ_t $(= \tilde{\Delta}_t)$ is a symmetric matrix. Now we take the limit for $t \to 0$ and conclude that also Δ is a symmetric matrix.

Altogether, we now proved: if for $[b\vdots A] \in \mathcal{O}$ the relations (9.3.2.a,b) hold, then we have $[b\vdots A] \in V_i$, and vice versa. Now, we may proceed as in the case rank $\hat{A}$ = i and observe that (eliminating σ_2 from (9.3.2.a) by means of (9.3.2.b)) the functions in the l.h.s. of (9.3.2) constitute a system of $\frac{1}{2}(n-i)(n-i+1) + (n-i) = \frac{1}{2}(n-i)(n-i+3)$ defining functions for $V_i \cap \mathcal{O}$. This completes the proof of Lemma 9.3.1. □

<u>Exercise 9.3.1.</u> Let the subsets $\tilde{V}_i$ and $\tilde{\tilde{V}}_i$ be defined as follows

$$\tilde{V}_i = \{[A\vdots b] \in V_i \mid \text{rank } A = i\}$$

$$\tilde{\tilde{V}}_i = \{[A\vdots b] \in V_i \mid \text{rank } A = i-1\}$$

Prove that $\tilde{V}_i$ is <u>open</u> in V_i, and that $\tilde{\tilde{V}}_i$ is a <u>closed</u> submanifold of V_i of codimension one.

We introduce our next transversality condition on smooth functions f.

Condition (***) $\widetilde{j^2 f} \pitchfork V_i \qquad i = 0,1,\dots,n-1.$

The subset of F_* of all functions f which fulfil Condition (***) is denoted by F_{***}.

The following result characterizes - generically - the structure of the set Ext(f).

Theorem 9.3.2.

a_1. The set F_{***} is a C^∞-generic subset of $C^\infty(\mathbb{R}^n,\mathbb{R})$.

a_2. If $f \in F_{***}$, then Ext(f) is a closed, stratified subset of $\mathbb{R}^n$ of codimension ≥ 2, with strata $(\widetilde{j^2 f})^{-1}(V_i)$, $i = 0,1,\dots,n-1$.

In the special case where n = 2, 3 or 4, we have:

b_1. If $f \in F_{***}$, then Ext(f) is a closed submanifold of $\mathbb{R}^n$ of codimension two (or empty).

b_2. The set F_{***} is C^3-open in $C^\infty(\mathbb{R}^n,\mathbb{R})$.

Proof.

a_1. This assertion follows directly from Theorem 7.4.1, Remark 7.4.2, Lemma 9.3.1.b and the fact that F_* is C^3-open and C^3-dense in $C^\infty(\mathbb{R}^n,\mathbb{R})$.

a_2. Note that - for $f \in F_{***}$ - the set Ext(f) just equals $(\widetilde{j^2 f})^{-1}[\bigcup_{i=0}^{n-1} V_i]$; cf. Lemma 9.1.2. The assertion follows from Theorem 7.3.1, Lemma 9.3.1.b, and the fact that $\bigcup_{i=0}^{n-1} V_i$ is closed in $\mathbb{R}^m$, $m = \frac{1}{2}n(n+3)$.

b_1. Let $x \in \text{Ext}(f)$ and thus $\widetilde{j^2 f}(x) \in V_{i_0}$ for some index i_0 with $0 \leq i_0 \leq n-1$. From Condition (***) it follows: codim $V_{i_0} \leq n$. On the other hand, Lemma 9.3.1.b yields that codim $V_{i_0} = 2$ if $i_0 = n-1$ and codim $V_{i_0} \geq 5$ if $i_0 \leq n_0-2$, where the equality occurs iff $i_0 = n-2$. From these observations, we may conclude that in the cases where $n \leq 4$, we have $x \in \text{Ext}(f)$ iff $\widetilde{j^2 f}(x) \in V_{n-1}$ and hence, $\text{Ext}(f) = (\widetilde{j^2 f})^{-1}(V_{n-1})$. Now, the assertion is a consequence of Part a_2 of this theorem (and of Lemma 9.3.1.b).

b_2. The proof of this assertion is based on the Openess-Principle, cf. Theorem 7.3.7. Firstly, we give a reformulation of this principle which fits our purpose. In order to simplify the notations, we denote the mapping $\widetilde{j^2 f}: \mathbb{R}^n \to \mathbb{R}^m$, $m = \frac{1}{2}n(n+3)$, by h. Moreover, we put $M_i = \mathbb{R}^n \times V_i$, $i = 0,\dots,n-1$. Let $f \in F_{***}$, i.e. $\mathrm{Graph}(h) \pitchfork M_i$, $i = 0,\dots,n-1$. <u>In this case, the Openess-Principle reads:</u>

There exists a C^1-neighborhood V of h having the property: $\mathrm{Graph}(g) \pitchfork M_i$, $i = 0,\dots,n-1$, $g \in V$

iff the following condition is fulfilled:

<u>Condition $\mathcal{O}$</u>: For every $i = 0,\dots,n-1$ and every compact subset $K \subset \mathbb{R}^{n+m}$ with $K \cap M_i \neq \emptyset$ we have: $\inf_{z \in K \cap M_i} \phi_{h,M_i}(z) > 0$,

where $z = (x,y)$, $x \in \mathbb{R}^n$, $y \in \mathbb{R}^m$ and the <u>continuous</u> functions ϕ_{h,M_i} on M_i are defined by Formula (7.3.8), i.e. as a sum of two terms: $\|y-h(x)\|$ and a term consisting of the sum of certain determinants.

We are done if we are able to prove that in our case Condition $\mathcal{O}$ is fulfilled (For then, application of the Openess-Principle yields the existence of a C^3-neighborhood of f, say U, such that $u \in U$ implies $\widetilde{j^2 u} \pitchfork V_i$, $i = 0,\dots,n-1$, hence, F_{***} is C^3-open in $C^\infty(\mathbb{R}^n,\mathbb{R})$).
Firstly, we show that Condition $\mathcal{O}$ holds in the case of the highest dimensional manifold M_{n-1} $(= \mathbb{R}^n \times V_{n-1})$. Note that (cf. Property 7.3.7) from h $(= \widetilde{j^2 f}) \pitchfork V_{n-1}$ it follows:

$$\phi_{h,M_{n-1}}(z) > 0 \quad \text{for all} \quad z \in M_{n-1}. \tag{I}$$

Now, <u>suppose</u> that $\inf_{z \in K \cap M_{n-1}} \phi_{h,M_{n-1}}(z) = 0$. We lead this assumption to a contradiction in the following five steps:

1. $K \cap \bigcup_{i=0}^{n-1} M_i$ is compact;

2. In view of our assumption and in view of Step 1 there is a sequence $z_k = (x_k,y_k) \in M_{n-1}$ which tends, if $k \to \infty$, to $\hat{z} = (\hat{x},\hat{y}) \in M_{i_0}$, for some i_0, $0 \le i_0 \le n-1$, with $\phi_{h,M_{i_0}}(\hat{z}) = 0$;

3. In view of the continuity of $\phi_{h,M_{n-1}}$ and in view of the inequality (I) we have $i_0 < n-1$;

4. Since $n \le 4$, the transversality Condition (***) yields $h(\mathbb{R}^n) \cap V_i = \emptyset$ if $i < n-1$, cf. Part b_1 of this theorem. Thus, $\|\hat{y}-h(\hat{x})\| =: \delta > 0$.
5. From Step 4 it follows that for k sufficiently large, the following inequality holds: $\phi_{h,M_{n-1}}(\bar{z}_k) \ge \|y_k-h(x_k)\| > \frac{1}{2}\delta\ (> 0)$.

   This is in contradiction with our assumption.

Finally, we have to show that Condition O also holds in the case of manifolds M_i, $i < n-1$. However, this proof will be omitted since it is similar to the above proof in case M_{n-1}. (Note that, if $i < n-1$, then $\phi_{h,M_i}(x,y)$ reduces to $\|y-h(x)\|$). □

Remark 9.3.1. (The case $n = 1$).
Let f be a smooth function from $\mathbb{R}$ to $\mathbb{R}$. Then, we may define the Gradient Newton-flow $N(f)$ in accordance with Remark 9.2.2 (put $F = f'$). In the one-dimensional case, the sets A_i and V_i, $i = 0,1$, can be defined as in the case $n \ge 2$. So, for the function f, the Conditions (**) and (***) do have a meaning. One easily verifies that Condition (*) (i.e."f is Morse function") is equivalent with Condition (***). So, in the one-dimensional case, we have $F_* = F_{***}$.
If $f \in F_*$ then $\mathrm{Ext}(f) = \emptyset$. (Note that this property is formally in accordance with the assertion of Theorem 9.3.2.b_1). Moreover, Condition (**) yields that f' is a Morse function and hence Crit(f) is a discrete subset of $\mathbb{R}$. (So, if - in Fig. 9.2.2 - we put $F = f'$, then $f \notin F_{**}$). □

For functions $f \in C^\infty(\mathbb{R}^n,\mathbb{R})$, $n \ge 2$, we introduce the following notations.

$$\partial_i(x) = \frac{\partial f}{\partial x_i}(x)\ ;\ \partial_{ij}(x) = \frac{\partial^2 f}{\partial x_i \partial x_j}(x)\ ;\ \partial_{ijk}(x) = \frac{\partial^3 f}{\partial x_i \partial x_j \partial x_k}(x)\ ;$$
$$i,j,k=1,\ldots,n.$$

If no confusion is possible, then we shall delete the argument from this notations; thus $\partial_i = \partial_i(x)$ etc.

In the case of a Gradient Newton-flow, the equation (9.1.3) takes the following form:

$$\frac{dx}{dt} = -\widetilde{D^2 f(x)} \cdot D^T f(x). \tag{9.3.3}$$

Our next step is to describe the structure of the local phase-portrait of $N(f)$ in the case where $f \in F_{***}$ and $n = 2$. To this aim we need the following lemma.

Lemma 9.3.2. For $f \in C^{\infty}(\mathbb{R}^2,\mathbb{R})$ we have:

Condition (***) implies Condition (*).

Proof. Suppose that Condition (***) holds for f. Let $\hat{x}$ be a zero for $D^T f$, i.e. $(\partial_1,\partial_2)_{\hat{x}} = (0,0)$. In view of Condition (***) the intersection $\widetilde{j^2f}(\mathbb{R}^2) \cap V_0$ is empty. So, there remain two possibilities: either $\widetilde{j^2f}(\hat{x}) \in V_2$ (and thus Condition (*) holds) or $\widetilde{j^2f}(\hat{x}) \in V_1$. We are done, if we can prove that the latter possibility is in contradiction with Condition (***). In order to prove this, let us suppose that $\widetilde{j^2f}(\hat{x}) \in V_1$. We identify the matrix

$$\begin{pmatrix} y_1 & y_3 & y_4 \\ y_2 & y_4 & y_5 \end{pmatrix}$$

with $y = (y_1,y_2,y_3,y_4,y_5)^T \in \mathbb{R}^5$. According to this identification, we have

$$\widetilde{j^2f}(x) = (\partial_1,\partial_2,\partial_{11},\partial_{12},\partial_{22})^T.$$

We put $\hat{\partial} = \widetilde{j^2f}(\hat{x})$. Then, we may write $\hat{\partial} = (0,0,\hat{\partial}_{11},\hat{\partial}_{12},\hat{\partial}_{22})^T$. The set V_1 is a 3-dimensional submanifold of $\mathbb{R}^5$, cf. Lemma 9.3.1.b, and Condition (***) reads in this case:

$$D\widetilde{j^2f}(\hat{x})(\mathbb{R}^2) + T_{\hat{\partial}}V_1 = \mathbb{R}^5, \tag{9.3.4}$$

where $T_{\hat{\partial}}V_1$ stands for the tangent space at $\hat{\partial}$ to V_1.
Since $\hat{\partial} = (0,0,\hat{\partial}_{11},\hat{\partial}_{12},\hat{\partial}_{22})$ is supposed to be an element of V_1, we may assume (no loss of generality) that $(\hat{\partial}_{11},\hat{\partial}_{12}) \neq (0,0)$. Now, one easily verifies that, locally around $\hat{\partial}$, the manifold V_1 may be given by the defining system (cf. Section 7.2) of functions $h_1(y) = h_2(y) = 0$, where

$$h_1(y) = \det\begin{pmatrix} y_1 & y_3 \\ y_2 & y_4 \end{pmatrix}, \quad h_2(y) = \det\begin{pmatrix} y_3 & y_4 \\ y_4 & y_5 \end{pmatrix}.$$

Thus, the normal space $(N_{\hat{\partial}}V_1)$ at $\hat{\partial}$ to V_1 is spanned by the vectors $D^T h_1(\hat{\partial}) = (\hat{\partial}_{12},-\hat{\partial}_{11},0,0,0)^T$, $D^T h_2(\hat{\partial}) = (0,0,\hat{\partial}_{22},-\hat{\partial}_{12},\hat{\partial}_{11})^T$.
The linear space $D\widetilde{j^2 f}(\hat{x})\ \mathbb{R}^2)$ is spanned, cf. (9.3.4), by the vectors $w_1(\hat{\partial}) = (\hat{\partial}_{11},\hat{\partial}_{12},\hat{\partial}_{111},\hat{\partial}_{112},\hat{\partial}_{122})^T$, $w_2(\hat{\partial}) = (\hat{\partial}_{12},\hat{\partial}_{22},\hat{\partial}_{112},\hat{\partial}_{122},\hat{\partial}_{222})^T$.
Now, let the 2×2 matrix Δ be defined by

$$\Delta = (< D^T h_i(\hat{\partial}),\ w_j(\hat{\partial}))_{i,j=1,2}, \tag{9.3.5}$$

where <.,.> denotes the standard inner product in $\mathbb{R}^5$. In view of Exercise 7.2.2, the transversality condition (9.3.4), i.e. Condition (***) in the point $\hat{x}$, is equivalent with

$$\det \Delta \neq 0. \tag{9.3.6}$$

However, a straightforward calculation (taking into account that $\hat{\partial}_{11}\hat{\partial}_{22} - \hat{\partial}_{12}^2 = 0$) learns that $\det \Delta = 0$. Hence, the condition (9.3.4), and thus Condition (***) is violated. □

In order to give our next result we need the concept of <u>elementary equilibrium state of a smooth vectorfield</u>: let x_0 be equilibrium state of the smooth vectorfield F on $\mathbb{R}^n$ (i.e. $F(x_0) = 0$). Then, we call x_0 <u>elementary</u> if the matrix $DF(x_0)$ is nonsingular. In the <u>special</u> case where F is defined on $\mathbb{R}^2$ and both eigenvalues of $DF(x_0)$ are equal and positive (negative), then x_0 is called a <u>dicritical unstable (stable) node</u> for F. One can prove (cf. Andronov et al [2] page 171) that such x_0 enjoys the following property: each semi-trajectory of F tending to x_0 does so in a definite direction, and to each direction there corresponds exactly one such semi-trajectory.
Now, let x_1 be an (eventually non-elementary) equilibrium state for a vector field on $\mathbb{R}^2$, then x_1 is called a <u>center</u>, if there exists a neighnorhood V of x_1 such that the trajectory through any point of $V\setminus\{x_1\}$ is periodic and contains x_1 in its interior.

By E, we denote the set of all functions $f \in C^\infty(\mathbb{R}^2,\mathbb{R})$ which fulfil Condition (***).

The following theorem describes - for a generic class of functions - the possible local phase-portraits of 2-dimensional Gradient Newton-flows around the equilibrium states (see also Fig. 9.3.1).

Theorem 9.3.3.

a. The set E is C^3-open and -dense in $C^\infty(\mathbb{R}^2,\mathbb{R})$.

Let $f \in E$ and let $\hat{x}$ be an equilibrium state for $N(f)$. Then,

b. The point $\hat{x}$, being either an essential or an extraneous singularity, is an elementary equilibrium state for $N(f)$.

Moreover,

(i) If $\hat{x} \in \mathrm{Ess}(f)$, then $\hat{x}$ is either a stable or an unstable dicritical node for $N(f)$.

(ii) If $\hat{x} \in \mathrm{Ext}(f)$, then $\hat{x}$ is either a saddle point or a center for $N(f)$.

Proof.

a. This is a direct consequence of Theorem 9.3.2.b.2 and Lemma 9.3.2.

b. In view of Lemma 9.3.2, the function f fulfils Condition (*). Hence, the sets Ess(f) and Ext(f) are well defined and disjoint, cf. Section 9.1.

(i) Let $\hat{x} \in \mathrm{Ess}(f)$.

In view of (9.1.4) we have:

$$D[-\widetilde{D^2 f}(x)\cdot D^T f(x)]_{\hat{x}} = -\det D^2 f(\hat{x})\cdot I_2 ,$$

where $\det D^2 f(\hat{x}) \neq 0$. Thus, depending on sign $\det D^2 f(\hat{x})$, point $\hat{x}$ is either a stable or an unstable <u>dicritical node</u> for $N(f)$, compare also Fig. 9.1.2.b,c.

(ii) Let $\hat{x} \in \mathrm{Ext}(f)$.

We adopt the notations as used in the proof of Lemma 9.3.2. In view of Theorem 9.3.2.b.1 (proof), we have $\hat{\partial}$ (= $\widetilde{j^2 f}(\hat{x})$)$\in V_1$. From the definition of extraneous singularity, it follows that $(\partial_1(\hat{x}),\partial_2(\hat{x})) \neq (0,0)$. Thus, locally around $\hat{\partial}$, the 3-dimensional manifold V_1 is given by the defining system of functions $h_1(y) = h_3(y) = 0$, where

$$h_1(y) = \det\begin{pmatrix} y_1 & y_3 \\ y_2 & y_4 \end{pmatrix} , \quad h_3(y) = \det\begin{pmatrix} y_1 & y_4 \\ y_2 & y_5 \end{pmatrix} .$$

Thus, the normal space $(N_{\hat{\partial}}V_1)$ at $\hat{\partial}$ to V_1 is spanned by the vectors $D^T h_1(\hat{\partial}) = (\hat{\partial}_{12}, -\hat{\partial}_{11}, -\hat{\partial}_2, \hat{\partial}_1, 0)^T$, $D^T h_3(\hat{\partial}) = (\hat{\partial}_{22}, -\hat{\partial}_{12}, 0, -\hat{\partial}_2, \hat{\partial}_1)^T$. The vectors $w_1(\hat{\partial})$ and $w_2(\hat{\partial})$ are defined as in the proof of Lemma 9.3.2. As an analogue of the 2×2 matrix Δ (cf. (9.3.5) we define the matrix $\tilde{\Delta}$.

$$\tilde{\Delta} = \begin{pmatrix} \langle D^T h_1(\hat{\partial}),\ w_1(\hat{\partial})\rangle & \langle D^T h_3(\hat{\partial}),\ w_1(\hat{\partial})\rangle \\ \langle D^T h_1(\hat{\partial}),\ w_2(\hat{\partial})\rangle & \langle D^T h_3(\hat{\partial}),\ w_2(\hat{\partial})\rangle \end{pmatrix}$$

Condition (***) is - at $\hat{x}$ - equivalent with (cf. (9.3.6)):

$$\det \tilde{\Delta} \neq 0. \tag{9.3.7}$$

Taking into account that $\det D^2 f(\hat{x}) = (\hat{\partial}_{11}\hat{\partial}_{22} - \hat{\partial}_{12}^2) = 0$, a straightforward calculation shows that

$$\det \tilde{\Delta} = \det D[-\widetilde{D^2 f(x)} \cdot D^T f(x)]_{\hat{x}} .$$

Consequently, the point $\hat{x}$ is an elementary equilibrium state for $N(f)$. Now, we assume (no loss of generality) that at $\hat{x}$ we have $\partial_2(\hat{x}) \neq 0$. Hence, the trajectory map (cf. Section 9.1) associated with $N(f)$ is - locally around $\hat{x}$ - given (cf. also (9.1.7)) as the <u>function</u>:

$$Tf(x) = \frac{\partial_1(x)}{\partial_2(x)} .$$

A straightforward calculation (taking into account that $\hat{x}$ is a critical point for Tf and that $\hat{\partial}_{11}\hat{\partial}_{22} - \hat{\partial}_{12}^2 = 0$) yields:

$$D^2 Tf(\hat{x}) = \tilde{\Delta}/\partial_2^2(\hat{x}) .$$

So - in view of (9.3.7) - the point $\hat{x}$ is a nondegenerate critical point for Tf. Consequently, the Morse-lemma (cf. Theorem 2.7.2) together with the fact that - locally - the trajectories of $N(f)$ are just the level lines of Tf (cf. Lemma 9.1.4 and Relation (9.1.7)) yields Assertion (ii). □

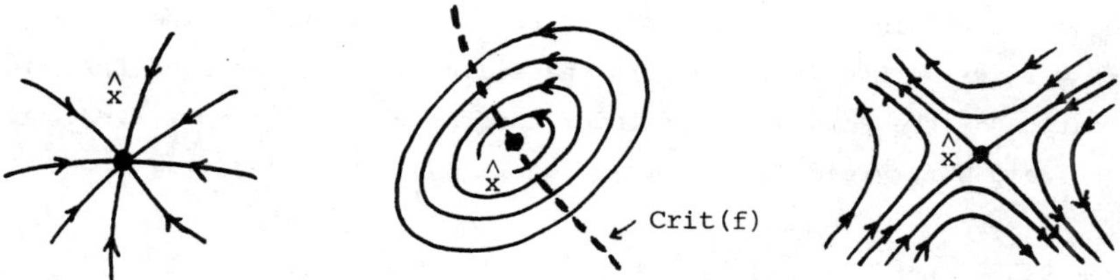

$\hat{x} \in \mathrm{Ess}(f)$ $\hat{x} \in \mathrm{Ext}(f)$ $\hat{x} \in \mathrm{Ext}(f)$

$\det D^2 f(\hat{x}) > 0$ $\det \tilde{\Delta} > 0$ $\det \tilde{\Delta} < 0$

Fig. 9.3.1

Note that, in view of Theorem 9.3.1 and 9.3.3 the set $F_{**} \cap E$ is C^3-open and dense in $C^\infty(\mathbb{R}^2,\mathbb{R})$. So, from Corollary 9.3.1 it follows that - generically - not only Theorem 9.3.3 holds, but also: the set of critical points for $D^T f$ is a 1-dimensional manifold (without boundary) or empty. In the case where the extraneous singularity $\hat{x}$ is a center for $N(f)$, on each (periodic) trajectory around $\hat{x}$, there are at least two critical points for $D^T f$ (cf. (9.1.2) and Fig. 9.3.1).

Let $f \in E$ be given. Motivated by the above Theorem 9.3.3 we distinguish between extraneous singularities of "saddlepoint" type and of "center" type; the corresponding subsets of Ext(f) are denoted by Sadd(f) resp. Cent(f).

Let γ be a Global Boundary (cf. Section 9.2) or a periodic trajectory for $N(f)$. Note that in both cases, γ is a closed, smooth Jordan curve in $\mathbb{R}^2$ which does not contain $N(f)$-equilibrium states. In view of Theorem 9.3.3 there are only finitely many essential and extraneous singularities for f in the interior (γ_{int}) of γ. We put $e_\gamma(f) = \# \,\mathrm{Ess}(f) \cap \gamma_{int}$, $s_\gamma(f) = \# \,\mathrm{Sadd}(f) \cap \gamma_{int}$ and $c_\gamma(f) = \# \,\mathrm{Cent}(f) \cap \gamma_{int}$. Then we have the following result:

<u>Lemma 9.3.3</u>. Let $f \in E$ and let γ be a Global Boundary or a periodic trajectory for $N(f)$. Then,

$$s_\gamma(f) - c_\gamma(f) = e_\gamma(f) - 1 \,. \qquad (9.3.8)$$

Proof. All equilibrium states of $N(f)$ are elementary. Let $\hat{x} \in Ess(f) \cup Ext(f)$ and denote its Poincaré-index ([22]) w.r.t. $N(f)$ by $Ind(\hat{x})$. Then we have: $Ind(\hat{x}) = 1$ if $\hat{x} \in Ess(f) \cup Cent(f)$, and $Ind(\hat{x}) = -1$ if $\hat{x} \in Sadd(f)$. Now, application of the Poincaré-Hopf index theorem (cf. [55]) to γ w.r.t. the vector field $N(f)$ yields the desired result (Note that $Ind(\gamma) = 1$). □

Theorem 9.3.4. Let $f \in C^{\infty}(\mathbb{R}^2, \mathbb{R})$ be a function for which Condition (*) holds. Then, we have:
If there are no extraneous singularities for f, then $N(f)$ does not exhibit periodic trajectories.

Proof. Note that in the proof of Lemma 9.3.3, the Condition (***) is used to get information on (the Poincaré-index of) the extraneous singularities. So, in case $Ext(f) = \emptyset$, the statement of Lemma 9.3.3 remains true if we only ask that f fulfils Condition (*). In particular, under the assumption that γ is a periodic trajectory for $N(f)$, we have: $e_{\gamma}(f) = 1$. Hence, on $\bar{\gamma}_{int}$ there is exactly one equilibrium state -say $\hat{x}$- for the vector field $D^T f$. In view of Condition (*), the Poincaré index of $\hat{x}$ w.r.t. $D^T f$ equals ± 1 On the other hand, since along γ the vector $D^T f$ has constant direction, the index of γ w.r.t. $D^T f$ equals zero. Now, application of the Poincaré-Hopf theorem to γ w.r.t. the vector field $D^T f$ yields a contradiction. □

Note that the above theorem is in the same spirit as Theorem 9.2.2.

Remark 9.3.2. The relation (9.3.8) could be derived by means of appropriately adapted techniques from Morse-theory (cf. Chapter 5) as well.

The Branin-conjectures.

In this intermezzo, we discuss some conjectures due to F. Branin as stated in e.g. [6] and [19]:

C.1. "An extraneous singularity cannot be of focus type (with $N(f)$-trajectories spiralling to it)", and

C.2. "If there are no extraneous singularities, one can find all zeros for $D^T f$ be following (for increasing or decreasing t) a $N(f)$-trajectory until one reaches a zero for $D^T f$ and following another trajectory from this zero until one reaches another zero, and so on" (In other words: in the absence of extraneous singularities, system (9.3.3) provides a "globally convergent method for finding all zeros for $D^T f$".

Apparently, Theorem 9.3.3 justifies Conjecture C.1 in two dimensions for a C^3-open and dense class of functions f.
In Section 9.4, Example 9.4.1, we shall reject Conjecture C.2 by giving a counterexample by means of a function f <u>satisfying Condition (***)</u>.

The following result, which holds under fairly weak conditions, is related to Conjecture C.2.

<u>Theorem 9.3.5</u>. Let $F \in C^\infty(\mathbb{R}^n,\mathbb{R}^n)$, <u>$n \geq 2$</u>, be a <u>mapping</u> which fulfils Condition (*) and suppose that Ext(F) is <u>empty</u>. Then, for each $\hat{x} \in$ Ess(F) there exists a maximal N(F)-trajectory, say $\phi_x(t)$, $\phi_x(0) = x$, with $\lim_{t\to\infty} \phi_x(t) = \hat{x}$ (or $\lim_{t\to-\infty} \phi_x(t) = \hat{x}$) which is <u>unbounded</u> (as a point set).

<u>Proof</u>. We adopt the notations as used in Section 9.2 (especially in (the proofs of) Corollary 9.2.1 and Theorem 9.2.2). Let $\beta \in$ Ess(F) be a repellor and $\alpha_i \in$ Ess(F) be the attractors for N(F), i = 1,2,... . Let $B(\alpha_i)$ be small balls around α_i. Let $B(\beta)$ be a small ball around β such that $B(\beta) \cap B(\alpha_i) = \emptyset$, all i. Suppose that $\phi_x(t)$ is bounded for every $x \in \partial B(\beta)$. Define $A_i = \{x \mid x \in \partial B(\beta) \cap \text{basin } \alpha_i\}$. It follows, cf. Lemma 9.2.2 (proof) that $\partial B(\beta)$, as a connected set, is the disjoint union of <u>open</u> sets A_i. So there is only one non-empty A_i, say A_1. Now, we construct a diffeomorphism ψ between $\partial B(\beta)$ and $\partial B(\alpha_1)$ in a similar way as in the proof of Theorem 9.2.2 and this leads to a contradiction (Use also a homotopy/index argument). □

<u>Remark 9.3.3</u>. Global convergence aspects of the continuous Newton-method for finding the zeros of a mapping $F \in C^2(\mathbb{R}^n,\mathbb{R}^n)$ are discussed by several authors (see e.g. [13], [32] and [49]).
I. Diener ([9], [10]) has given a sufficient condition for the inverse image of a point under a C^2-mapping $\mathbb{R}^n \to \mathbb{R}^k$ to be connected. Applying this result, he obtains the following theorem:

Let $F(x_0) \neq 0$ and let A be an $(n-1)\times n$ matrix of rank n-1 such that $F(x_0) \in$ Ker A. If

$$\sup_{x\in\mathbb{R}^n} \| (A\cdot DF(x)\cdot D^TF(x)\cdot A^T)^{-1}\| < \infty$$

then, the union of all the maximal $N(F)$-trajectories on which $F(x) = \lambda F(x_0)$, some $\lambda \in \mathbb{R}$, constitutes a connected submanifold of dimension one, containing all zeros for F. (Here, $\|\cdot\|$ stands for the Euclidean norm). □

For two-dimensional Gradient Newton-flows we consider structural stability under the assumption of a global boundary.
To this aim let $M \subset \mathbb{R}^2$ be a fixed smooth submanifold, diffeomorphic to a circle. The manifold M will serve as a global boundary. We define:

$$E(M) = \{f \in E \mid N(f) \pitchfork M \text{ and } N(f) \text{ points inwards (to } M_{int})\}.$$

The set $E(M)$ is easily seen to be a C^3-open subset of E (but not a dense one). We show that structural stability of the corresponding Newton-flows holds for a C^3-open and <u>dense</u> subset $E_s(M)$ of $E(M)$, where

$$E_s(M) = \left\{ f \in E(M) \;\middle|\; \begin{array}{l} \text{distinct critical points of the trajectory} \\ \text{map Tf in the set } M_{int}\backslash \text{Ess}(f) \text{ do have} \\ \text{distinct critical values.} \end{array} \right\}$$

In the above definition the symbol Tf is an abbreviation for TD^Tf. The subscript "s" in $E_s(M)$ stands for "separating" (cf. also Theorem 6.3.1, where separating functions play an analogous crucial role).

<u>Lemma 9.3.4</u>. The set $E_s(M)$ is C^3-open and dense in $E(M)$.

<u>Proof</u>. The open part follows from the fact that there are only a finite number of critical points of Tf in $M_{int}\backslash\text{Ess}(f)$, all of them being non-degenerate as critical points. The density part follows if we show how to move a critical value by means of an arbitrarily small local perturbation of the function f. In fact, let $\bar{x} \in M_{int}\backslash\text{Ess}(f)$ be a critical point for Tf. Without loss of generality, we may assume that $\frac{\partial f}{\partial x_2}(\bar{x}) \neq 0$. Then, Tf can be (locally) represented by means of the map $F\colon x \to \frac{\partial f}{\partial x_1} \Big/ \frac{\partial f}{\partial x_2}(x)$. So, $DF(x) = 0$ and $D^2F(\bar{x})$ is nonsingular. Put $f_\varepsilon(x_1,x_2) = f(x_1,x_2) + \varepsilon x_1$. This gives rise to the map $F_\varepsilon\colon x \mapsto (\frac{\partial f}{\partial x_1}(x) + \varepsilon)/\frac{\partial f}{\partial x_2}(x)$. For ε near zero, the critical

point $x(\varepsilon)$ of F_ε near $\bar{x}$ depends smoothly on ε (use the Implicit Function Theorem), and for the critical value map $\phi(\varepsilon) := F_\varepsilon(x(\varepsilon))$ we obtain (cf. also Formula (6.1.16)):

$$\phi'(0) = \frac{\partial F_\varepsilon}{\partial \varepsilon}(\bar{x})\Big|_{\varepsilon=0} = 1/\frac{\partial f}{\partial x_2}(\bar{x}) \neq 0.$$

From this we see that the critical value $\phi(\varepsilon)$ changes with a linear rate in ε. In order to change the critical values of all critical points of the trajectory map within M_{int} <u>independently</u>, we have to perform the perturbations locally, say

$$\hat{f}_\varepsilon(x_1,x_2) = f(x_1,x_2) + \varepsilon \cdot x_1 \cdot \rho(x_1,x_2),$$

where ρ is a nonnegative smooth function with support in a small disc around $\bar{x}$, and being equal to one throughout some smaller disc around $\bar{x}$. This completes the proof of the lemma. □

<u>Theorem 9.3.6.</u> (Structural Stability). Let $f \in E_s(M)$. Then, there exists a C^3-neighborhood $\mathcal{O}$ of f with the property:
For each $g \in \mathcal{O}$ there exists a homeomorphism $\Phi_g: \overline{M_{int}} \to \overline{M_{int}}$ such that Φ_g sends trajectories of $N(g)$ onto trajectories of $N(f)$, thereby preserving the orientation.

<u>Proof.</u> The proof consists of a combination of the ideas in the proof of Theorem 6.3.1 and the fact that the essential singularities are attractors, resp. repellors for $N(f)$. Therefore, we can be short with the present outline. Let $f \in E_s(M)$ and let $\{p_1,\dots,p_s\}$ denote the set of essential singularities of f (= critical points of f) in M_{int}. For each p_i we choose a small disc $B_i \subset M_{int}\backslash \mathrm{Ext}(f)$ with p_i as its center (the discs being pairwise disjoint) such that $N(f)$ is transversal to the boundary ∂B_i, $i = 1,\dots,s$. For a function g which is C^3-close to f, the essential singularities within M_{int} are close to those of f, hence lying in the discs B_i, and $N(g)$ will also be transversal to the boundary ∂B_i, $i = 1,\dots,s$. Put $X = \overline{M_{int}} \backslash \bigcup_{i=1}^{s} \overline{B_i}$. Then, X is a compact smooth manifold with boundary ∂X, where $\partial X = M \cup \bigcup_{i=1}^{s} \partial B_i$. For g being C^3-close to f, we can shift the extraneous singularities of g within X to those of f (diffeomorphism on X) and,

moreover, we can shift the corresponding critical values of the trajectory map by means of a diffeomorphism on its image space, namely S^1. For the latter shift on S^1 we need a nonvanishing tangential vector field on S^1, given for example, by means of the vector field $(x_1,x_2) \mapsto (-x_2,x_1)$ restricted to S^1; the latter vector field generalizes the unit vector field on $\mathbb{R}^1$. Now, suppose that g is C^3-close to f with the same critical points and critical values for the trajectory map with respect to X.

Note that $N(g) \pitchfork \partial X$ and hence, Tg has no critical points on ∂X. Using the ideas in the proof of Theorem 6.3.1, we can construct a diffeomorphism ψ_g: $X \to X$ with $Tg = Tf \circ \psi_g$. The final step consists in extending the diffeomorphism ψ_g on X to the required homeomorphism Φ_g on $\overline{M_{int}}$.

Let p_i be an attractor for $N(f)$ and let p_i' be the corresponding attractor for $N(g)$ in B_i. Let Φ, resp. Φ' be the flows of $N(f)$, resp. $N(g)$ on B_i. Then, we define Φ_g as follows:

$$\begin{cases} \Phi_g(\Phi'(x,t)) = \Phi(\psi_g(x),t) & \underline{\text{for } x \in \partial B_i \text{ and } t \geq 0}, \\ \Phi_g(p_i') = p_i & . \end{cases}$$

In case that p_i is a repellor, the construction of Φ_g is analogous. Now, it is not difficult to see that Φ_g is the required homeomorphism. (see also the proof of Theorem 8.1.2). □

Remark 9.3.4. Up till now, we have analyzed the phase protraits of Gradient Newton-flows around extraneous singularities only in the two-dimensional case, cf. Theorem 9.3.3. However, similar results do hold in dimensions 3 and 4. Here, we merely give a short outline on the subject; full details will be published separately. We restrict ourselves to the 3-dimensional case.

Let the function $f \in C^\infty(\mathbb{R}^3,\mathbb{R})$ be contained in F_{***}. On Ext(f), we may assume (no loss of generality) that $\partial_3 \neq 0$. Hence, the functions $\phi = \frac{\partial_1}{\partial_3}$ and $\psi = \frac{\partial_2}{\partial_3}$ are well-defined and smooth on the 1-dim. manifold Ext(f), cf. Theorem 9.3.2. We consider a connected component of Ext(f), which is smoothly parametrized by $x(s) = (x_1(s),x_2(s),x_3(s))$; s = arc-length. In view of relation (9.1.6) we have:

$$\text{rank}\begin{pmatrix} D\phi \\ D\psi \end{pmatrix}_{x=x(s)} = 1. \tag{9.3.8}$$

Consequently, we may assume (no loss of generality) that $D\psi_{|x(s)} \neq 0$, and moreover, $D\phi_{|x(s)} = \lambda(s)\cdot D\psi_{|x(s)}$, where the function $\lambda(s)$ is uniquely determined. The local manifold M_s at $x(s)$ is given by the equation $\psi(x) = \psi(x(s))$. We consider the one-parameter family of Lagrange functions L_s, where

$$L_s(x) = \phi(x) - \lambda(s)\cdot\psi(x).$$

In view of Exercise 3.2.3, the "restricted" Hessian-matrix $D_x^2 L_s|_{TM_s}(x(s))$ is nonsingular if and only if

$$\det \left(\begin{array}{c|c} D_x^2 L_s & D^T\psi \\ \hline D\psi & 0 \end{array}\right)_{x(s)} \neq 0 . \tag{9.3.9}$$

Let y_1, y_2, y_3 be new coordinates such that $y_1 = \psi$. Then, the condition (9.3.9) is equivalent with

$$\det \left(\begin{array}{c|c} D_y^2\phi & \begin{matrix}1\\0\\0\end{matrix} \\ \hline 1\ 0\ 0 & 0 \end{array}\right)_{y(s)} = \det \begin{pmatrix} \phi_{y_2y_2} & \phi_{y_2y_3} \\ \phi_{y_3y_2} & \phi_{y_3y_3} \end{pmatrix}_{y(s)} \neq 0 \tag{9.3.10}$$

On the other hand, the 1-manifold Ext(f) intersects M_s transversally (in $y(s)$) if and only if $y_1'(s) \neq 0$. (Here $y(s) = (y_1(s), y_2(s), y_3(s))$ is the parametrization of Ext(f) w.r.t. the y-coordinates).
Moreover, from (9.3.8) it follows that $\phi_{y_2}(y(s)) = \phi_{y_3}(y(s)) = 0$, all s, and thus

$$\begin{pmatrix} \phi_{y_1y_2} & \phi_{y_2y_2} & \phi_{y_3y_2} \\ \phi_{y_1y_3} & \phi_{y_2y_3} & \phi_{y_3y_3} \end{pmatrix}_{y(s)} \cdot \begin{pmatrix} y_1'(s) \\ y_2'(s) \\ y_3'(s) \end{pmatrix} = 0 \tag{9.3.11}$$

So we may conclude (cf. (9.3.10) and (9.3.11)) that if $D_y^2 L_s|_{TM_s}(y(s))$ is nonsingular, then: $M_s \pitchfork \mathrm{Ext}(f)$ at $y(s)$. The converse is also true. This follows from the fact that (as a tough but <u>straightforward</u> calculation based on Exercise 7.2.2 learns), at $y(s)$, Condition (***) implies:

$$\text{rank}\begin{pmatrix} \phi_{y_1y_2} & \phi_{y_2y_2} & \phi_{y_3y_2} \\ \phi_{y_1y_3} & \phi_{y_2y_3} & \phi_{y_3y_3} \end{pmatrix}_{y(s)} = 2$$

Note that, locally around a regular point in M_s, the $N(f)$-trajectories are just the level lines of the restriction $\phi_{|M_s}$ (cf. Lemma 9.1.4.b). Hence -w.r.t. the y-coordinates- these trajectories are (locally) contained in the planes y_1 = const. Moreover, if condition (9.3.10) holds at y(s), then the local phase portrait, restricted to $y_1 = y_1(s)$, is either of the "center type" or of the "saddle point type", cf. Fig. 9.3.2.

A next step should be the investigation of those extraneous singularities where (9.3.10) does not hold (or, equivalently, where Ext(f) intersects the manifold M_s non-transversally). The corresponding subset of Ext(f) is denoted by $\text{Ext}_d(f)$. Then, the following result holds:
"Under an additional genericity condition (similar to Conditions (*), (***)) the set $\text{Ext}_d(f)$ is a discrete subset of Ext(f). Moreover, if $y(s_0) \in \text{Ext}_d(f)$, then $y_1''(s_0) \neq 0$ and the phase portrait of $N(f)$, restricted to $y_1 = y(s_0)$, is of a "cusp type" (cf. Fig. 9.3.2)." □

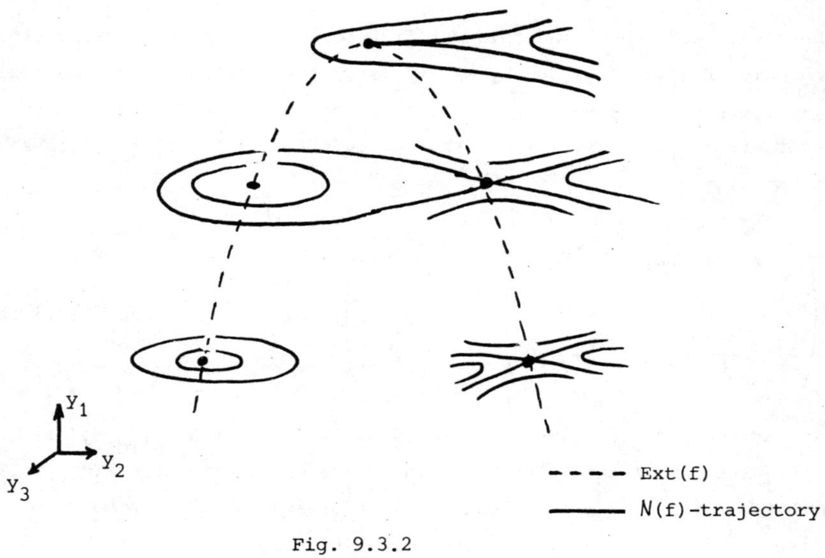

Fig. 9.3.2

Generalization to Newton-flows w.r.t. mappings from $\mathbb{R}^n$ to $\mathbb{R}^n$.

One may ask whether the results on Gradient Newton-flows which we have obtained above (cf. Theorems 9.3.2 and 9.3.3) do have their analogues in the case of the more general Newton-flows $N(F)$, $F \in C^\infty(\mathbb{R}^n,\mathbb{R}^n)$. The answer is positive. In fact, the required proofs are similar and -occasionally- more straightforward. This will be explained below.

Let B be an n×n-matrix and let b be an element of $\mathbb{R}^n$. Then, the augmented matrix $[B\vdots b]$ can be identified as an element of $\mathbb{R}^{n+n^2}$. The subset of $\mathbb{R}^{n+n^2}$ of all matrices $[B\vdots b]$ of rank equal to i is denoted by U_i $(0 \le i \le n)$. As a straightforward consequence of Example 7.3.3 we have:

The set U_i constitute C^∞-submanifolds of $\mathbb{R}^{n+n^2}$ of codimension $(n-i)(n-i+1)$, $i = 0,\dots,n-1$.

Note that the proof of the analogue statement in Lemma 9.3.1 is much more complicated, since in that case we must take into account the symmetry of the matrix A).

For the mapping $F \in C^\infty(\mathbb{R}^n,\mathbb{R}^n)$ we define $\widetilde{j^1F}(x) = [DF(x) \vdots F(x)]$, i.e. $\widetilde{j^1F}$ is a reduced 1-jet of F. Then, we propose the following transversality condition:

Condition (****). $\widetilde{j^1F} \pitchfork U_i$, $i = 0,\dots,n-1$.

The set of all mappings $F \in F$ which fulfil Condition (****) is denoted by F_{****}.

The proofs of the following results are similar to their analogues in the case of Gradient Newton-flows and are left as exercises to the reader.

Exercise 9.3.2. Prove the following statement (compare also Theorem 9.3.2).

a.1. The set F_{****} is a C^∞-generic subset of $C^\infty(\mathbb{R}^n,\mathbb{R}^n)$.

a.2. If $F \in F_{****}$, then Ext(F) is a closed stratified subset of $\mathbb{R}^n$ of co-dimension ≥ 2.

In the special case where n = 2,3,4 or 5, we have:

b. If $F \in F_{****}$, then Ext(F) is a closed submanifold of $\mathbb{R}^n$ of codimension two (or empty). □

It is possible to prove that the family $\{U_i\}_{i=0,\ldots,n}$ constitutes already a Whitney-regular stratification for $\mathbb{R}^{n+n^2}$ (cf. Section 7.5). Using this fact as well as Theorem 7.5.1, the following refinement of the result of Exercise 9.3.2 becomes transparent.

Exercise 9.3.3. Prove the following statement (compare also Theorem 9.3.3).

a.1. The set F_{****} is C^2-open in $C^\infty(\mathbb{R}^n,\mathbb{R}^n)$.

a.2. If $F \in F_{****}$, then Ext(F) is a Whitney-regular stratified subset of $\mathbb{R}^n$ of codimension ≥ 2.

Exercise 9.3.4. (Compare also Lemma 9.3.2).

Let F be a smooth mapping from $\mathbb{R}^2$ to $\mathbb{R}^2$.

Prove that Condition (****) implies Condition (*).

Finally, we present the analogue of our result on the local phase-portraits of 2-dimensional Gradient Newton-flows:

Exercise 9.3.5. Prove the following statement (compare also Theorem 9.3.3).

a. The set of mappings $F \in C^\infty(\mathbb{R}^2,\mathbb{R}^2)$ which fulfil Condition (****) is C^2-open and -dense.

Suppose that Condition (****) holds for F and let $\hat{x}$ be an equilibrium state for $N(F)$. Then,

b. The assertion of Theorem 9.3.3.b holds for $\hat{x}$ (if in this assertion f is replaced by F). □

Exercise 9.3.6. Prove that the statements of Lemma 9.3.3 and Theorem 9.3.4 do hold in the case where $F \in C^\infty(\mathbb{R}^2,\mathbb{R}^2)$ fulfils Condition (****) resp. Condition (*). □

9.4. Meromorphic Newton-flows.

Let $\mathbb{C}$ be the complex plane and let F be a complex valued function on $\mathbb{C}$. A point in $\mathbb{C}$ where F fails to be complex analytic [58] is called a singularity for F. The function F is called meromorphic if all its (finite) singularities are poles. An entire function is a meromorphic function without (finite) poles.

This section is concerned with the study of Newton-flows w.r.t. meromorphic functions (the so-called Meromorphic Newton-flows). As important subcases, we treat Newton-flows w.r.t. entire functions and Newton-flows w.r.t. rational functions. We merely give a concise survey on the subject and shall delete most of the proofs. For these proofs as well as for more details we refer to [37], [42] and [75].

First of all, we make the concept of Meromorphic Newton-flow more precise. To this aim, let F be a <u>non-constant</u>, meromorphic function. By N(F) resp. P(F) we denote the set of zeros resp. of poles for F. We define $C(F) = N(F')\backslash N(F)$, where F' stands for the usual derivative of F. An element of C(F) is called a <u>critical point</u> for F. Note that N(F), P(F) and C(F) are <u>discrete subsets</u> of $\mathbb{C}$.

Now, we consider the complex analogue of system (9.1.1), i.e. the system

$$\frac{dz}{dt} = -\frac{F(z)}{F'(z)} \tag{9.4.1}$$

The singularities for the r.h.s. of (9.4.1) are just the points of $N(F) \cup P(F) \cup C(F)$. The singularity for $\frac{F}{F'}$ at a point $z_0 \in N(F) \cup P(F)$ can be removed by defining $\frac{F(z_0)}{F'(z_0)} := 0$. On the other hand, the singularity for $\frac{F}{F'}$ at a critical point can not be removed. Consequently, from now on, we shall regard the r.h.s. of (9.4.1) as a <u>complex analytic vector field</u> on $\mathbb{C}\backslash C(F)$.

It is possible to "desingularize" system (9.4.1) in such a way that the resulting system is real analytic, in x_1 (= Re z) and x_2 (= Im z), on the whole plane. We perform this desingularization procedure in two steps (compare the change over from (9.1.1) to (9.1.3)).

<u>First Desingularization Step.</u>

On $\mathbb{C}\backslash P(F)$ we consider the vector field $\tilde{N}(F)$ defined by

$$\tilde{N}(F)_{|z} = -\overline{F'}(z)F(z) ,$$

where $\overline{F'}(z)$ stands for the complex conjugate of $F'(z)$. Obviously, $\tilde{N}(F)$ depends real analytic on x_1 and x_2; moreover, the equilibrium-states are the elements of N(F) ∪ C(F). Note that, outside P(F) ∪ C(F), the vector field $\tilde{N}(F)$ is obtained by multiplying $-\frac{F(z)}{F'(z)}$ (= r.h.s. of (9.4.1)) with the non-negative factor $|F'(z)|^2$. Consequently, on $\mathbb{C}\setminus(P(F) \cup C(F))$, the phase-portraits of $\tilde{N}(F)$ and of system (9.4.1) are the same (included the orientation).

In case F is an entire function, (i.e. $P(F) = \emptyset$) then $\tilde{N}(F)$ is globally defined on $\mathbb{C}$, whereas this is not true in the case of strictly meromorphic functions (i.e. $P(F) \neq \emptyset$). So, we need another desingularization step:

Second Desingularization Step.

On $\mathbb{C}$, we consider the vector field $\bar{N}(F)$ which is defined as follows:

$$\bar{N}(F)_{|z} = \begin{cases} (1 + |F(z)|^4)^{-1}\tilde{N}(F)_{|z} & \text{if } z \in \mathbb{C}\setminus P(F) \\ 0 & \text{if } z \in P(F) \end{cases}$$

Since -on $\mathbb{C}\setminus P(F)$- both $(1+|F(z)|^4)^{-1}$ and $\tilde{N}(F)_{|z}$ depend real analytic on x_1 and x_2, this is also true for $\bar{N}(F)_{|z}$. Obviously, the equilibrium states for $\bar{N}(F)$ are the points of N(F) ∪ P(F) ∪ C(F). Since, the factor $(1+|F(z)|^4)^{-1}$ is strictly positive outside the latter set, the phase-portraits of $\tilde{N}(F)$ and of $\bar{N}(F)$ are the same on $\mathbb{C}\setminus P(F)$, including the orientations. By inspection, one easily verifies that

$$\bar{N}(F) = -\bar{N}(\tfrac{1}{F}). \tag{9.4.2}$$

From this relation, it follows that $\bar{N}(F)$ is also real analytic in P(F).

The flow, $\bar{N}(F)$ -as defined above- will be referred to as to the Meromorphic Newton-flow w.r.t. the function F.
In the special case where F is entire, both $\tilde{N}(F)$ as well as $\bar{N}(F)$ will be referred to as to the Entire Newton-flow w.r.t. the function F (Note that the latter convention is reasonable, since in case F is entire the flows $\tilde{N}(F)$ and $\bar{N}(F)$ have equal phase portraits).

The relation between Entire Newton-flows and Gradient Newton-flows.

In this intermezzo, we explain how the concept of Entire Newton-flow fits into the framework of Gradient Newton-flows as introduced in Section 9.3.

To this aim, we assume, for the moment, that F is a (non-constant) entire function. We write F = u+iv, where u and v are real valued functions on $\mathbb{C}$. Then, for all $z = x_1+ix_2$, the Cauchy-Riemann relations hold:

$$u_{x_1} = v_{x_2} \qquad (9.4.3.a)$$

$$u_{x_2} = -v_{x_1} \qquad (9.4.3.b)$$

where u_{x_1}, v_{x_2} etc. denote partial derivatives.

Identifying $\mathbb{C}$ with $\mathbb{R}^2$ (i.e. $z = x_1+ix_2 \leftrightarrow (x_1,x_2)$) we may regard F as a smooth mapping from $\mathbb{R}^2$ to $\mathbb{R}^2$ given by $F(x,y) = (u,v)^T$.
From (9.4.3) it follows that

$$\det DF(x) = |F'(z)|^2. \qquad (9.4.4)$$

One easily verifies -using (9.4.3) and (9.4.4)- that, in terms of $F = (u,v)^T$, system (9.4.1) takes the form of system (9.1.1), whereas $\tilde{N}(F) = N(F)$, cf. (9.1.3).

Now, we consider the mapping $\bar{F}: \mathbb{R}^2 \to \mathbb{R}^2$, $\bar{F} = (u,-v)^T$. Then, in view of (9.4.3.b), this mapping may be written in the form

$$\bar{F} = D^T f, \qquad (9.4.5)$$

where f is a smooth function on $\mathbb{R}^2$. So, we may consider the <u>Gradient Newton-flow</u> (cf. Section 9.3) $N(f)$ $(= N(\bar{F}))$. (Note that - although the function f is determined by $\bar{F}$ up to a constant, the Gradient Newton-flow corresponding with $\bar{F}$ is uniquely determined).
The phase-portraits of $\tilde{N}(F)$ and $N(f)$ are the same, <u>up to opposite orientation.</u> This contention can be verified by writing down -explicitly in terms of u,v and their partial derivatives- the equations for $\tilde{N}(F)$ and $N(f)$ $(= N(\bar{F}))$ and by using the Cauchy-Riemann relations (9.4.3). So, we may conclude that <u>the study of Entire Newton-flows implies the study of a certain class of Gradient Newton-flows on the plane.</u>
A natural question is: how "special" is the class of Gradient Newton-flows $N(f)$, where f is induced by the entire function F? In order to answer this question, we briefly inspect the Conditions (*), (**) and (***) in the case of these special functions f.

Apparently, cf. (9.4.4) and (9.4.5), we have

<u>Condition (*) holds for f if and only if all zeros for F are simple.</u>

We adopt the notations as used in Section 9.3 (cf. Lemma 9.3.2 and Theorem 9.3.3). Then, we may write $Df(x) = (\partial_1, \partial_2)$ and thus $\partial_1 = u$, $\partial_2 = -v$. In view of the Cauchy-Riemann relations, we find:

$$\operatorname{rank}(D^2 f(x)) = \operatorname{rank}\begin{pmatrix} +u_{x_1} & -v_{x_1} \\ -v_{x_1} & -u_{x_1} \end{pmatrix} = \begin{cases} 2 & \text{if} \quad F'(z) \neq 0 \\ 0 & \text{if} \quad F'(z) = 0 \end{cases} \tag{9.4.6}$$

In particular, $\operatorname{rank}(D^2 f(x)) \neq 1$ for all $x \in \mathbb{R}^2$. Hence, a dimension argument yields:

<u>Condition (**) holds for f if and only if $\mathbb{N}(F') = \emptyset$.</u>

We proceed by investigating Condition (***) and consider a point $\hat{z} = \hat{x}_1 + i\hat{x}_2$, i.e. $\hat{x} = (\hat{x}_1, \hat{x}_2)$. If $F'(\hat{z}) \neq 0$, then -at $\hat{x}$- the Condition (***) is always fulfilled, cf. (9.4.6). If $F'(\hat{z}) = 0$, then we distinguish between two possibilities: $\hat{z} \in N(F)$ and $\hat{z} \in C(F)$.
In the case where $\hat{z} \in N(F)$, we obtain a contradiction since Condition (***) is -at $\hat{z}$- equivalent with: $\hat{z}$ is a <u>simple</u> zero for F (Use (9.4.6) and recall that Condition (***) implies: $\operatorname{rank}[D^2 f(\hat{x}) \vdots D^T f(\hat{x})] > 0$). In the case where $\hat{z} \in C(F)$, i.e. $F(\hat{z}) \neq 0$ and $F'(\hat{z}) = 0$, we write: $Df(\hat{x}) = (\hat{\partial}_1, \hat{\partial}_2)$ $(\neq (0,0))$, compare the proof of Theorem 9.3.3. From the proof of this latter theorem it also follows (cf. (9.3.7)) that Condition (***) is -at $\hat{x}$- equivalent with <u>$\det \tilde{\Delta} \neq 0$.</u>
In view of the Cauchy-Riemann relations we have $\partial_{11} = -\partial_{22}$ and thus $\partial_{111} = -\partial_{122}$, $\partial_{112} = -\partial_{222}$. Using these relations, a straightforward calculation shows:

$$\tilde{\Delta} = \left(\begin{array}{c|c} -\hat{\partial}_2 \cdot \hat{\partial}_{111} + \hat{\partial}_1 \cdot \hat{\partial}_{112} & -\hat{\partial}_2 \cdot \hat{\partial}_{112} - \hat{\partial}_1 \cdot \hat{\partial}_{111} \\ \hline -\hat{\partial}_2 \cdot \hat{\partial}_{112} - \hat{\partial}_1 \cdot \hat{\partial}_{111} & +\hat{\partial}_2 \cdot \hat{\partial}_{111} - \hat{\partial}_1 \cdot \hat{\partial}_{112} \end{array}\right)$$

Thus, since $(\hat{\partial}_1, \hat{\partial}_2) \neq (0,0)$, we find: $\det \tilde{\Delta} = 0$ iff $\hat{\partial}_{112} = \hat{\partial}_{111} = 0$.
From this, it easily follows that Condition (***) is -at $\hat{x}$- equivalent

with: $\hat{z}$ is a simple zero for F'.

Altogether, we have proved:

Condition (***) holds for f iff all zeros for F and F' are simple.

We conclude this intermezzo on the relation between Entire Newton-flows and Gradient Newton-flows with the following observation (which is easily verified):

Under the assumption of Condition (*), we have: Ess(f) = N(F) and Crit(f) = Ext(f) = C(F).

So, if we restrict ourselves to Gradient Newton-flows which are induced by entire functions, the transversality Conditions (*), (**), (***) as well as the sets of essential and extraneous singularities take an extremely simple form. □

Now, we return to the general case of (non-constant) meromorphic functions F. In order to describe the global features of the corresponding Meromorphic Newton-flows, we have to take into account the occurrance of poles and must pass on to the "desingularized" flow $\bar{N}(F)$. As hard tools, we have at our disposal the theory of (real analytic) autonomous differential equations in the plane as well as the theory of complex variables in (one) complex variable.

We begin with the description of the possible local phase-portraits of a Meromorphic Newton-flow $\bar{N}(F)$. To this aim, let $\gamma(z_0)$ be the maximal trajectory of (9.4.1) through a point $z_0 \notin N(F) \cup P(F) \cup C(F)$. Note, in view of the above desingularization procedure, that the maximal $\bar{N}(F)$-trajectory through z_0 just coincides with $\gamma(z_0)$. On the other hand, the maximal $\bar{N}(F)$-trajectory through a point $z_1 \in N(F) \cup P(f) \cup C(F)$ reduces to the singleton $\{z_1\}$. If $\gamma(z_0)$ is given by

$$z(t)\ ,\ z(0) = z_0\ ,\ t \in (a,b)\ ,\ \text{eventually } a = -\infty\ ,\ b = +\infty,$$

then, direct integration of (9.4.1) yields

$$F(z(t)) = e^{-t}F(z_0). \tag{9.4.7}$$

It follows that the maximal trajectory $\gamma(z_0)$ is contained in the inverse image under F of the line $\arg w = \arg F(z_0)$, see also Section 9.1, especially (9.1.2). Application of the elementary properties of (multifold) conformal mappings (cf. [59]) yields:

Lemma 9.4.1. (The local phase-portraits of $\bar{N}(F)$).
The local phase-portrait of $\bar{N}(F)$ around a point, say $\hat{z}$, is of the types as depicted in Fig. 9.4.1. □

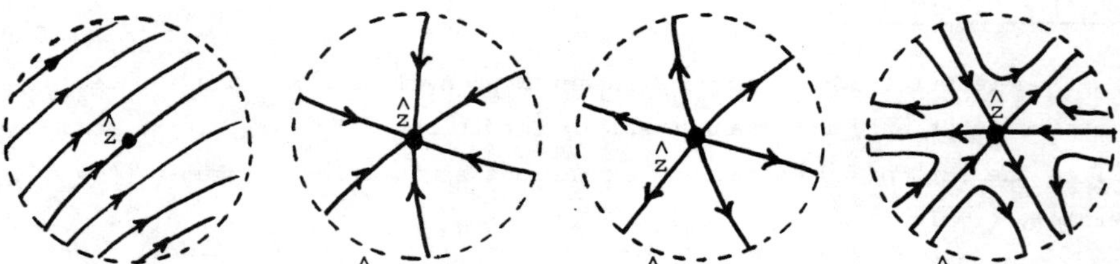

	$\hat{z} \in N(F)$	$\hat{z} \in P(F)$	$\hat{z} \in C(F)$
regular point	stable dicritical node	unstable dicritical node	k-fold saddlepoint
a	b	c	d

Fig. 9.4.1

For the definition of dicritical (un-)stable node we refer to Section 9.3.

In the case where $\hat{z} \in C(F)$, the integer k stands for the multiplicity of $\hat{z}$ as a zero for F'; in Fig. 9.4.1.d we have $k = 2$.

Now, we focus our attention to the global aspects of the phase partrait of $\bar{N}(F)$.

The limiting set of the maximal trajectory $\gamma(z_0)$ is extremely simple as it is pointed out in the following lemma.

Lemma 9.4.2.

1. Either $\lim_{t\uparrow b} (z(t)) = \infty$ or $\lim_{t\uparrow b} (z(t)) = z_* \in \mathbb{C}$; in the latter case, we have $z_* \in N(F)$ if $b = \infty$ and $z_* \in C(F)$ if $b < \infty$.
2. Either $\lim_{t\downarrow a} z(t) = \infty$ or $\lim_{t\downarrow a} z(t) = z_* \in \mathbb{C}$; in the latter case, we have $z_* \in P(F)$ if $a = -\infty$ and $z_* \in C(F)$ if $a > -\infty$. □

We emphasize that, although the above lemma is in the same spirit as Bendixon's theorem on limiting sets of maximal trajectories for 2-dimensional vector fields (cf. [55]), our result neither requires a Global Boundary Condition nor a compact support for the vector field $\bar{N}(f)$. Therefore, the lemma needs an independent proof (cf. [37]).

Let $\hat{z}$ be a zero or a pole for F. Then, we define

Definition 9.4.1. The basin $B(\hat{z})$ of $\hat{z}$ is the set

$$B(\hat{z}) = \{\hat{z}\} \cup \{z_0 \in \mathbb{C} \mid \lim z(t) = \hat{z};\ z(0) = z_0\},$$

where the limit is taken $t \to +\infty$ if $\hat{z} \in N(F)$, respectively $t \to -\infty$ if $\hat{z} \in P(F)$.

Let $\partial B(\hat{z})$ stand for the boundary of the basin $B(\hat{z})$. Then, we have the following result:

Lemma 9.4.3. Let $\hat{z}$ be a zero or a pole for the non-constant meromorphic function F. Then,

a. $\partial B(\hat{z}) = \emptyset$ iff F is of the form $F(z) = \alpha(z-\hat{z})^n$, where $\alpha \in \mathbb{C}$ is a constant and n is a positive (negative) natural number if $\hat{z}$ is a zero (pole) for F.
b. If the boundary $\partial B(\hat{z})$ of the basin $B(\hat{z})$ is non-empty, then it is the union of the (topological) closures of maximal $\bar{N}(F)$-trajectories.
c. If $B(\hat{z})$ is bounded (and thus $\partial B(\hat{z}) \neq \emptyset$), then there lies at least one pole (zero) on $\partial B(\hat{z})$ if $\hat{z}$ is a zero (pole) for F. □

The proof of this lemma, which is based on (9.4.7), the continuous dependence of the solutions of (9.4.1) on the initial conditions as well as on some elementary complex function theory (the Casorati-Weierstrass theorem) is to be found in [37].

As a direct consequence of Lemma 9.4.3.c we have

Corollary 9.4.1. The basin of a zero for a (non-constant) entire function in unbounded. □

D. Braess [5] has obtained the latter result -in a different way- for the polynomial case. Compare also Section 9.3, where we derived similar results on Gradient Newton-flows (Theorem 9.3.5).

In the following example, some of the features of the above lemma are illustrated. Moreover, the example provides a counter example for Branin's conjecture C.1 on the global convergence of Newton's method in the absence of extraneous singularities, cf. Section 9.3.

Example 9.4.1. Let $F(z) = e^{z}-1$. Then,

$$N(F) = \{2k\pi i \mid k \in \mathbb{Z}\};\ C(F) = P(F) = \emptyset.$$

In view of Lemma 9.4.3 (and Corollary 9.4.1) the basins of the zeros for F are unbounded and their boundaries are non-empty. Moreover, since $P(F) = C(F) = \emptyset$, each component contains one single maximal $\bar{N}(F)$-trajectory. In fact, the phase portrait of $\bar{N}(F)$ is of the form as depicted in Fig. 9.4.2; the basins of the zeros for F are separated by the trajectories $\{z \mid \text{Im } z = (2k+1)\pi\}$. Since all zeros for F are simple it follows that the function f, $\bar{F} = D^{T}f$, fulfils Condition (***), and thus also Condition (*). Moreover, from $N(F') = \emptyset$, it follows that also Condition (**) holds. □

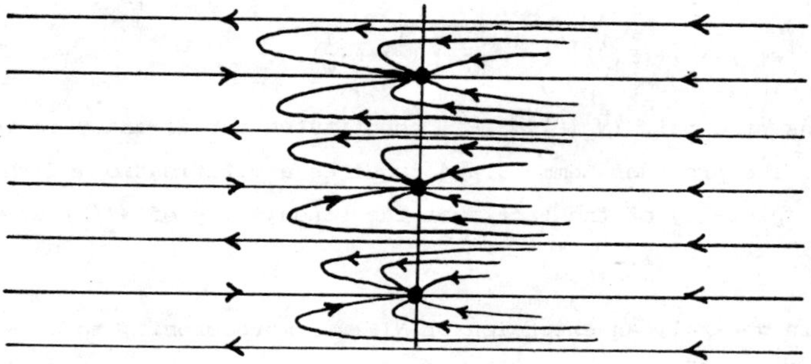

Fig. 9.4.2

For special choices of the function F, it is possible to extend $\bar{N}(F)$ -by means of the transformation $w = \frac{1}{z}$ -to a real analytic vector field on the 2-sphere S^2. Here, we only treat the case where F is a rational function.

Let R be the set of all non-constant rational functions F with $F = \frac{p_n}{q_m}$ and p_n (resp. q_m) polynomials of degree n (m) which are relatively prime. Then, we have the following result.

<u>Lemma 9.4.4</u>. <u>(Extension to S^2)</u>
For each $F \in R$, there exists a real analytic vector field $\bar{\bar{N}}(F)$ defined on the whole S^2, such that on the z-chart of S^2 (i.e. on $S^2\backslash\{z = \infty\}$) the phase-portrait of $\bar{\bar{N}}(F)$ equals the phase-portrait of $\bar{N}(F)$ (up to a stereographic projection).
Moreover, we have $\bar{\bar{N}}(F) = -\bar{\bar{N}}(\frac{1}{F})$. □

Here, we merely give the explicit expressions of $\bar{\bar{N}}(F)$ w.r.t. the z-chart and the w-chart: $\bar{\bar{N}}_z(F)$ resp. $\bar{\bar{N}}_w(F)$; compare also the change over from (9.4.1) to $\bar{N}(F)$ via $\tilde{N}(F)$. <u>If $n \neq m$</u>, then

$$\bar{\bar{N}}_z(F) = -(1+|z|^2)^{|n-m|+1}\cdot(1+|F(z)|^4)^{-1}\cdot\bar{F}'(z)\cdot F(z) \left(= (1+|z|^2)^{|n-m|+1}\cdot\bar{N}(F)_{|z}\right)$$

$$\bar{\bar{N}}_w(F) = -(1+|w|^2)^{|n-m|+1}\cdot|w|^{-2|n-m|+2}\cdot(1+|F(\tfrac{1}{w})|^4)^{-1}\cdot\overline{\frac{d}{dw}F(\tfrac{1}{w})}\cdot F(\tfrac{1}{w}).$$

<u>If $n = m$</u>, then

$\bar{\bar{N}}_z(F) = -(1+|z|^2)^2 \cdot (1+|F(z)|^4)^{-1} \cdot \overline{F'(z) \cdot F(z)}\ (= (1+|z|^2)^2 \bar{N}(F)_{|z})$.

$\bar{\bar{N}}_w(F) = -(1+|w|^2)^2 \cdot (1+|F(\frac{1}{w})|^4)^{-1} \cdot \overline{\frac{d}{dw} F(\frac{1}{w}) \cdot F(\frac{1}{w})}$.

In both cases, the pair $(\bar{\bar{N}}_z(F), \bar{\bar{N}}_w(F))$ constitutes a real analytic vector field on S^2. The proof of Lemma 9.4.4 requires a verification of this statement (especially of the fact that the singularity of $\bar{\bar{N}}_w(F)$ at $w = 0$ is removable, cf. [37]).

Note that, in general, an extension of $\bar{N}(F)$, F meromorphic, to a real analytic $\bar{\bar{N}}(F)$ on S^2 (as presented above) is <u>not</u> possible. This follows from the existence of meromorphic functions whose finite zeros do accumulate at $z = \infty$ (e.g. sin z, or tan z).

In order to describe the local phase portrait of $\bar{\bar{N}}(F)$, $F \in \mathcal{R}$, around $z = \infty$, we introduce the following integers:

Let $F \in \mathcal{R}$, $F = \frac{p_n}{q_m}$. Then,

> s = number of critical points for F, each counted a number of times equal to its multiplicity as a zero for F'

and, in the case where n = m,

$$k = 2n - \deg(p_n' q_n - p_n q_n') - 2.$$

<u>Lemma 9.4.5</u>. For $F \in \mathcal{R}$, $F = \frac{p_n}{q_m}$, we have:

a. If $n < m$ (resp. $n > m$), then $z = \infty$ is a <u>stable</u> (resp. <u>unstable</u>) <u>dicritical node</u> for $\bar{\bar{N}}(F)$.

b. If $n = m$, then

   (i) The point $z = \infty$ is either a regular point for $\bar{\bar{N}}(F)$ (in case $k = 0$) or it is a k-fold saddlepoint (in case $k \geq 1$).

   (ii) Moreover, $k = \# N(F) + \# P(F) - s - 2$, where # stands for cardinality.

<u>Remark 9.4.1</u>. The <u>Fundamental Theorem of Algebra</u> follows directly -by considering $F(z) = \frac{p_n(z)}{z^n}$ - from Lemma 9.4.5.b, (ii) and the facts that $k \geq 0$ and $s \geq 0$.

It is possible (cf. [37]) to endow the set $\mathcal{R}$ with a topology τ which is natural in the following sense:
Let $F \in \mathcal{R}$ be represented by $\frac{p_n}{q_m}$. Given $\varepsilon > 0$ sufficiently small, then there exists a τ-neighborhood Ω of F such that for each $G \in \Omega$, the function G can be represented by $\frac{\tilde{p}_n}{\tilde{q}_m}$ such that the coefficients of $\tilde{p}_n$, $\tilde{q}_m$ are in ε-neighborhoods of the corresponding coefficients of p_n, q_m.

Roughly speaking, $\bar{\bar{N}}(F)$ is called structurally stable if small perturbations of the coefficients of F (not changing the degree of numerator and denominator) do not alter the qualitative features of the resulting phase portraits of the corresponding Rational Newton-flows. The precise definition is as follows (cf. also Section 8.1 for the concept of topological equivalency between vector fields):

Definition 9.4.2.
The Rational Newton-flow $\bar{\bar{N}}(F)$ is called structurally stable if a τ-neighborhood -say Ω- of F exists, such that for each $G \in \Omega$, the vector fields $\bar{\bar{N}}(F)$ and $\bar{\bar{N}}(G)$ are topologically equivalent.

We introduce the concept of non-degenerate rational function. (In the below definition, $z = \infty$ is resp. a zero, pole or critical point for F if $w = 0$ is a zero, pole or critical point for $F(\frac{1}{w})$).

Definition 9.4.3. The function $F \in \mathcal{R}$ is called non-degenerate if
1. All finite zeros and poles for F are simple.
2. All critical points for F -eventually including $z = \infty$- are simple (as zeros for F').
3. No two critical points for F are "connected" by a $\bar{\bar{N}}(F)$-trajectory.

The subset of $\mathcal{R}$ consisting of all non-degenerate rational functions is denoted by $\tilde{\mathcal{R}}$.
As a main result we have:

Theorem 9.4.1.
(i) $\tilde{\mathcal{R}}$ is τ-open and τ-dense in $\mathcal{R}$.
(ii) $\bar{\bar{N}}(F)$ is structurally stable iff $F \in \tilde{\mathcal{R}}$. □

Remark 9.4.2. The proof of Theorem 9.4.1 (ii) (cf. [37]) is essentially based on the three following steps:

1. All equilibrium states of $\overline{N}(F)$, $F \in \tilde{R}$, are hyperbolic;
2. The mapping $T: F(\in R) \mapsto \overline{\overline{N}}(F)$ is (τ,c)-continuous, where c stands for the C^1-Whitney-topology on the set of smooth vector fields on S^2;
3. Application of the de Baggis-Peixoto Characterization Theorem for structurally stable vector fields on compact 2-manifolds (cf. [64]) yields the desired result.

We emphasize, that Theorem 9.4.1 is in the same spirit as the result, we obtained on Gradient Newton-flows in Theorem 9.3.6. In fact, let $F \in \tilde{R}$ be entire, i.e. F is a polynomial of degree ≥ 1. Then, the non-degeneracy conditions 1 and 2 in Definition 9.4.3 reduce to: all zeros for F and F' are simple. Let f be a function on $\mathbb{R}^2$ given by $\bar{F} = D^T f$, cf. (9.4.5). Then, Condition (***) holds for f. Moreover, the non-degeneracy condition 3 holds if the trajectory map TF (= $TD^T f$) is separating. From Lemma 9.4.5.a it follows that the circle $|z| = R$ provides a Global Boundary for $N(f)$, if R is chosen sufficiently large. Altogether we have shown now, that if $\tilde{T f}$ is separating, then the function f belongs to a set $E_s(M)$.

Remark 9.4.3. There is a physical interpretation of Theorem 9.4.1. One easily sees that system (9.4.1) yields the differential equation for the stream lines of a steady stream with complex potential - log F(z), cf.[59]. The case where $F \in R$, corresponds to steady streams which have -regarded on the Riemann Sphere- only finitely many "sinks" and "sources". We may expect that -generically- the phase portrait of such a stream behaves extremely regular under small perturbations of the coefficients of the underlying rational function. This is expressed in Theorem 9.4.1.

The following simple example illustrates some of the features of the above theory.

Example 9.4.2. Let $F(z) = (\frac{az+b}{cz+d})^n$, with ad ≠ bc and n = 1,2,... In case where ac ≠ 0 we have $N(F) = \{-\frac{b}{a}\}$, $P(F) = \{-\frac{d}{c}\}$, $C(F) = \emptyset$, and $z = \infty$ is a regular point for $\overline{N}(F)$. If a = 0 then, the point $z = \infty$ is the only zero for

F, $P(F) = \{-\frac{d}{c}\}$ and $C(F) = \emptyset$. If $c = 0$, then $z = \infty$ is the only pole, $N(F) = \{-\frac{b}{c}\}$ and $C(F) = \emptyset$. Only in the case $n = 1$, the flow $\bar{\bar{N}}(F)$ is structurally stable (although, for fixed a, b, c, d, the phase-portraits are the same for all n).

Note that the cases $a = 0$ resp. $c = 0$ provide the only examples of Meromorphic Newton-flows with the property that the boundary of the basin of one of its nodes is empty (cf. Lemma 9.4.3.a).

The systems $\bar{\bar{N}}(F)$ are called "north-south flows", cf. Fig. 9.4.3.

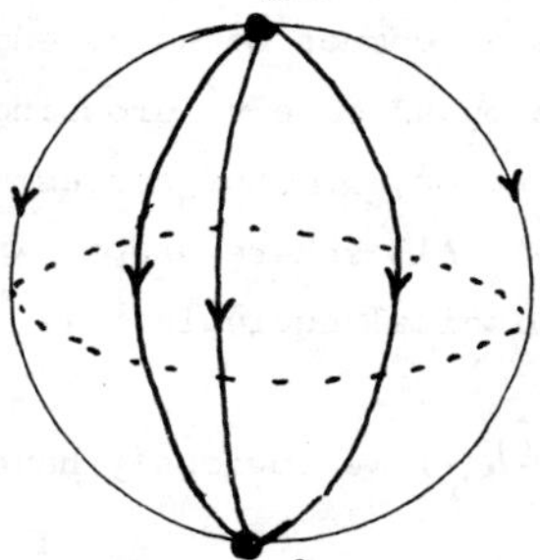

$\bar{\bar{N}}(F)$, $F = (\frac{az+b}{cz+d})^n$, $b = c = 0$

Fig. 9.4.3

With $F \in \tilde{R}$, we may associate a sphere graph $\bar{\bar{G}}(F)$, i.e. a realization of an abstract graph -say $G(F)$- on the sphere S^2.

<u>Definition 9.4.4.</u> For $F \in \tilde{R}$, the sphere graph $\bar{\bar{G}}(F)$ is defined as follows:

- The <u>vertices of</u> $\bar{\bar{G}}(F)$ are the <u>zeros</u> for F (including eventually $z = \infty$).
- The <u>edges</u> of $\bar{\bar{G}}(F)$ are the (topological) closures of the <u>unstable manifolds</u> at the saddle points for $\bar{\bar{N}}(F)$.

Note that this definition makes sense in view of Lemma 9.4.2 and Definition 9.4.3, and that in the case of a north-south flow (cf. Example 9.4.2) the graph reduces to a graph with only one vertex and no edges

As a direct consequence of Lemmas 9.4.1, 9.4.4, 9.4.5, the Poincaré-Hopf index-theorem on the sphere [22] and Euler's polyhedron formula for graphs embedded in S^2 [15], we have:

Lemma 9.4.5. Let $F \in \tilde{R}$, then $\bar{\bar{G}}(F)$ is a connected sphere-graph.

Given a function $F \in \tilde{R}$. If we delete from S^2 all edges and vertices of $\bar{\bar{G}}(F)$, then the connected components of the resulting set are called the regions of $\bar{\bar{G}}(F)$. One easily verifies that the regions of $\bar{\bar{G}}(F)$ are just the basins of the poles for F (these basins being defined as in the case of $\bar{N}(F)$, cf. Definition 9.4.1).

A geometrical dual of $\bar{\bar{G}}(F)$ is constructed as follows: place a vertex in each region of $\bar{\bar{G}}(F)$ and if v is a common boundary edge of two $\bar{\bar{G}}(F)$-regions, join the corresponding vertices by an egde v^* crossing only v; if v is in the boundary of only one $\bar{\bar{G}}(F)$-region, join the corresponding vertex to itself (by a loop crossing only v). All sphere graphs which can be obtained in this way from $\bar{\bar{G}}(F)$, are topological equivalent (i.e. are mapped onto each other by a homeomorphism $S^2 \to S^2$).

In view of the relation $\bar{\bar{N}}(F) = -\bar{\bar{N}}(\frac{1}{F})$ we obviously have:

The graph $\bar{\bar{G}}(\frac{1}{F})$ is a geometrical dual of $\bar{\bar{G}}(F)$.

Now, let F $(= \frac{p_n}{q_m})$ be a rational function in $\tilde{R}$ such that $n > m$. The plane-graph $\bar{G}(F)$ can be defined w.r.t. $\bar{N}(F)$, as we defined the sphere-graph $\bar{\bar{G}}(F)$ w.r.t. $\bar{\bar{N}}(F)$, cf. Definition 9.4.4. Due to the construction of $\bar{\bar{N}}(F)$ from $\bar{N}(F)$, cf. Lemma 9.4.4, it will be clear that -essentially up to a stereographic projection- the sphere graph $\bar{\bar{G}}(F)$ equals $\bar{G}(F)$; in particular it follows that $\bar{G}(F)$ is connected.

Let C be a cycle of the plane-graph $\bar{G}(F)$ and let Int(C) stand for the interior of C. The integers n_C, r_C and ℓ_C are defined as follows:

n_C = number of $\bar{G}(F)$-vertices in Int(C) (= # N(F) ∩ Int(C)).

r_C = number of $\bar{G}(F)$-regions in Int(C) (= # P(F) ∩ Int(C)).

ℓ_C = number of $\bar{G}(F)$-vertices on C (= # N(F) ∩ C).

We have the following lemma.

Lemma 9.4.6. Let F $(= \frac{p_n}{q_m}) \in \tilde{R}$, n > m. Then for each cycle C in $\bar{G}(F)$ the following inequality holds:

$$n_C < r_C < n_C + \ell_C \; . \tag{9.4.8}$$

Proof. The vertices of C are denoted by $\omega_{C(\ell)}$, $\ell = 1,\ldots,\ell_C$. The angles between two consecutive edges of C spanning a sector of Int(C) at $\omega_{C(\ell)}$ are given by $2\pi\phi_{C(\ell)}$. Since the $\bar{G}(F)$-vertices are the dicritical stable nodes for $\bar{N}(F)$ and the $\bar{G}(F)$-edges are built up by trajectories of $\bar{N}(F)$ we have: $0 < \phi_{C(\ell)} < 1$, $\ell = 1,\ldots,\ell_C$. See also Fig. 9.4.4.

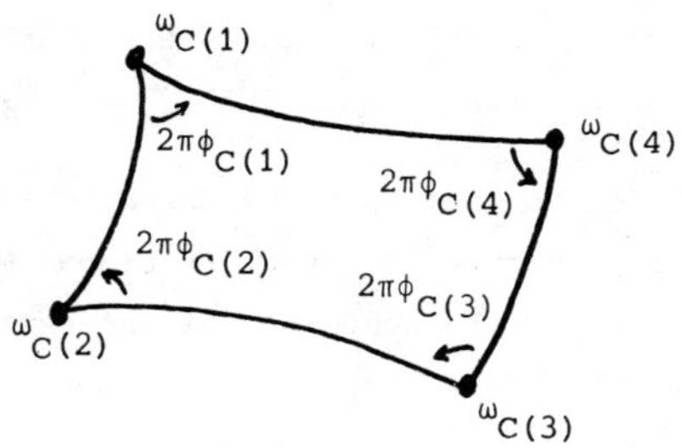

Fig. 9.4.4

The Cauchy-principal value, cf.[59], of $\int_C \frac{F'(z)}{F(z)}\,dz$ is well-defined, and since the zeros and finite poles for F are simple, we find for it

$$\int_C \frac{F'(z)}{F(z)}\,dz = 2\pi i \left[\left(\sum_{\ell=1}^{\ell_C} \phi_{C(\ell)}\right) + n_C - r_C\right] \tag{1}$$

Since C is built up by trajectories of system (9.4.1) one easily derives that l.h.s. of (1) is a real value and consequently:

$$\sum_{\ell=1}^{\ell_C} \phi_{C(\ell)} = r_C - n_C .$$

In view of the inequalities $0 < \phi_{C(\ell)} < 1$, the desired result follows immediately. □

Apparently, the inequalities (9.4.8) make sense for any connected plane-graph $\bar{G}$. Moreover, if (9.4.8) holds for each cycle in $\bar{G}$, then this is also true for any graph $\bar{G}$, which is topologically equivalent with $\bar{G}$.

Let $\bar{\bar{G}}$ be a connected sphere graph, and let r_0 be one of its regions. The image of $\bar{\bar{G}}$, under the stereographic projection ϕ_P w.r.t. a point P in r_0, is a plane-graph $(\tilde{\phi}_P(\bar{\bar{G}}))$. If Q is another point in r_0, then we obviously have: $\tilde{\phi}_P(\bar{\bar{G}})$ and $\tilde{\phi}_Q(\bar{\bar{G}})$ are topological equivalent.

Now, the following definition makes sense:

Definition 9.4.5. The connected graph $\bar{\bar{G}}$ is called admissible (w.r.t. a region r_0) if the plane graph $\tilde{\phi}_P(\bar{\bar{G}})$ fulfils the inequalities (9.4.8) for each of its cycles. (Here, ϕ_P stands for the stereographic projection w.r.t. an arbitrary point $P \in r_0$).

Now, we are in the position to formulate the Characterization and Classification Theorem on structurally stable Rational Newton-flows. For the proof (which is far beyond the scope of this presentation) we refer to [42], [75].

Theorem 9.4.2.

a. Let $\bar{\bar{G}}$ be a connected sphere graph. Then, there exists a rational function $F \in \tilde{R}$ with the property that $\bar{\bar{G}}(F)$ is topologically equivalent with $\bar{\bar{G}}$ iff either $\bar{\bar{G}}$ is admissible or its geometrical dual $\bar{\bar{G}}^*$ is admissible.
   Moreover, if $\bar{\bar{G}}$ resp. $\bar{\bar{G}}^*$ is admissible, then F is of the form $F = \frac{p_n}{q_m}$ with $n \geq m-1$ resp. $m \geq n-1$.
b. Given the struturally stable rational Newton-flows $\bar{\bar{N}}(F_1)$ and $\bar{\bar{N}}(F_2)$, i.e. $F_1, F_2 \in \tilde{R}$. Then,
   $\bar{\bar{N}}(F_1)$ and $\bar{\bar{N}}(F_2)$ are topologically equivalent, if and only if $\bar{\bar{G}}(F_1)$ and $\bar{\bar{G}}(F_2)$ are topologically equivalent.

Corollary 9.4.2. Let $\bar{\bar{G}}$ be an arbitrary sphere-tree (i.e. $\bar{\bar{G}}$ is connected and does not exhibit cycles). Then, $\bar{\bar{G}}$ is -in a trivial way- admissible. Consequently, an arbitrary sphere-tree can be "realized" -up to topological equivalence- as the graph $\bar{\bar{G}}(F)$, where $F \in \tilde{R}$ is a polynomial.

Remark 9.4.4. For a proof of Corollary 9.4.2 which is not based on Theorem 9.4.2, we refer to our paper [37]. Independently, M. Shub and B. Williams [68] have found a proof. See also S. Smale [72], who posed the result of Corollary 9.4.2 as an open problem.

The following example gives us the opportunity to see Theorem 9.4.2 in action.

Example 9.4.3. Due to Theorem 9.4.2, the sphere graphs, as depicted in Fig. 9.4.5, can be realized -up to topological equivalence- as the graphs $\bar{\bar{G}}(F)$, $F \in \tilde{R}$. This is not true for the sphere graphs as depicted in Fig. 9.4.6. Note that the graph in Fig. 9.4.5.e has the property that its geometrical dual is admissible as well, hence the underlying rational functions are of the form $\frac{p_n}{q_m}$ with $|n-m| \leqq 1$ (cf. Theorem 9.4.2.a). In Figures 9.4.5.d and 9.4.6 the dotted lines stand for the geometrical duals.

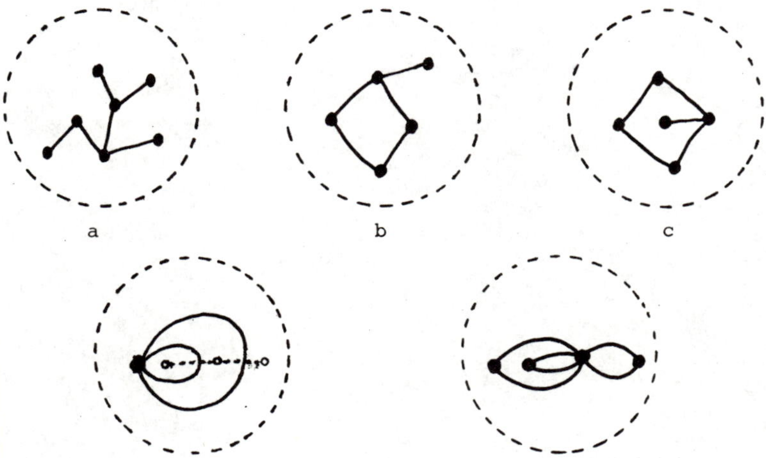

Fig. 9.4.5

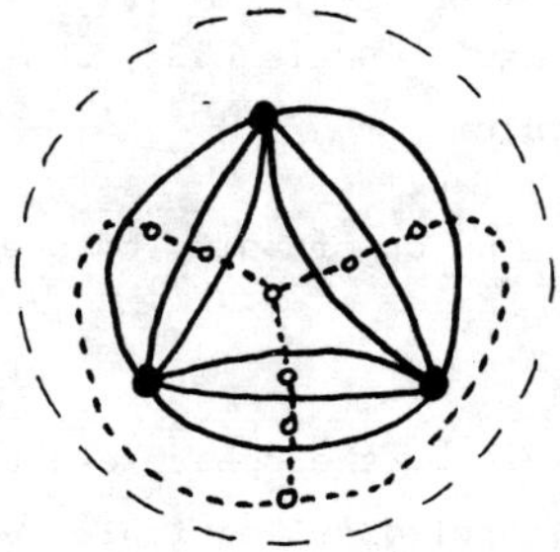

Fig. 9.4.6

10. PARAMETRIC ASPECTS

10.1. Critical manifolds.

In the present chapter we consider (smooth) optimization problems depending on parameters. We start with some general information about unconstrained r-parameter-families.

A smooth r-parameter family can be represented by means of a function $F \in C^\infty(\mathbb{R}^n \times \mathbb{R}^r, \mathbb{R})$, where $(x,t) \mapsto F(x,t)$, $x \in \mathbb{R}^n$ and $t \in \mathbb{R}^r$. Indeed, for each fixed $t \in \mathbb{R}^r$ (the parameter space) we have a function $F(\cdot,t) \in C^\infty(\mathbb{R}^n,\mathbb{R})$ and, as the parameter t varies from $\bar{t}_1$ to $\bar{t}_2$, the function $F(\cdot,\bar{t}_1)$ is deformed into the function $F(\cdot,\bar{t}_2)$.

For an r-parameter-family F we define the critical point set $\Sigma(F)$ by

$$\Sigma(F) = \{(x,t) \in \mathbb{R}^n \times \mathbb{R}^r \mid D_x F(x,t) = 0\}. \qquad (10.1.1)$$

So, the set $\Sigma(F) \cap (\mathbb{R}^n \times \{\bar{t}\})$ is precisely the set of critical points for the fixed function $F(\cdot,\bar{t})$. Recall that generically an $f \in C^\infty(\mathbb{R}^n,\mathbb{R})$ is non-degenerate (cf. Example 7.4.3). However, we cannot expect that $F(\cdot,t)$ is nondegenerate for all values of the parameter $t \in \mathbb{R}^r$. As a consequence, for certain values of t the equations $D_x F(x,t) = 0$ and $\det D_x^2 F(x,t) = 0$ will be satisfied simultaneously.

Note that the set $\Sigma(F)$ is determined by n equations on $\mathbb{R}^n \times \mathbb{R}^r$:
$\frac{\partial}{\partial x_i} F(x,t) = 0$, $i = 1,\dots,n$.

Condition A. An $F \in C^\infty(\mathbb{R}^n \times \mathbb{R}^r, \mathbb{R})$ is said to fulfil Condition A if the set of vectors $\{D^T[\frac{\partial}{\partial x_i} F(x,t)]$, $i = 1,\dots,n\}$ in $\mathbb{R}^n \times \mathbb{R}^r$ is linearly independent for all $(x,t) \in \Sigma(F)$.

If F satisfies Condition A, then, obviously, $\Sigma(F)$ is an r-dimensional C^∞-manifold in $\mathbb{R}^n \times \mathbb{R}^r$ (if not empty), and we call $\Sigma(F)$ the critical manifold for F.

Theorem 10.1.1. For fixed n,r, let $\mathcal{F}(A)$ be defined by

$$\mathcal{F}(A) = \{F \in C^\infty(\mathbb{R}^n \times \mathbb{R}^r, \mathbb{R}) \mid F \text{ fulfils Condition A}\}.$$

Then, $\mathcal{F}(A)$ is C^k-dense for all k; moreover, $\mathcal{F}(A)$ is C^k-open for all $k \geq 2$.

Proof. Consider the reduced 1-jet extension $\widetilde{j^1F}$: $(x,t) \mapsto D_xF(x,t)$. Then, F satisfies Condition A iff $\widetilde{j^1F} \pitchfork \{0_n\}$. Now, the statement of the theorem follows from the Jet Transversality Theorem (Theorem 7.4.1) and Remark 7.4.3. □

We define:

$$\Sigma_d(F) = \{(x,t) \in \Sigma(F) \,|\, \det D_x^2F(x,t) = 0\}. \tag{10.1.2}$$

The set $\Sigma_d(F)$ is the set of points $(\bar{x},\bar{t})$ for which $\bar{x}$ is a degenerate critical point for $F(\cdot,\bar{t})$; in other words, at the points of $\Sigma_d(F)$ we cannot apply the Implicit Function Theorem in order to parametrize locally the set $\Sigma(F)$ by means of the parameter t.

We proceed with a study of the local structure of $\Sigma_d(F)$. To this aim we consider the following reduced 2-jet extension $\widetilde{j^2F}$:

$$\widetilde{j^2F}(x,t) = [\underbrace{D_xF}_{n}\, , \underbrace{D_x^2F}_{\frac{1}{2}n(n+1)}]|_{(x,t)} \tag{10.1.3}$$

Let $\mathcal{B}_k$ denote the set of all symmetric n×n matrices for which exactly k eigenvalues vanish. Put $\mathcal{B}^* = \bigcup_{i=1}^{n} \mathcal{B}_k$.

From Example 7.3.4 we see that $\mathcal{B}^*$ is a stratified set in $\mathbb{R}^{\frac{1}{2}n(n+1)}$ with strata $\mathcal{B}_1,\ldots,\mathcal{B}_n$, and the codimension of $\mathcal{B}_k$ in $\mathbb{R}^{\frac{1}{2}n(n+1)}$ equals $\frac{1}{2}k(k+1)$. In Section 7.5 we mentioned that the latter stratification is also Whitney-regular. Furthermore, it is easily verified that $\mathcal{B}_i \subset \bar{\mathcal{B}}_j$ for $i > j$, and that $\bigcup_{j=i}^{n} \mathcal{B}_j$ is a closed set.

Recall that $(\bar{x},\bar{t}) \in \Sigma_d(F)$ iff $D_xF(\bar{x},\bar{t}) = 0$ and $D_x^2F(\bar{x},\bar{t}) \in \mathcal{B}^*$. In the target space of $\widetilde{j^2F}$ we consider the set M^*:

$$M^* = \{0_n\} \times \mathcal{B}^*. \tag{10.1.4}$$

The set M^* is closed and it has a Whitney-regular stratification, the strata being $\{0_n\} \times \mathcal{B}_k$, $k = 1,\ldots,n$.

Moreover, we have

$$\Sigma_d(F) = (\widetilde{j^2F})^{-1}[M^*] \tag{10.1.5}$$

Condition B. An $F \in C^\infty(\mathbb{R}^n \times \mathbb{R}^r, \mathbb{R})$ is said to fulfil Condition B if $\widetilde{j^2F} \pitchfork M^*$.

Remark 10.1.1. If $F \in C^\infty(\mathbb{R}^n \times \mathbb{R}^r, \mathbb{R})$ fulfils Condition B, then F fulfils automatically Condition A. To see this, note that $\mathcal{B}_0$, the set of all non-singular symmetric n×n matrices, is open in $\mathbb{R}^{\frac{1}{2}n(n+1)}$.

Put $\mathcal{B} = \bigcup_{i=0}^{n} \mathcal{B}_i$ and $\mathcal{M} = \{0_n\} \times \mathcal{B}$. Note that $\mathcal{M}$ is a Whitney-regular stratified set with strata $\{0_n\} \times \mathcal{B}_k$, k = 0,1,...,n, that $\widetilde{j^2F} \pitchfork M^*$ iff $\widetilde{j^2F} \pitchfork \mathcal{M}$ and that $\mathcal{B}$ fills up the whole space $\mathbb{R}^{\frac{1}{2}n(n+1)}$.

Let $F(B)$ be defined as follows (n,r fixed):

$$F(B) = \{F \in C^\infty(\mathbb{R}^n \times \mathbb{R}^r, \mathbb{R}) \,|\, F \text{ fulfils Condition B}\}.$$

Theorem 10.1.2. The set $F(B)$ is C^ℓ-dense for all ℓ, and $F(B)$ is C^ℓ-open for all $\ell \geq 3$.

If $F \in F(B)$, then $\Sigma_d(F)$ is a Whitney-regular stratified set with strata $(\widetilde{j^2F})^{-1}[\{0_n\} \times \mathcal{B}_k]$, k = 1,...,n (each of them being a submanifold of $\Sigma(F)$) and dim $\Sigma_d(F)$ = r-1 if $\Sigma_d(F) \neq \emptyset$. In particular we have:

Case r = 1: $\Sigma_d(F)$ is a discrete point set

Case r = 2: $\Sigma_d(F)$ is a one-dimensional manifold

Proof. The open/dense part of the theorem as well as the Whitney-regularity of the given stratification for $\Sigma_d(F)$ is an immediate consequence of Theorem 7.5.1 (and Remark 7.4.3).

The codimension of $\{0_n\} \times \mathcal{B}_k$ in $\mathbb{R}^n \times \mathbb{R}^{\frac{1}{2}n(n+1)}$ (= the target space of $\widetilde{j^2F}$) equals $n + \frac{1}{2}k(k+1)$. Next, let M be a manifold which intersects $\{0_n\} \times \mathcal{B}_k$ transversally and for which $M \cap (\{0_n\} \times \mathcal{B}_k) \neq \emptyset$. Then, it is not difficult to see that $M \cap (\{0_n\} \cap \mathcal{B}_1) \neq \emptyset$. Consequently, if $F \in F(B)$ and $\Sigma_d(F) \neq \emptyset$, we obtain: dim $\Sigma_d(F) = (n+r) - (n+\frac{1}{2}\cdot 1(1+1)) = r-1$; in particular, in case r = 1 we have dim $\Sigma_d(F) = 0$.

Furthermore, if $F \in F(B)$ and $(\widetilde{j^2F})^{-1}(\{0_n\} \times \mathcal{B}_k) \neq \emptyset$ for a $k \in \{1,\ldots,n\}$, then codim$(\{0_n\} \times \mathcal{B}_k) \leq n+r$ (= the number of available variables); thus, it holds $n + \frac{1}{2}k(k+1) \leq n+r$, or

$$r \geq \frac{1}{2}k(k+1). \tag{10.1.6}$$

In particular, in case $r = 2$, Formula (10.1.6) implies $k \le 1$. From the latter inequality it follows, in case $r = 2$, that $\Sigma_d(F) = (\widetilde{j^2F})^{-1}(\{0_n\} \times \mathcal{B}_1)$; this implies the last assertion of the theorem. □

Example 10.1.2. Put $n = r = 1$, $F_1(x,t) = \frac{1}{3}x^3 - t^2x$, $F_2(x,t) = \frac{1}{4}x^4 + tx$, $F_3(x,t) = \frac{1}{3}x^3 + tx$. Then, $F_1 \notin \mathcal{F}(A)$, $F_2 \in \mathcal{F}(A)$, $F_2 \notin \mathcal{F}(B)$, $F_3 \in \mathcal{F}(B)$, $\Sigma_d(F_i) = \{0\}$, $i = 1,2,3$ (see Fig. 10.1.1).

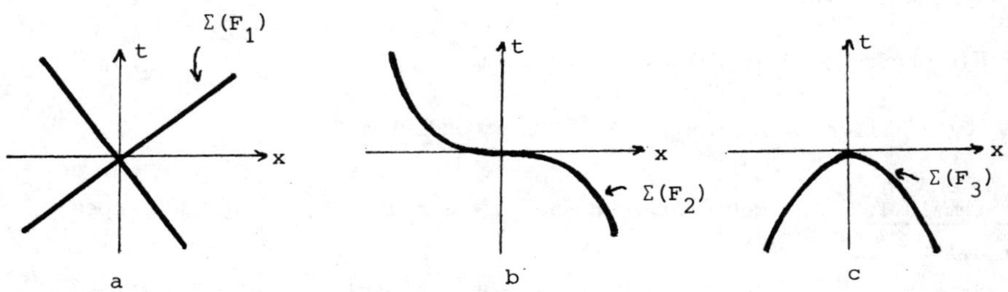

Fig. 10.1.1

Now, let us turn to the easiest constrained case. To this aim let $h_i, g_j \in C^\infty(\mathbb{R}^n \times \mathbb{R}^r, \mathbb{R})$, $i \in I, j \in J$ where I and J are finite index sets. These functions define for each parameter $t \in \mathbb{R}^r$ a feasible set $M(t)$,

$$M(t) = \{x \in \mathbb{R}^n \mid h_i(x,t) = 0,\ i \in I,\ g_j(x,t) \ge 0,\ j \in J\} \tag{10.1.7}$$

As in Chapter 3 we put

$$J_0(x,t) = \{j \in J \mid g_j(x,t) = 0\} \tag{10.1.8}$$

Let $\bar{x}$ belong to $M(\bar{t})$ and suppose that $M(\bar{t})$ is regular at $\bar{x}$, i.e. the set

$$\{D_x h_i,\ i \in I,\ D_x g_j,\ j \in J_0(\bar{x},\bar{t})\}\big|_{(\bar{x},\bar{t})}$$

is linearly independent.

In this "regular case" we can construct a local C^∞-coordinate transformation $(y,u) = \Phi(x,t)$ sending t-hyperplanes (of $\mathbb{R}^{n+1}$) onto u-hyperplanes such that (locally) $M(t)$ transforms to the constant set $\mathbb{H}^p \times \mathbb{R}^q$; here, $p = |J_0(\bar{x},\bar{t})|$ and $q = n - |I| - |J_0(\bar{x},\bar{t})|$.

In fact (cf. proof of Lemma 3.1.2), assume that $I = \{1,\dots,m\}$, $J_0(\bar{x},\bar{t}) = \{1,\dots,p\}$ and choose vectors $\xi_{m+p+1},\dots,\xi_n$, such that the set $\{D_x h_i, i \in I, D_x g_j, j \in J_0(\bar{x},\bar{t})\}_{|(\bar{x},\bar{t})} \cup \{\xi_{m+p+1},\dots,\xi_n\}$ forms a basis for $\mathbb{R}^n$. Then, put

$$\Phi(x,t) = (h_1,\dots,h_m,g_1,\dots,g_p{}_{|(x,t)},\xi^T_{m+p+1}(x-\bar{x}),\dots,\xi^T_n(x-\bar{x}),t-\bar{t}). \tag{10.1.9}$$

From the very construction it follows that there exist open neighborhoods $\mathcal{O}$ and $\mathcal{V}$ of $(\bar{x},\bar{t})$ and 0_{n+1} respectively such that $\Phi\colon \mathcal{O} \to \mathcal{V}$ is a diffeomorphism and moreover:

$$\Phi[\mathcal{O} \cap (\underline{\underline{M(t)}}\times\{t\})] = \mathcal{V} \cap (\{0_m\} \times \underline{\underline{\mathbb{H}^p \times \mathbb{R}^q}} \times \{t-\bar{t}\}) \tag{10.1.10}$$

Let $(\tilde{x},\tilde{t}) \in \mathcal{O}$ and $\tilde{x} \in M(\tilde{t})$; moreover, let $\tilde{x}$ be a critical point for $F(\cdot,\tilde{t})_{|M(\tilde{t})}$. Then (cf. Lemma 3.2.2), there exist (unique) real numbers $\tilde{\lambda}_i$, $\tilde{\mu}_j$, $i \in I$, $j \in J_0(\tilde{x},\tilde{t}) \subset J_0(\bar{x},\bar{t})$ such that

$$D_x F = \sum_{i\in I} \tilde{\lambda}_i D_x h_i + \sum_{j\in J_0(\tilde{x},\tilde{t})} \tilde{\mu}_j D_x g_j{}_{|(\tilde{x},\tilde{t})} \tag{10.1.11}$$

If we consider F in the new local coordinates around $(\bar{x},\bar{t})$, i.e. if we look at $\tilde{F} := F\circ\Phi^{-1}{}_{|\mathcal{V}}$, the we have with $(\tilde{y},\tilde{u}) = \Phi(\bar{x},\bar{t})$:

> the point $\tilde{y}$ is a critical point for $\tilde{F}(\cdot,\tilde{u})_{|\{0_m\} \times \mathbb{H}^p \times \mathbb{R}^q}$,
> and each Lagrange parameter $\tilde{\mu}_j$ is equal to the corresponding
> <u>partial derivative</u> $\frac{\partial \tilde{F}}{\partial y_j}(\tilde{y},\tilde{u})$, $j \in J_0(\tilde{x},\tilde{t})$ (exercise).

Altogether, by means of the local coordinate transformation Φ, we obtain a <u>constant</u> and <u>simple feasible set</u> $\mathbb{H}^p \times \mathbb{R}^q$ and, moreover, <u>we get rid of Lagrange parameters</u> since they just become partial derivatives corresponding to the $\mathbb{H}^p$-directions. In particular, if $p = 0$, we are, in the new coordinates, in the unconstrained situation.

In the unconstrained case we saw that the critical point set $\Sigma(F)$ is generically an r-dimensional smooth manifold. Motivated by the above discussion we now study analogously the (+)-Kuhn-Tucker set $\Sigma_+(F)$ with respect to the constant feasible set $\mathbb{H}^p \times \mathbb{R}^q$, where $p+q = n$:

$$\Sigma_+(F) = \left\{ (x,t) \in (\mathbb{H}^p \times \mathbb{R}^q) \times \mathbb{R}^r \;\middle|\; \begin{array}{l} x \text{ is a } (+)\text{-Kuhn-Tucker point} \\ \text{for } F(\cdot,t)_{|\mathbb{H}^p \times \mathbb{R}^q} \end{array} \right\} \tag{10.1.12}$$

As an abbreviation we denote $\partial_i F = \frac{\partial}{\partial x_i} F$. A point $(\bar{x},\bar{t})$ belongs to $\Sigma_+(F)$ iff the following conditions hold:

c_1. $x_i \geq 0$ and $\partial_i F \geq 0$, $i = 1,\ldots,p$

c_2. $x_i \cdot \partial_i F = 0$, $i = 1,\ldots,p$ (complementarity)

c_3. $\partial_i F = 0$, $i = p+1,\ldots,n$.

Having the conditions c_1-c_3 in mind, we introduce the following (specially ordered) reduced 1-jet extension $\widetilde{j^1}F$:

$$\widetilde{j^1}F(x,t) = (x_1,\partial_1 F,\ldots,x_p,\partial_p F,\partial_{p+1}F,\ldots,\partial_n F). \tag{10.1.13}$$

The corresponding (reduced) 1-jet space becomes the space $\mathbb{R}^{2p} \times \mathbb{R}^q$. Let us denote a typical point from the 1-jet space by $(u_1,v_1,\ldots,u_p,v_p,v_{p+1},\ldots,v_n)$, and consider the (u_i,v_i)-subspace for some $i \in \{1,\ldots,p\}$. The subset defined by the equation $u_i v_i = 0$ (compare condition c_2) is the union of the u_i-axis and the v_i-axis (Fig. 10.1.2.a). The obvious stratification for this set consists of the origin and remaining half-rays (Fig. 10.1.2.b).

This stratification is of course Whitney-regular. Let Σ_+^1 denote the subset $\{(u_i,v_i) \mid u_i v_i = 0 \text{ and } u_i \geq 0,\ v_i \geq 0\}$. Then Σ_+^1 has the stratification as in Fig. 10.1.2.c. Note that Σ_+^1 is a 1-dimensional topological manifold in $\mathbb{R}^2$ which is not differentiable at the origin; by a topological manifold in $\mathbb{R}^n$ of dimension m we mean a subset $N \subset \mathbb{R}^n$ having the property: for every $\bar{x} \in N$ there exists an open $\mathbb{R}^n$-neighborhood $\mathcal{O}_{\bar{x}}$ such that $N \cap \mathcal{O}_{\bar{x}}$ is homeomorphic with an open subset of $\mathbb{R}^m$. In Fig. 10.1.2.d a possible parametrization of Σ_+^1 is sketched.

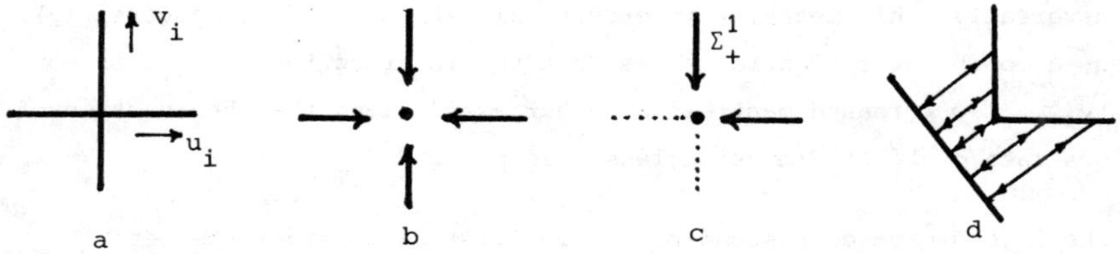

Fig. 10.1.2

The conditions c_1 and c_2 together give rise to the product set $\Sigma^p_+ := \Sigma^1_+ \times \ldots \times \Sigma^1_+$ in $\mathbb{R}^{2p}$. The stratification of the set Σ^p_+ will be the product-stratification: take a stratum of each Σ^1_+ and form the product; all possible products together define the product-stratification (cf. Fig. 10.1.3)

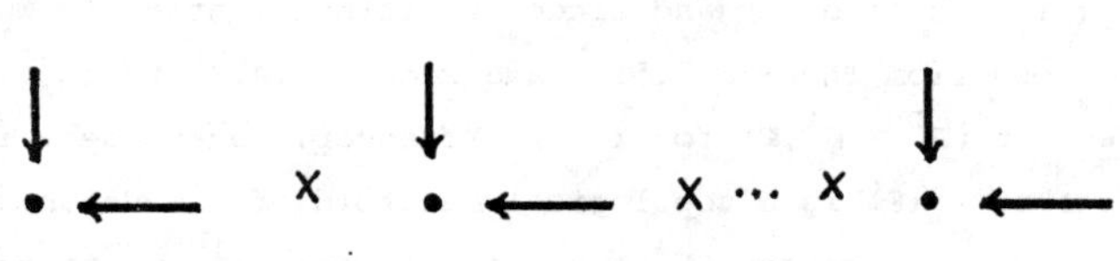

Fig. 10.1.3

Note that Σ^p_+ is a p-dimensional topological manifold in $\mathbb{R}^{2p}$ which fails to be smooth along the strata of dimension less than p. On the other hand, Σ^p_+ can be regarded as being pieced together from 2^p orthants from $\mathbb{R}^p$. Such manifolds are called "creased" manifolds by S. Schecter. For a precise definition we refer to the excellent reference [67]. Roughly speaking, a creased manifold (of dimension p) is a special type of a piecewise smooth manifold: each piece is locally diffeomorphic with an orthant of $\mathbb{R}^p$ and they are fitted together, combinatorially, in the same way as the orthants of $\mathbb{R}^p$. In particular, a creased manifold is a topological manifold. In particular, the "creases" in Σ^p_+ occur exactly along the strata of dimension less than p.

Now, suppose that $M \subset \mathbb{R}^{2p}$ is a smooth manifold which intersects Σ_+^p transversally (this means, M intersects all strata of Σ_+^p transversally). Then, a moment of reflection shows that the intersection $M \cap \Sigma_+^p$, if not empty, is also a creased manifold with "creases" along the intersection of M with strata of Σ_+^p of dimension less than p.

In the 1-jet space corresponding to (10.1.13) we consider the set $\Sigma_+^p \times \{0_q\}$ with the obvious product stratification.
It is important to note:

$$(x,t) \in \Sigma_+(F) \quad \text{iff} \quad \widetilde{j^1} F(x,t) \in \Sigma_+^p \times \{0_q\} .$$

We define

$$F^\star = \{F \in C^\infty(\mathbb{R}^n \times \mathbb{R}^r, \mathbb{R}) \mid \widetilde{j^1} F \pitchfork \Sigma_+^p \times \{0_q\}\} \quad . \tag{10.1.14}$$

Since $\Sigma_+^p \times \{0_q\}$ is a closed set and since its stratification is Whitney-regular, it follows from Theorem 7.5.1 (and Remark 7.4.3) that $F^\star$ is C^k-open and dense in $C^\infty(\mathbb{R}^n \times \mathbb{R}^r, \mathbb{R})$ for $k \geq 2$. Moreover, (exercise), if $F \in F^\star$ and $\Sigma_+(F) \neq \emptyset$, then $\Sigma_+(F)$ is a topological manifold of dimension r (even a creased manifold with creases at those points where $\widetilde{j^1} F$ meets points of strata of $\Sigma_+^p \times \{0_q\}$ of dimension less than p). In particular, in case r = 1 the set $\Sigma_+(F)$ is a piecewise smooth curve. Compare also Remark 10.3.2 in Section 10.3.

So far the regular case. At points where $M(t)$ is not regular the situation becomes much more complicated. A partial result is obtained by S. Schecter in [67] under the assumption of special rank conditions; we will not go into details here and refer to the interesting paper of S. Schecter [67] which also contains a stratification idea. For an explanation of the difficulties appearing in the general r-parameter case we refer to our study in [43].

10.2. One-parameter families of functions ($\mathbb{R}^n$).

We proceed with the discussion of Section 10.1 for the special case r = 1. The notations will be as in Section 10.1.
Suppose that $F \in C^\infty(\mathbb{R}^n \times \mathbb{R}, \mathbb{R})$ satisfies Condition A. Then $\Sigma(F)$ is a

1-dimensional C^∞-manifold (without boundary) in $\mathbb{R}^n \times \mathbb{R}$, consisting of a countable number of connected components. Obviously, $\Sigma(F)$ is a closed subset of $\mathbb{R}^n \times \mathbb{R}$. Let C be a component of $\Sigma(F)$. If C is bounded, then C is diffeomorphic with the circle S^1; if C is unbounded, then C is diffeomorphic with $\mathbb{R}$.

Example 10.2.1. Put $F(x,t) = \frac{1}{3}x_1^3 + (t^2-1)x_1 + \sum_{i=2}^{n} x_i^2$. Then, F satisfies Condition B (and thus Condition A), and

$$\Sigma(F) = \{(x,t) \in \mathbb{R}^n \times \mathbb{R} \mid x_1^2+t^2 = 1,\ x_i = 0,\ i = 2,\ldots,n\},$$

i.e. $\Sigma(F)$ is a circle.

Theorem 10.2.1. Suppose that $F \in C^\infty(\mathbb{R}^n \times \mathbb{R},\mathbb{R})$ satisfies Condition A and that C_d is a connected component of $\Sigma_d(F)$. Then, there exist two non-negative integers k_1, k_2, $k_1+k_2 = n-1$, such that for all $(\bar{x},\bar{t}) \in C_d$, k_1 (k_2) is equal to the number of negative (positive) eigenvalues of $D_x^2F(\bar{x},\bar{t})$.

Proof. Suppose that F satisfies Condition A. The linear independence of the system $\{D^T \frac{\partial}{\partial x_i} F(x,t),\ i = 1,\ldots,n\}$ implies that rank $D_x^2F(x,t) \geq n-1$. So, if $(\bar{x},\bar{t}) \in \Sigma_d(F)$, then rank $D_x^2F(\bar{x},\bar{t}) = n-1$. The assertion of the theorem follows from the facts that C_d is connected and that eigenvalues depend continuously on the matrix elements. □

Theorem 10.2.2. Suppose that $F \in C^\infty(\mathbb{R}^n \times \mathbb{R},\mathbb{R})$ satisfies Condition A. Put $\mathrm{Det}(x,t) = \det D_x^2F(x,t)$ and $\phi(x,t) = t$. Then, the following statements are equivalent:

(i) F satifies Condition B,

(ii) $\phi_{|\Sigma(F)}$ is nondegenerate,

(iii) If $\mathrm{Det}_{|\Sigma(F)}$ (locally viewed as function of 1-variable) vanishes, then its derivative does not vanish.

Proof. Firstly, we show: $(\bar{x},\bar{t}) \in \Sigma(F)$ is a critical point for $\phi_{|\Sigma(F)}$ iff $(\bar{x},\bar{t}) \in \Sigma_d(F)$.
In fact, let $(\bar{x},\bar{t}) \in \Sigma(F)$ be a critical point for $\phi_{|\Sigma(F)}$. So, there exist unique numbers $\bar{\lambda}_1,\dots,\bar{\lambda}_n \in \mathbb{R}$ such that

$$(0,\dots,0,1) = \sum_{i=1}^{n} \bar{\lambda}_i D \frac{\partial}{\partial x_i} F(\bar{x},\bar{t}) \quad . \tag{10.2.1}$$

Formula (10.2.1) implies that $D_x^2F(\bar{x},\bar{t})$ is singular and thus $(\bar{x},\bar{t}) \in \Sigma_d(F)$.
On the other hand, suppose that $(\bar{x},\bar{t}) \in \Sigma_d(F)$. Then, by Theorem 10.2.1, exactly one eigenvalue of $D_x^2F(\bar{x},\bar{t})$ vanishes. Let $(v_1,\dots,v_n)^T$ generate the 1-dimensional eigenspace belonging to this vanishing eigenvalue. Since the system $\{D^T \frac{\partial}{\partial x_i} F(\bar{x},\bar{t}),\ i = 1,\dots,n\}$ is linearly independent, we have $\sum_{i=1}^{n} v_i D \frac{\partial}{\partial x_i} F(\bar{x},\bar{t}) \neq 0$. Put $\alpha = \sum_{i=1}^{n} v_i \frac{\partial}{\partial t}(\frac{\partial}{\partial x_i} F(\bar{x},\bar{t}))$. Then, $\alpha \neq 0$ and $\bar{\lambda}_i := \frac{1}{\alpha} v_i$, $i = 1,\dots,n$, satisfy (10.2.1), i.e. $(\bar{x},\bar{t})$ is a critical point for $\phi_{|\Sigma(F)}$.

Suppose that $(\bar{x},\bar{t}) \in \Sigma(F)$ is a critical point for $\phi_{|\Sigma(F)}$. After a linear coordinate transformation on the x-variables we may assume that

$$D_x^2F(\bar{x},\bar{t}) = \mathrm{diag}(0,\varepsilon_2,\varepsilon_3,\dots,\varepsilon_n), \varepsilon_i \neq 0,\ i = 2,\dots,n. \tag{10.2.2}$$

Then, in Formula (10.2.1) we have $\bar{\lambda}_1 \neq 0$, $\bar{\lambda}_i = 0$, $i = 2,\dots,n$ and thus $\frac{\partial^2}{\partial x_1 \partial t} F(\bar{x},\bar{t}) \neq 0$.

Furthermore, the tangentspace $T_{(\bar{x},\bar{t})}\Sigma(F)$ becomes:

$$T_{(\bar{x},\bar{t})}\Sigma(F) = \{(x,t) \in \mathbb{R}^{n+1} \mid t = 0,\ x_i = 0,\ i = 2,\dots,n\}. \tag{10.2.3}$$

Consequently, since $\bar{\lambda}_1 \neq 0$, the critical point $(\bar{x},\bar{t})$ for $\phi_{|\Sigma(F)}$ is non-degenerate iff $\frac{\partial^3}{\partial x_1^3} F(\bar{x},\bar{t}) \neq 0$.

In the next step, we look closer at Condition B in this case. Since, by assumption, F satisfies Condition A, we merely have to fix our attention to the set $\Sigma_d(F)$. Let $(\bar{x},\bar{t}) \in \Sigma_d(F)$. Without loss of generality we may assume again that (10.2.2) holds.
With the abbreviation $F_i = \frac{\partial}{\partial x_i} F$, $F_t = \frac{\partial}{\partial t} F$, etc., we put:

$$j^2F(x,t) = (\underbrace{F_1,\ldots,F_n}_{n}, \underbrace{F_{11},F_{12},\ldots,F_{nn}}_{\frac{1}{2}n(n+1)-1}, F, F_t,\ldots,F_{tt}, x, t).$$

Now, F satisfies Condition B <u>at the point $(\bar{x},\bar{t})$</u> iff $j^2F[\mathbb{R}^{n+1}]$ intersects $\{0\} \times \mathcal{B}_1 \times \mathbb{R}^p$, $0 \in \mathbb{R}^n$, transversally in $\mathcal{J}(n+1,1,2)$ at the point $j^2F(\bar{x},\bar{t})$ and we contend that this is equivalent with $F_{1t}(\bar{x},\bar{t}) \neq 0$, $F_{111}(\bar{x},\bar{t}) \neq 0$. In fact, this implies the equivalence of (i), (ii). To this aim we have to calculate the tangent space to $\mathcal{B}_1$ at the point $\bar{A} = \mathrm{diag}(0,\varepsilon_2,\ldots,\varepsilon_n)$. From Example 7.3.4 it follows that a symmetric n×n matrix A -in a small neighborhood of $\mathrm{diag}(0,\varepsilon_2,\ldots,\varepsilon_n)$- is in $\mathcal{B}_1$ iff A satisfies the equation $h(a_{11},\ldots,a_{nn}) = 0$, where

$$\left.\begin{aligned} &h(a_{11},\ldots,a_{nn}) = a_{11} - (a_{12},\ldots,a_{1n})\tilde{A}^{-1}(a_{12},\ldots,a_{1n})^T, \\ &\text{and} \\ &A = \left(\begin{array}{c|c} a_{11} & a_{12} \ \cdots \ a_{1n} \\ \hline a_{21} & \\ \vdots & \tilde{A} \\ a_{n1} & \end{array}\right). \end{aligned}\right\} \qquad (10.2.4)$$

From (10.2.4) we obtain $\dfrac{\partial h(\bar{A})}{\partial a_{11}} = 1$, $\dfrac{\partial h(\bar{A})}{\partial a_{ij}} = 0$ for $(i,j) \neq (1,1)$.

So, the tangentspace $T_{\bar{A}}\mathcal{B}_1$ is equal to the space of all symmetric n×n matrices $C = (c_{ij})$ with $c_{11} = 0$.

In order that $j^2F[\mathbb{R}^{n+1}]$ and $\{0\} \times \mathcal{B}_1 \times \mathbb{R}^p$ intersect transversally in $\mathcal{J}(n+1,1,2)$ at the point $j^2F(\bar{x},\bar{t})$, the tangentspace at $j^2F(\bar{x},\bar{t})$ of both manifolds should span the whole space $\mathcal{J}(n+1,1,2)$ (note that $j^2F(\bar{x},\bar{t})$ lies in the intersection of both manifolds since $(\bar{x},\bar{t}) \in \Sigma_d(F)$). Consequently, the following $N(n+1,1,2) \times N(n+1,1,2)$ matrix M must be nonsingular:

$$M = \left(\begin{array}{cccc|c} F_{11} & \cdots & F_{n1} & F_{111} & \\ \vdots & & \vdots & \vdots & * \\ F_{1n} & \cdots & F_{nn} & F_{11n} & \\ F_{1t} & \cdots & F_{nt} & F_{11t} & \\ \hline & & 0 & & I \end{array}\right)\Bigg|_{(\bar{x},\bar{t})}, \qquad (10.2.5)$$

I being the identity matrix of appropriate dimension.

Recalling Formula (10.2.2), it becomes obvious that M is nonsingular iff $F_{1t} \cdot F_{111}|_{(\bar{x},\bar{t})} \neq 0$. This proves the equivalence of (i), (ii).

Finally, we prove the equivalence of (ii), (iii). Again, we may assume that $(\bar{x},\bar{t}) \in \Sigma_d(F)$ and that (10.2.2) holds. In view of (10.2.3) we have to show that $\frac{\partial}{\partial x_1}$ Det$(\bar{x},\bar{t}) \neq 0$ iff $F_{111}(\bar{x},\bar{t}) \neq 0$. Let us denote the determinant by the symbol $|\cdot|$. Then,

$$\frac{\partial}{\partial x_1} \operatorname{Det}(\bar{x},\bar{t}) = \frac{\partial}{\partial x_1} \begin{vmatrix} F_{11} & \cdots & F_{1n} \\ \vdots & & \vdots \\ F_{n1} & \cdots & F_{nn} \end{vmatrix}_{|(\bar{x},\bar{t})} =$$

$$= \sum_{i=1}^{n} \begin{vmatrix} F_{11} & & \begin{pmatrix} F_{1i} \\ \vdots \\ F_{ni} \end{pmatrix} & & F_{1n} \\ \vdots & \cdots & \frac{\partial}{\partial x_i} & \cdots & \vdots \\ F_{n1} & & & & F_{nn} \end{vmatrix}_{|(\bar{x},\bar{t})} =$$

$$= \begin{vmatrix} F_{111} & 0 & \cdots & 0 \\ \vdots & & \varepsilon_2 & 0 \\ F_{n11} & 0 & \ddots & \varepsilon_n \end{vmatrix}_{|(\bar{x},\bar{t})} = F_{111}(\bar{x},\bar{t}) \cdot \varepsilon_2 \cdots \varepsilon_n .$$

Since $\varepsilon_i \neq 0$, $i = 2,\ldots,n$, this completes the proof of the theorem. □

Define $\Sigma_r(F)$ as follows:

$$\Sigma_r(F) = \Sigma(F) \backslash \Sigma_d(F). \tag{10.2.6}$$

Corollary 10.2.1. Suppose that $F \in C^\infty(\mathbb{R}^n \times \mathbb{R}, \mathbb{R})$ satisfies Condition B. Then, each $(\bar{x},\bar{t}) \in \Sigma_d(F)$ is an isolated point in $\Sigma(F)$. The 1-dimensional manifold $\Sigma_r(F)$ consists of a countable number of connected components. Let C be a connected component of $\Sigma_r(F)$. Then, the quadratic (co-)index of $F(\cdot,t)$ is constant on C. Furthermore, if we pass a point $(\bar{x},\bar{t}) \in \Sigma_d(F)$ on $\Sigma(F)$, then the quadratic index of $F(\cdot,t)$ changes exactly by one.

Corollary 10.2.2. Suppose that $F \in C^{\infty}(\mathbb{R}^n \times \mathbb{R},\mathbb{R})$ satisfies Condition B. Let $a < b$, (x_a,a), $(x_b,b) \in \Sigma_r(F)$ and suppose that (x_a,a), (x_b,b) lie on the same connected component of $\Sigma(F) \cap (\mathbb{R}^n \times [a,b])$. Then,

$$QI(F(\cdot,a)_{|x=x_a}) \equiv QI(F(\cdot,b)_{|x=x_b}) \pmod{2}.$$

If $(x_a,a),(y_a,a) \in \Sigma_r(F)$, $x_a \neq y_a$ and if (x_a,a), (y_a,a) lie in the same connected component of $\Sigma(F) \cap (\mathbb{R}^n \times [a,\infty))$ resp. $\Sigma(F) \cap (\mathbb{R}^n \times (-\infty,a])$. Then,

$$QI(F(\cdot,a)_{|x=x_a}) \equiv QI(F(\cdot,a)_{|x=y_a}) + 1 \pmod{2}. \qquad \square$$

We end this section with a discussion on the realizability of the statements in Corollary 10.2.2.

To this aim, let us introduce the following family of functions:

$$F(n) = \{F \in C^{\infty}(\mathbb{R}^n \times \mathbb{R},\mathbb{R})\,|\,F(x,t) = \|x\|^2 \text{ outside } B(0,1) \times \mathbb{R}\},$$

where $B(0,1)$ stands for the Euclidean ball in $\mathbb{R}^n$ with center 0 and radius 1. The following theorem shows that the statements in Corollary 10.2.2 are completely realizable within the family $F(n)$. We have chosen the family $F(n)$ just in order to show that "all action" may take place in a compact subset of $\mathbb{R}^n$.

Let $n \geq 1$ be arbitrary, but fixed. A finite sequence $\alpha_0,\alpha_1,\ldots,\alpha_k$, $k \geq 0$, is called an index sequence if $\alpha_i \in \{0,1,\ldots,n\}$, $i = 1,\ldots,k$ and $|\alpha_j-\alpha_{j+1}| = 1$, $j = 0,1,\ldots,k-1$.

Theorem 10.2.3. Let $\alpha_0,\alpha_1,\ldots,\alpha_k$ be an index sequence.

(1) If k is even, then there exists an $F_1 \in F(n)$ such that:

(i) F_1 satisfies Condition B,

(ii) there exist $x_0,x_1 \in B(0,1)$ such that $(x_0,0)$, $(x_1,1) \in \Sigma_r$ and $QI(F_1(\cdot,0)_{|x=x_0}) = \alpha_0$, $QI(F_1(\cdot,1)_{|x=x_1}) = \alpha_k$,

(iii) $F_1(\cdot,t) = F_1(\cdot,0)$ for all $t \leq 0$,

$F_1(\cdot,t) = F_1(\cdot,1)$ for all $t \geq 1$,

(iv) the points $(x_0,0)$, $(x_1,1)$ lie on the same connected component of $\Sigma(F)$. Moreover, walking from $(x_0,0)$ to $(x_1,1)$ on $\Sigma(F)$, the changes in the quadratic index QI for $F(\cdot,t)$ take place precisely according to the sequence $\alpha_0,\alpha_1,\ldots,\alpha_k$.

(2) If k is odd, then there exists an $F_2 \in F(n)$ such that:

(i) F_2 satisfies Condition B,

(ii) there exist $x_0, y_0 \in B(0,1)$, $x_0 \neq y_0$ such that $(x_0,0),(y_0,0) \in \Sigma_r$ and $QI(F_2(\cdot,0)_{|x=x_0}) = \alpha_0$, $QI(F_2(\cdot,0)_{|x=y_0}) = \alpha_k$,

(iii) $F_2(\cdot,t) = F_2(\cdot,0)$ for all $t \leq 0$,

$F_2(\cdot,t) = F_2(\cdot,1)$ for all $t \geq 1$,

(iv) the points $(x_0,0)$, $(y_0,0)$ lie on the same component of $\Sigma(F)$. Moreover, walking form $(x_0,0)$ to $(y_0,0)$ on $\Sigma(F)$, the changes in the quadratic index QI for $F(\cdot,t)$ take place precisely according to the sequence $\alpha_0,\alpha_1,\ldots,\alpha_k$.

Proof. At certain places in the proof we will restrict ourselves to the basic constructive topological ideas and omit the details. The main idea of using "local models" as below is contained in [41].

Step 1. First of all we note that the special parameter-choices 0 and 1 for t as well as the orientation of the t-axis is irrelevant. In fact, we may take t' as a new parameter instead of t by putting $t' = \alpha t+\beta$, $\alpha,\beta \in \mathbb{R}$, $\alpha \neq 0$.

Step 2. Local Model I. Let $f \in C^\infty(\mathbb{R}^n,\mathbb{R})$, $\bar{x} \in \mathbb{R}^n$, $Df(\bar{x}) \neq 0$ and $k \in \{1,2,\ldots,n\}$. Let U be an open neighborhood of $\bar{x}$ such that $Df(x) \neq 0$ for all $x \in U$. The Local Model I yields the creation of an $F \in C^\infty(\mathbb{R}^n \times \mathbb{R},\mathbb{R})$ having the following properties:
$F(x,t) = f(x)$ outside $K \times \mathbb{R}$, K a compact neighborhood of $\bar{x}$, $K \subset U$;
$F(x,t) = f(x)$ for all $t \leq 0$; $F(\cdot,t) = F(\cdot,1)$ for all $t \geq 1$; F satisfies Condition B on $U \times \mathbb{R}$; for all $t \in (\frac{1}{2},\infty)$, $F(\cdot,t)_{|U}$ has exactly two critical points, both of them nondegenerate and of quadratic index k, k-1 respectively; for all $t \in (-\infty,\frac{1}{2})$, $F(\cdot,t)_{|U}$ has no critical points.
Thus, the Local Model I provides the creation of a pair of nondegenerate

critical points of quadratic index k, k-1 respectively. In fact, $\Sigma(F)_{|U \times \mathbb{R}}$ has the form as in Fig. 10.2.1.

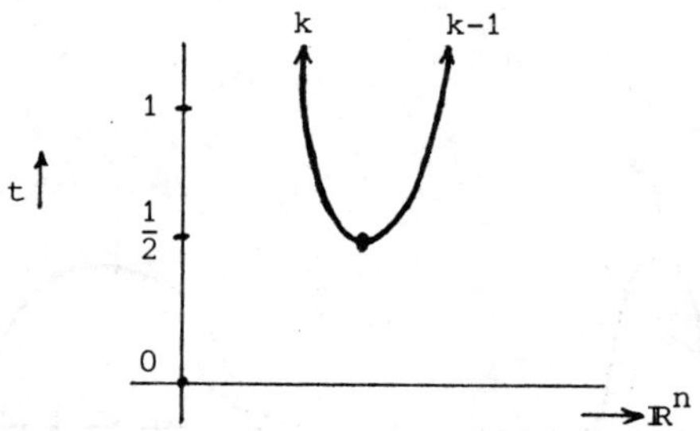

Fig. 10.2.1

The construction of Local Model I:

Our proof will be an appropriate modification of a construction made in [62, proof of Lemma 8.2].

Firstly we prove the following "global version":

Given $k \in \{1,\dots,n\}$. Then there exists a $G \in C^\infty(\mathbb{R}^n \times \mathbb{R}, \mathbb{R})$ with the following properties:

(1) G satisfies Condition B,

(2) $G(x_1,\dots,x_n,t) = x_1$ outside $K \times \mathbb{R}$, K a compact subset of $\mathbb{R}^n$,

(3) $G(x,t) = x_1$ for $t \le 0$; $G(\cdot,t) = G(\cdot,1)$ for $t \ge 1$,

(4) for $t < \frac{1}{2}$, $G(\cdot,t)$ has no critical points,

(5) for $t > \frac{1}{2}$, $G(\cdot,t)$ has exactly two critical points, both of them nondegenerate and of quadratic index k, k-1 respectively.

Proof of the "global version":

Let $\zeta \in C^\infty(\mathbb{R},\mathbb{R})$ be a function with the following properties (cf. Fig. 10.2.2.a):

(i) $\zeta = 0$ on $\mathbb{R}\setminus(-1,1)$, $\zeta > 0$ on $(-1,1)$,

(ii) ζ is strictly monotone on $(-1,0)$ and on $(0,1)$,

(iii) $\zeta(0) = 2$, $\zeta'(0) = 0$, $\zeta''(0) < 0$ (prime denoting differentiation).

Let $\xi \in C^\infty(\mathbb{R},\mathbb{R})$ be defined as follows (cf. Fig. 10.2.2.b):
$\xi(u) = \zeta(u)$ for $u \le 1$, $\xi(u) = -\zeta(2-u)$ for $u > 1$.

Define $\eta \in C^\infty(\mathbb{R},\mathbb{R})$ as follows (cf. Fig. 10.2.2.c):
$\eta(u) = 0$ for $u \le -1$, $\eta(u) = \int_{-1}^{u} \xi(\tau)d\tau$ for $u > -1$.

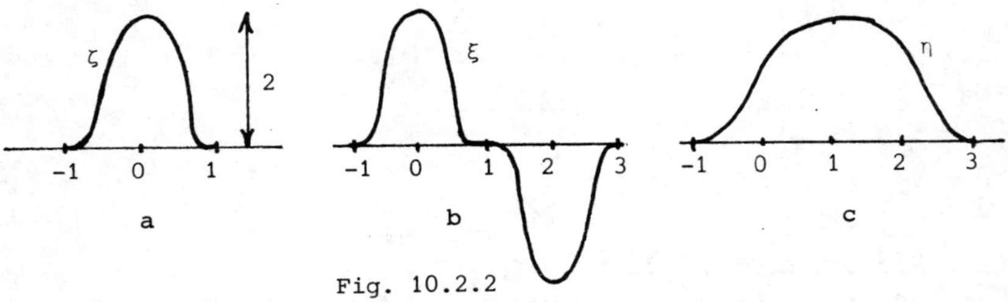

Fig. 10.2.2

We proceed by introducing $\phi \in C^\infty(\mathbb{R},\mathbb{R})$ with the following properties:
$\phi(u) = 0$ for $u \le 0$, $\phi(u) = 1$ for $u \ge 1$, ϕ strictly monotone on $(0,1)$,
$\phi(\frac{1}{2}) = \frac{1}{2}$ (cf. Fig. 10.2.3.a).

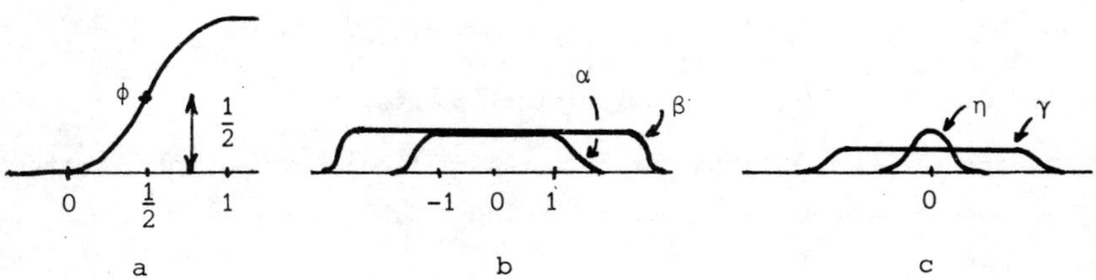

Fig. 10.2.3

We partition $\mathbb{R}^n$ as $\mathbb{R} \times \mathbb{R}^{k-1} \times \mathbb{R}^{n-k}$ and write the general point $x \in \mathbb{R}^n$ as
$x = (x_1,y,z)$, $x_1 \in \mathbb{R}$, $y \in \mathbb{R}^{k-1}$, $z \in \mathbb{R}^{n-k}$. As an abbreviation we put
$y^2 = \|y\|^2$, $z^2 = \|z\|^2$. Define:

$$G^*(x,t) = x_1 + \phi(t)[-\eta(x_1)-y^2+z^2]. \tag{10.2.7}$$

We contend that G^* satisfies the properties (1), (3), (4), (5). To see this, we calculate D_xG^*:

$$D_xG^*(x,t) = (1-\phi(t)\eta'(x_1), -2\phi(t)y, +2\phi(t)z). \tag{10.2.8}$$

Since $\eta'(x_1) = \xi(x_1)$ and $\xi \leq 2$, the equation $\phi(t)\eta'(x_1) = 1$ is solvable only if $\phi(t) \geq \frac{1}{2}$ and thus $t \geq \frac{1}{2}$. This shows property (4). Furthermore, it shows that $D_xG^*(x,t) = 0$ only if $y = 0$, $z = 0$.
From (10.2.4) we see that $D_xG^*(x,t) = 0$ and $\det D_x^2G^*(x,t) = 0$ iff $y = 0$, $z = 0$, $1-\phi(t)\eta'(x_1) = 0$ and $\phi(t)\eta''(x_1) = 0$. By the very construction (cf. definition of ξ), this is the case iff $y = 0$, $z = 0$, $x_1 = 0$, $t = \frac{1}{2}$. Note that $\eta'''(0) = \xi''(0) \neq 0$ and $\phi'(\frac{1}{2})\eta'(0) \neq 0$. This implies that G^* satisfies Condition B and so, property (1) holds for G^*. Note that the function $x_1 - \eta(x_1)$ has exactly two critical points, both of them nondegenerate (local minimum, local maximum). Consequently, $G^*(\cdot,1)$ has exactly two critical points, both of them nondegenerate, and of quadratic index k, k-1 respectively. Now, the validity of property (5) is easily verified. The validity of property (3) follows immediately from the definition of the function ϕ.

It remains to alter G^* into G such that property (2) is also valid. As in [62] we introduce the following <u>nonnegative</u> functions $\alpha,\beta,\gamma \in C^\infty(\mathbb{R},\mathbb{R})$, <u>all of them being indentically zero outside a compact subset of $\mathbb{R}$</u> with the following properties (cf. Fig. 10.2.3.b,c):

(a) $\alpha(u) = 1$ for $|u| \leq 1$,

(b) $|\alpha'(u)| < (\max_{v\in\mathbb{R}} |\eta(v)|)^{-1}$ for all $u \in \mathbb{R}$,

(c) $\beta(u) = 1$ whenever $\alpha(u) \neq 0$,

(d) $\gamma(u) = 1$ whenever $\eta'(u) \neq 0$,

(e) $|\gamma'(u)| < (\max_{v\in\mathbb{R}} v\beta(v))^{-1}$

We define G as follows:

$$G(x,t) = x_1 + \phi(t)[-\eta(x_1)\alpha(y^2+z^2) + \gamma(x_1)(-y^2+z^2)\beta(y^2+z^2)], \tag{10.2.9}$$

and we contend that G satisfies all the properties (1) - (5).

The validity of properties (2), (3) is obvious. Concerning the properties (1), (4), (5) we have to deal with $D_x G(x,t)$. Firstly we consider $\frac{\partial}{\partial x_1} G(x,t)$:

$$\frac{\partial}{\partial x_1} G(x,t) = 1+\phi(t)[-\eta'(x_1)\alpha(y^2+z^2) + \gamma'(x_1)(-y^2+z^2)\beta(y^2+z^2)]. \tag{10.2.10}$$

From the estimate

$$|\gamma'(x_1)(-y^2+z^2)\beta(y^2+z^2)| \le |\gamma'(x_1)|(y^2+z^2)\beta(y^2+z^2) < 1 \text{ (cf. (e))},$$

we obtain that $\frac{\partial}{\partial x_1} G(x,t) \neq 0$ whenever $\phi(t)\eta'(x_1)\alpha(y^2+z^2) = 0$.
Consequently we may restrict our attention to the case $t > 0$, $\eta'(x_1) \neq 0$ (and thus $\gamma = 1$), $\alpha(y^2+z^2) \neq 0$ (thus $\beta = 1$) in order to find points at which $D_x G(x,t) = 0$.

Within the maximal open region where $\beta = 1$, $\gamma = 1$ we have:

$$D_x G(x,t) = (1,0,0)-\phi(t)[\eta'(x_1)\alpha(y^2+z^2),\ 2y(\eta(x_1)\alpha'(y^2+z^2)+1),\ 2z(\eta(x_1)\alpha'(y^2+z^2)-1)].$$

From (b) we obtain $|\eta(x_1)|\,|\alpha'(y^2+z^2)| < 1$. Therefore, $\eta(x_1)\alpha'(y^2+z^2) \pm 1 \neq 0$. Since t may be assumed to be positive, we have $\phi(t) > 0$. Altogether we obtain: $D_x G(x,t) = 0 \rightarrow y = 0,\ z = 0$.
So, from (a), we may restrict our attention further to the maximal open regions where $\alpha = \beta = \gamma = 1$. But then, $G(x,t)$ reduces to $x_1 + \phi(t)[-\eta(x_1)-y^2+z^2]$. The latter function equals $G^*(x,t)$ and we are done.

Now we proceed with the construction of the Local Model I.

So, let $f \in C^\infty(\mathbb{R}^n,\mathbb{R})$, $\bar{x} \in \mathbb{R}^n$, $Df(\bar{x}) \neq 0$, $k \in \{1,2,\ldots,n\}$ and $Df(x) \neq 0$ for all $x \in U$, U an open neighborhood of $\bar{x}$.
Without loss of generality we assume: $\bar{x} = 0$, $f(\bar{x}) = 0$, the open cube $(-\varepsilon,\varepsilon)^n \subset U$ ($\varepsilon > 0$ sufficiently small) and $f(x) = x_1$ for all $x \in (-\varepsilon,\varepsilon)^n$ (cf. Theorem 2.7.1).
Outside $(-\varepsilon,\varepsilon)^n \times \mathbb{R}$ we define $F(x,t) = f(x)$. Next, we put $g(u) = \tan(\frac{\pi u}{2\varepsilon})$ and define the diffeomorphism $\Phi: (-\varepsilon,\varepsilon)^n \rightarrow \mathbb{R}^n$, $y = \Phi(x)$ as follows: $y_i = g(x_i), i = 1,\ldots,n$. Finally, on $(-\varepsilon,\varepsilon)^n \times \mathbb{R}$ we define F as follows:

$$F(x,t) = g^{-1} \circ G(\Phi(x),t),$$

where G is the function constructed in the "global version". The verification of the fact that F satisfies all the properties as stated in the beginning of Step 2 is not difficult and will be omitted.

Step 3. Local Model II a,b.

Local Model II a defines a 1-parameter family $F(\cdot,t)$ of perturbations of $f(x) = x^2$ (function of one variable) in a neighborhood of $x = 0$ such that a pair of critical points of quadratic index 0, 1 resp. is created and such that a similar pair is destroyed, $F(\cdot,t)$ satisfying condition B, $F(x,t) = x^2$ for $t \leq 0$, $F(\cdot,t) = F(\cdot,1)$ for $t \geq 1$. See Fig. 10.2.4.a.

The Local Model II b is analogous to Local Model II a, but now the function $\bar{f}(x) = -x^2$ is perturbed. See Fig. 10.2.4.b (take $\bar{F} = -F$).

The Local Model II a (and thus II b) is easily constructed by a suitable combination of two copies of Local Model I, case n = 1, k = 1. See Fig. 10.2.4.c.

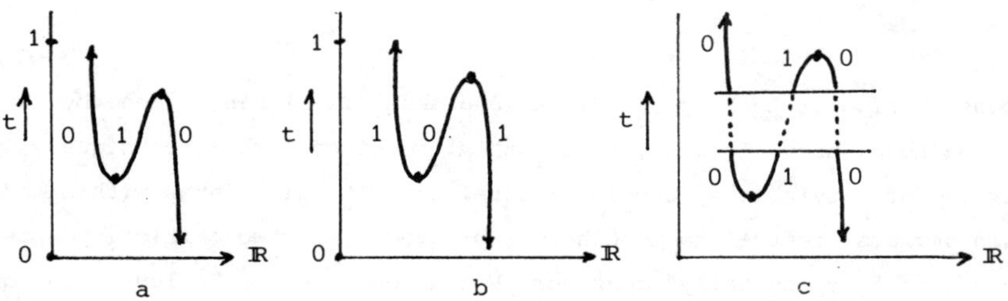

Fig. 10.2.4

Step 4. Local Model III a,b.

The Local Model III a defines a 1-parameter family $F(\cdot,t)$ of perturbations of $f(x_1,x_2) = x_1^2+x_2^2$ (function of two variables) in a neighborhood of $x = 0$ such that: $F(\cdot,t)$ satisfies Condition B, $\Sigma(F)$ consists of two connected components, along one of these components, say C_1, a critical point of quadratic index 0 changes via quadratic index 1 into a critical point of quadratic index 2. (increase of quadratic index by 2). Furthermore,

$F(x,t) = f(x)$ for $t \le 0$ and $F(\cdot,t) = F(\cdot,1)$ for $t \ge 1$. In fact, the idea is that $\Sigma(F)$ has the form as sketched in Fig. 10.2.5.a.
The Local Model III b provides an analogous 1-parameter perturbation family $\bar{F}(\cdot,t)$ for the function $\bar{f}(x) = -x_1^2-x_2^2$. The manifold $\Sigma(\bar{F})$ has the form as sketched in Fig. 10.2.5.b (take $\bar{F} = -F$).
Thus, Local Model III b provides a decrease of quadratic index by 2.

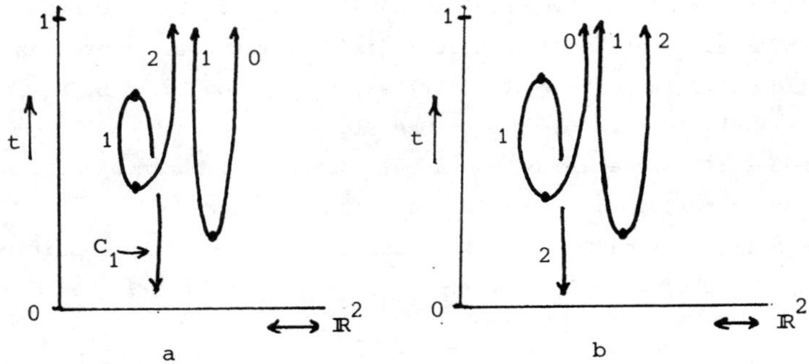

Fig. 10.2.5

The construction of Local Model III a (and thus III b) can be done by inserting Local Model I two times suitably, combined with a technique consisting of varying the functional level of critical points -within certain obvious limits- and a technique of cancelling two critical points which belong "topologically" together. For an outline of the last technique we refer to the cancellation theorem of M. Morse [62, Theorem 5.4]. In order to keep the constructions short and to visualize the geometrical idea on the other hand, we restrict ourselves to an intuitive proof by sketching the relevant level lines of $F(\cdot,t)$ for several values of t (Fig. 10.2.7). In order to understand the pictures completely, in Fig. 10.2.6 we sketched the relevant level lines of $\bar{F}(x_1,x_2,0)$, where $\bar{F}(x_1,x_2,t) = x_1^3+tx_1+x_2^2$. (standard example of a function satisfying Condition B).

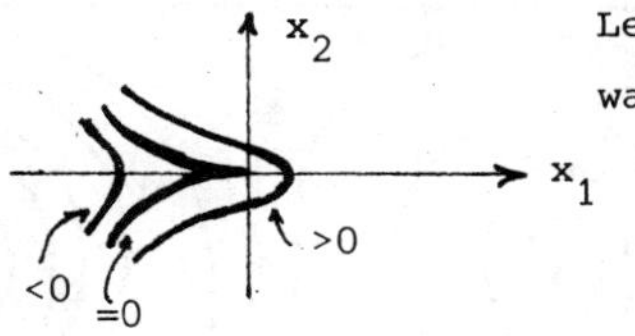

Level lines of $\phi(x) = x_1^3+x_2^2$ watch the 0-level!

Fig. 10.2.6

In Fig. 10.2.7.c, at the *, a pair of critical points of quadratic index 0, 1 resp. is born (thus, here we meet a point of $\Sigma_d(F)$). In Fig. 10.2.7.f a pair of critical points of quadratic index 1, 2 resp. is born. In passing Fig. 10.2.7.h, the "topological role" of the critical points of quadratic index 1 is interchanged. Finally, in Fig. 10.2.7.k, a pair of critical points of quadratic index 0, 1 resp. is cancelled.

Step 5. Now we come to the actual proof of our theorem. The basic work is already done in the foregoing steps.

Firstly we note that the proof in case $n = 1$ consists of an obvious combination of Local Model I ($k = 1$), IIa, IIb and therefore it will be deleted. So, now we assume $n \geq 2$.

Let $\alpha_0,\alpha_1,\ldots,\alpha_k$ be an index sequence.

Case I: k is even.

Case I.a: $k = 0$. If $\alpha_0 = 0$, put $F_1(x,t) = \|x\|^2$ and we are done.

If $\alpha_0 \neq 0$, let $\phi \in C^\infty(\mathbb{R}^n,\mathbb{R})$, ϕ having exactly 3 critical points, all of them nondegenerate with QI 0, α_0,α_0-1 resp. and $\phi(x) = \|x\|^2$ outside $B(0,1)$. The construction of ϕ can be done by means of Local Model I.

Now, put $F_1(x,t) = \phi(x)$ and we are done again.

Case I.b: $k \neq 0$. Start with a function $\phi \in C^\infty(\mathbb{R}^n,\mathbb{R})$, $\phi(x) = \|x\|^2$ outside $B(0,1)$, ϕ nondegenerate, having at least one critical point, say $\bar{x}$, of quadratic index α_0. The construction of ϕ can be done again by means of Local Model I. Put $F^{(1)}(x,t) = \phi(x)$ for $t \leq 0$.

There exists an open neighborhood U of $\bar{x}$, $U \subset \mathring{B}(0,1)$ such that $\phi_{|U}$ has the following form in suitable local coordinates (cf. Theorem 2.7.2):

$$\phi(y) = \phi(\bar{x}) - \sum_{i=1}^{\alpha_0} y_i^2 + \sum_{j=\alpha_0+1}^{n} y_j^2. \qquad (10.2.11)$$

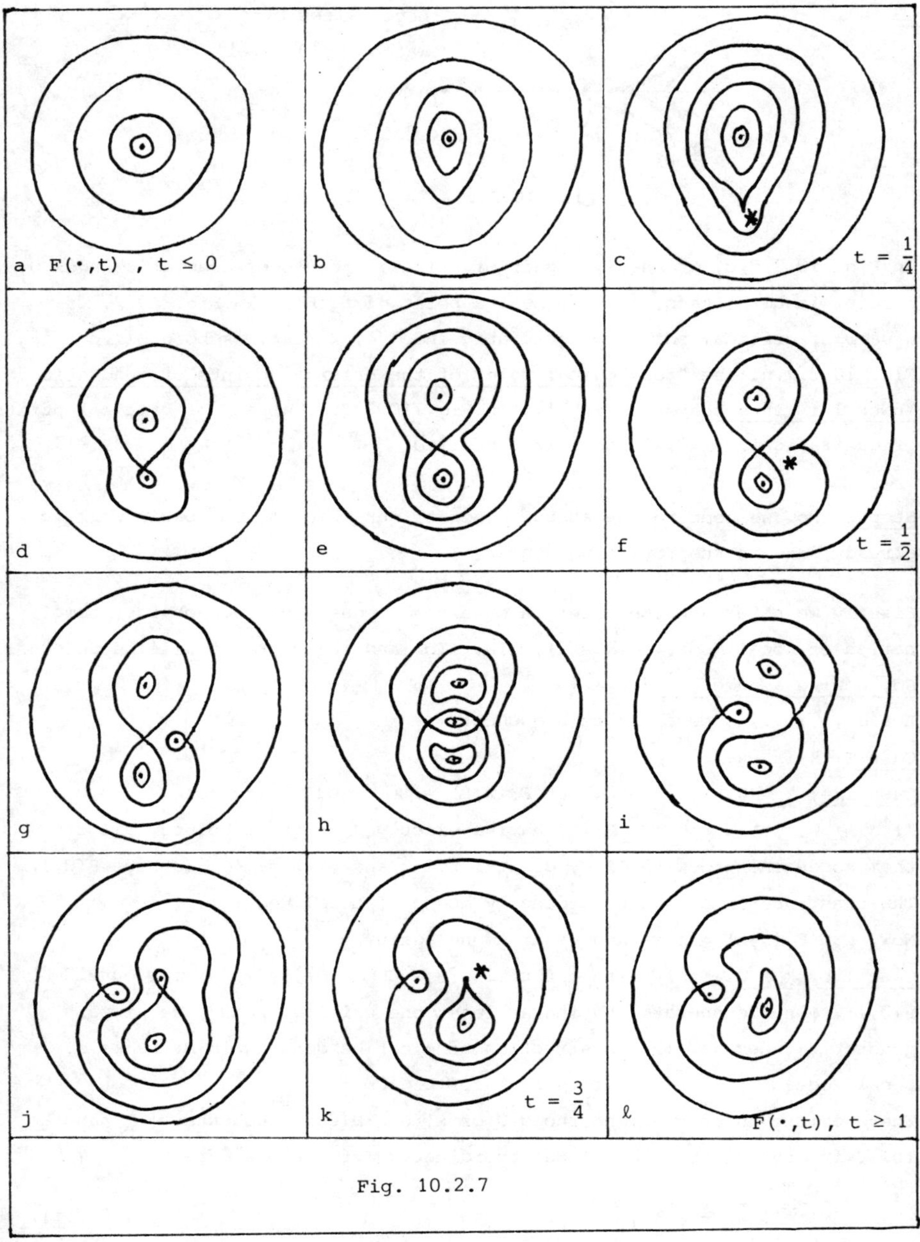

Fig. 10.2.7

On $(\mathbb{R}^n \setminus U) \times [0,1]$ we define $F^{(2)}(x,t) = \phi(x)$.
On $U \times [0,1]$ we define $F^{(2)}$ according to the following 4 possible subcases:

Subcase	1	2	3	4
$\alpha_1 =$	α_0+1	α_0-1	α_0+1	α_0-1
$\alpha_2 =$	α_0	α_0	α_0+2	α_0-2

Subcase 1. $\alpha_1 = \alpha_0+1$, $\alpha_2 = \alpha_0$. Note that $\alpha_0 < n$ in this subcase. We "plug in" Local Model II a on the y_n-coordinate (cf. (10.2.11)). Thus, $F^{(2)}$ -in the local coordinates $(y_1,\ldots,y_n,t)$- will have the form:

$$F^{(2)}(y_1,\ldots,y_n,t) = \phi(\bar{x}) - \sum_{i=1}^{\alpha_0} y_i^2 + \sum_{j=\alpha_0+1}^{n-1} y_j^2 + F(y_n,t),$$

$F(y_n,t)$ according to Local Model II a.

Subcase 2. $\alpha_1 = \alpha_0-1$, $\alpha_2 = \alpha_0$. Note that $\alpha_0 > 0$ in this subcase. Now we "plug in" Local IIb on the y_1-coordinate.

Subcase 3. $\alpha_1 = \alpha_0+1$, $\alpha_2 = \alpha_0+2$. Note that $\alpha_0 < n-1$ in this subcase. Insert Local Model III a on the coordinates y_{n-1}, y_n.
Thus, $F^{(2)}$ -in the local coordinates $(y_1,\ldots,y_n,t)$- will have the form:

$$F^{(2)}(y_1,\ldots,y_n,t) = \phi(\bar{x}) - \sum_{i=1}^{\alpha_0} y_i^2 + \sum_{j=\alpha_0+1}^{n-2} y_j^2 + F(y_{n-1},y_n,t),$$

$F(y_{n-1},y_n,t)$ according to Local Model III a.

Subcase 4. $\alpha_1 = \alpha_0-1$, $\alpha_2 = \alpha_0-2$. Note that $\alpha_0 > 1$ in this subcase. Insert Local Model III b on the coordinates y_1, y_2.

Taking $F^{(1)}$, $F^{(2)}$ together, we have constructed $F_1(x,t)$ for $t \le 1$. If $k = 2$, we put $F_1(\cdot,t) = F_1(\cdot,1)$ for $t \ge 1$ and we are done. If $k > 2$, we proceed in an obvious way for $t > 1$, replacing the function ϕ in the above considerations by $F^{(2)}(\cdot,1)$ and taking the remark in Step 1 into account. In this way, we obtain a finite number of "nested tubes" in which the "action" takes place. This proves Case I.

Case II: k is odd.

Case II.a: k = 1. This case is essentially treated by inserting Local Model I.

Case II.b: k > 1. Put $F^{(1)}(x,t) = \|x\|^2$ for $t \le 0$.
Take a point $\bar{x} \in \mathring{B}(0,1)\backslash\{0\}$ and an open neighborhood U of $\bar{x}$ such that $U \subset \mathring{B}(0,1)\backslash\{0\}$. Outside $U \times [0,1]$ we put $F^{(2)}(x,t) = \|x\|^2$. In $U \times [0,1]$ we plug in Local Model I, thus creating a new pair of critical points of quadratic index α_0, α_1 respectively. In this way, taking $F^{(1)}$, $F^{(2)}$ together we obtain $F_2(x,t)$ for $t \le 1$. Let $\bar{\bar{x}} \in U$ be the critical point of $F_2(\cdot,1)$ with quadratic index α_1. Note that the reduced sequence $\alpha_1,\ldots,\alpha_k$ has an even length. We proceed in a suitable neighborhood of $\bar{\bar{x}}$ with the reasoning in Case I in order to define $F_2(\cdot,t)$ for $t > 1$, taking again the remark in Step 1 into account. This proves Case II and completes the proof of our theorem. □

Remark 10.2.1. The proof of Theorem 10.2.3 might be changed in such a way that -for example- $\Sigma(F_1)$ resp. $\Sigma(F_2)$ has a minimal number of connected components. We will not dwell on these possible refinements at this place, but restrict ourselves to an illustrative example. Let n = 2 and consider the index sequence 0,1,2,1,0. So, k = 4. The construction in the proof of Theorem 10.2.3 (Step 5, Case I.b) provides an F_1 such that $\Sigma(F_1)$ has the form as in Fig. 10.2.8.a (two nested "tubes"). However, an $\tilde{F}_1 \in \mathcal{F}(2)$ exists such that $\Sigma(\tilde{F}_1)$ for which we are looking, consists of exactly one component (Fig. 10.2.8.b). For the construction of $\tilde{F}_1$, extend Fig. 10.2.7 by cancelling in Fig. 10.2.7.ℓ the local maximum against the saddlepoint.

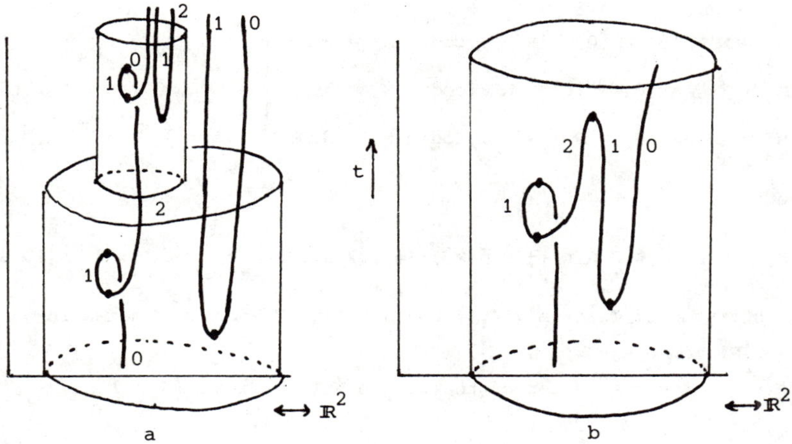

Fig. 10.2.8

10.3. One-parameter families of functions ($\mathbb{H}^n$).

In the foregoing section we discussed one-parameter families of functions on $\mathbb{R}^n$. Now, we focus our attention on the nonnegative orthant $\mathbb{H}^n$ of $\mathbb{R}^n$. As it is explained in Section 10.1, this set will locally be representative for the constrained case (with inequalities) as long as the regularity of the feasible set M(t) is satisfied. Let $F \in C^\infty(\mathbb{R}^n \times \mathbb{R}, \mathbb{R})$, $(x,t) \mapsto F(x,t)$, $x \in \mathbb{R}^n$ and $t \in \mathbb{R}$. The critical pointset $\Sigma^*(F)$ is defined in an analogous way as $\Sigma(F)$ is defined:

$$\Sigma^*(F) = \{(x,t) \in \mathbb{H}^n \times \mathbb{R} | x \text{ is a critical point for } F(\cdot,t)_{|\mathbb{H}^n}\}. \tag{10.3.1}$$

Firstly, we discuss the 1-dimensional case (n = 1) in detail. The general case (n > 1) will be treated by means of a "splitting-argument".

The case n = 1. Let $F \in C^\infty(\mathbb{R} \times \mathbb{R}, \mathbb{R})$. We write the 2-jet extension j^2F as follows:

$$j^2F(x,t) = [x, D_xF, D_x^2F, \underbrace{t, \text{ etc.}}_{\in \mathbb{R}^5}]. \tag{10.3.2}$$

In the 2-jet space $J(2,1,2)$ we consider the following three manifolds:

$$\left.\begin{aligned} M_1 &= \mathbb{R} \times \{0\} \times \{0\} \times \mathbb{R}^5 \ ; \ \text{codim } M_1 = 2, \\ M_2 &= \{0\} \times \{0\} \times \{0\} \times \mathbb{R}^5 ; \ \text{codim } M_2 = 3, \\ M_3 &= \{0\} \times \{0\} \times \mathbb{R} \times \mathbb{R}^5 \ ; \ \text{codim } M_3 = 2. \end{aligned}\right\} \tag{10.3.3}$$

Define

$$\mathcal{F} = \{F \in C^\infty(\mathbb{R} \times \mathbb{R},\mathbb{R}) \,|\, j^2F \pitchfork M_i, i = 1,2,3\}. \tag{10.3.4}$$

Then, in virtue of Theorem 7.4.1, the set $\mathcal{F}$ is C^k-open and dense in $C^\infty(\mathbb{R} \times \mathbb{R},\mathbb{R})$ for every $k \geq 3$.
In the remaining discussion of the case $n = 1$, we assume $F \in \mathcal{F}$.

Remark 10.3.1. If we compare the transversality condition (10.3.4) with (10.1.14) at points where $x = 0$, then the condition in (10.3.4) is stronger since the set M_2 is taken into account. Moreover, note that j^2F from (10.3.2) is automatically transversal to $\{0\} \times (\mathbb{R}\backslash\{0\}) \times \mathbb{R} \times \mathbb{R}^5$.

The set $\mathbb{H} \times \mathbb{R} := \mathbb{H}^1 \times \mathbb{R}^1$ will be stratified with two obvious strata, namely $\{0\} \times \mathbb{R}$ and $\{x|x > 0\} \times \mathbb{R}$.
The set $\Sigma^*(F) \cap (\{x|x > 0\} \times \mathbb{R})$ is discussed in detail in Section 10.2.
The set $\Sigma^*(F) \cap (\{0\} \times \mathbb{R})$ in a neighborhood of a point $(0,\bar{t}) \notin (j^2F)^{-1}(M_1 \cup M_2 \cup M_3)$ is obviously equal to $\{0\} \times (\bar{t}-\varepsilon,\bar{t}+\varepsilon)$, $\varepsilon > 0$ small enough.
Therefore, we restrict our attention to the case that $(0,\bar{t}) \in (j^2F)^{-1}(M_1 \cup M_2 \cup M_3)$ and discuss the set $\Sigma^*(F)$ in a neighborhood of $(0,\bar{t})$. Without loss of generality we may assume that $\bar{t} = 0$. Since $F \in \mathcal{F}$, we have at $(x,t) = (0,0)$:

$$\frac{\partial}{\partial x} F = 0 \ , \ \frac{\partial^2}{\partial x \partial t} F \neq 0 \ , \ \frac{\partial^2}{\partial x^2} F \neq 0 \ . \tag{10.3.5}$$

(For the first, resp. second inequality above use $j^2F \pitchfork M_3$ resp. $j^2F \pitchfork M_2$).
From (10.3.5) and the Implicit Function Theorem it follows that there exist two C^∞-functions η, ξ, each of one variable and both of them defined in an open neighborhood of 0 and satisfying:

$$D_xF(\eta(t),t) \equiv 0 \ , \ D_xF(x,\xi(x)) \equiv 0. \tag{10.3.6}$$

Consequently -in a neighborhood of (0,0)- the set $\{(x,t) \mid D_xF(x,t) = 0\}$ is a 1-dimensional manifold whose tangentspace at (0,0) does not coincide with the x-axis or the t-axis. Thus, it has locally the form as depicted in Fig. 10.3.1.a,b.

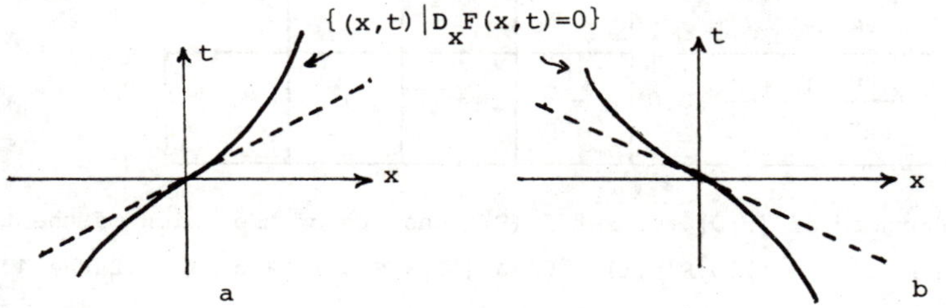

Fig. 10.3.1

From (10.3.6) it follows that -as t increases from negative to positive values- an (unconstrained) critical point enters the set $\mathbb{H}$, resp. leaves $\mathbb{H}$ according to $\frac{d\eta}{dt}(0) > 0$, resp. $\frac{d\eta}{dt}(0) < 0$ (cf. also Fig. 10.3.1.a,b). Note that, by the chain rule, we have $\frac{\partial\eta}{\partial t}(0) = -\left(\frac{\partial^2 F}{\partial x^2}\right)^{-1} \frac{\partial^2}{\partial x \partial t} F\Big|_{(0,0)} \neq 0$ (cf. (10.3.5)).

We put $\mu(t) = D_xF(0,t)$, i.e. $\mu(\tilde{t})$ is equal to the Lagrange parameter at the critical point $x = 0$ for $F(\cdot,\tilde{t})_{|\mathbb{H}}$ corresponding to the active linear constraint "$x \geq 0$". Obviously, $\mu(0) = 0$ and so, $x = 0$ is a <u>degenerated</u> critical point for $F(\cdot,0)_{|\mathbb{H}}$. As t increases from negative to positive values, the Lagrange parameter $\mu(t)$ changes its sign from - to +, resp. from + to - according to $\frac{d\mu}{dt}(0) > 0$, resp. < 0. Note that $\frac{d\mu}{dt}(0) = \frac{\partial^2}{\partial x \partial t} F(0,0) \neq 0$ (cf. (10.3.5)).

Consequently, for the description of $\Sigma^*(F)$ in a neighborhood of (0,0) we have to consider 4 cases:

	1	2	3	4
$\frac{\partial\mu}{\partial t}(0) = \frac{\partial^2 F}{\partial x \partial t}(0,0)$	+	+	−	−
$\frac{\partial^2 F}{\partial x^2}(0,0)$	+	−	+	−
$\frac{\partial\eta}{\partial t}(0) = -\left(\frac{\partial^2 F}{\partial x^2}\right)^{-1} \frac{\partial^2 F}{\partial x \partial t}\Big\vert(0,0)$	−	+	+	−

(10.3.7)

In a neighborhood of (0,0) the set $\Sigma^*(F)$ consists of the union of the t-axis and the set $\{(x,t) \mid D_x F(x,t) = 0\} \cap (\mathbb{H} \times \mathbb{R})$. In Fig. 10.3.2 the 4 cases according to (10.3.7) are sketched.

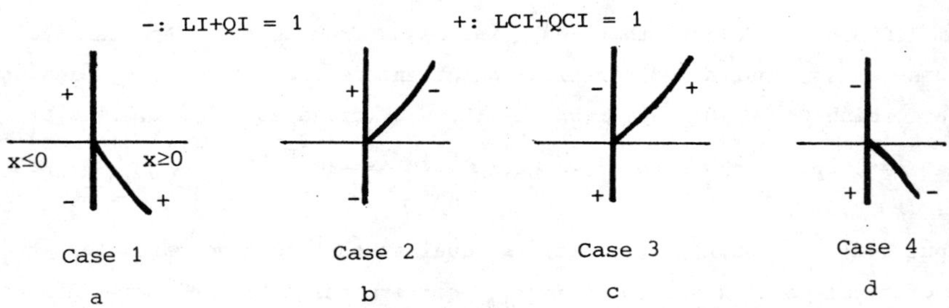

Fig. 10.3.2

Let us discuss Fig. 10.3.2.a (the others can be interpreted similarly). For $|\bar{t}|$ small, $\bar{t} < 0$, we have two nondegenerate critical points for $F(\cdot,\bar{t})_{|\mathbb{H}}$, namely the origin, with indices (LI,LCI,QI,QCI) = (1,0,0,0), and a point in the interior of $\mathbb{H}$, with indices (0,0,0,1). For $\bar{t} > 0$ we have exactly one nondegenerate critical point for $F(\cdot,\bar{t})_{|\mathbb{H}}$, namely the origin, with indices (0,1,0,0). So, walking along the "+" part of $\Sigma^*(F)$ in the increasing direction of t, an interior critical point becomes a critical point on the boundary and the positive square term changes into a positive linear term.

The general case (n > 1). Let us stratify the set $\mathbb{H}^n$ in the following natural way: the 0-dim. stratum will be the origin; the 1-dim. strata are of the type $\{(x_1,\ldots,x_n) \in \mathbb{H}^n | x_i > 0, x_j = 0, j \neq i\}$; the 2-dim. strata have the form $\{(x_1,\ldots,x_n) \in \mathbb{H}^n | x_i > 0, x_j > 0 (i \neq j), x_k = 0, k \neq i,j\}$; etc. (cf. also Definition 3.1.3). The set $\mathbb{H}^n \times \mathbb{R}$ is assumed to have the product stratification $\{W \times \mathbb{R} | W \text{ is a stratum of } \mathbb{H}^n\}$.

Condition B^*. An $F \in C^\infty(\mathbb{R}^n \times \mathbb{R}, \mathbb{R})$ is said to fulfil Condition B^* if the following holds for all $(\bar{x},\bar{t}) \in \Sigma^*(F)$ (where we assume without loss of generality: $\bar{x}_i > 0$, $i = 1,\ldots,k$ and $\bar{x}_j = 0$, $j = k+1,\ldots,n$):

either $\bar{x}$ is a nondegenerate critical point for $F(\cdot,\bar{t})_{|\mathbb{H}^n}$,

or $\bar{x}$ is a degenerate critical point for $F(\cdot,\bar{t})_{|\mathbb{H}^n}$ and exactly one of the following two cases holds:

Case I: $\dfrac{\partial F}{\partial x_j}(\bar{x},\bar{t}) \neq 0$ for all $j = k+1,\ldots,n$; (10.3.8)

put $G(x_1,\ldots,x_k,t) = F(x_1,\ldots,x_k,0,\ldots,0,t)$; (10.3.9)

then, G -viewed locally as an element of $C^\infty(\mathbb{R}^k \times \mathbb{R}, \mathbb{R})$- satisfies condition B in a neighborhood of $(\bar{x}_1,\ldots,\bar{x}_k,\bar{t})$;

Case II: As an abbreviation we put $F_i = \dfrac{\partial}{\partial x_i} F(\bar{x},\bar{t})$, $F_{ij} = \dfrac{\partial^2}{\partial x_i \partial x_j} F(\bar{x},\bar{t})$, $F_{it} = \dfrac{\partial^2}{\partial x_i \partial t} F(\bar{x},\bar{t})$;

a. the matrix $(F_{ij})_{i,j=1,\ldots,k}$ is nonsingular; (10.3.10)

b. for some $\ell \in \{k+1,\ldots,n\}$ we have

$F_\ell = 0$, $F_j \neq 0$, $j = \{k+1,\ldots,n\}\backslash\{\ell\}$; (10.3.11)

c. the matrix $\begin{pmatrix} F_{11} & \cdots & F_{1k} & F_{1t} \\ \vdots & & \vdots & \vdots \\ F_{k1} & \cdots & F_{kk} & F_{kt} \\ F_{\ell 1} & \cdots & F_{\ell k} & F_{\ell t} \end{pmatrix}$ is nonsingular; (10.3.12)

d. the matrix $(F_{ij})_{i,j \in \{1,\ldots,k\}\cup\{\ell\}}$ is nonsingular. (10.3.13)

□

We proceed with an explanation of Condition B^*. Suppose that $F \in C^\infty(\mathbb{R}^n \times \mathbb{R}, \mathbb{R})$ satisfies Condition B^*. Let $(\bar{x},\bar{t}) \in \Sigma^*(F)$ and assume again that $\bar{x}_i > 0$, $i = 1,\dots,k$, $\bar{x}_j = 0$, $j = k+1,\dots,n$. If $\bar{x}$ is a nondegenerate critical point for $F(\cdot,\bar{t})_{|\mathbb{H}^n}$, then there exists an $\mathbb{R}^{n+1}$-neighborhood $\mathcal{O}$ of $(\bar{x},\bar{t})$ such that $\Sigma^*(F) \cap \mathcal{O}$ lies entirely in the $(k+1)$-dim. stratum of $\mathbb{H}^n \times \mathbb{R}$ to which $(\bar{x},\bar{t})$ belongs (cf. Exercise 3.2.7). Furthermore, $\Sigma^*(F) \cap \mathcal{O}$ can be parametrized smoothly by t. Next, suppose that $\bar{x}$ is a degenerate critical point of $F(\cdot,\bar{t})_{|\mathbb{H}^n}$. If we are in Case I, then $\Sigma^*(F)$ lies again <u>locally</u> in the $(k+1)$-dim. stratum of $\mathbb{H}^n \times \mathbb{R}$ to which $(\bar{x},\bar{t})$ belongs and we are in fact in the situation as discussed in Section 10.2.

<u>Now suppose that we are in Case II.</u>

Note that the numbers F_j, $j = k+1,\dots,n$ are, in fact, the Lagrange-parameters at the critical point $\bar{x}$ for $F(\cdot,\bar{t})_{|\mathbb{H}^n}$ corresponding to the (active) linear inequality constraints "$x_j \geq 0$", $j = k+1,\dots,n$. From (10.3.10) and the Implicit Function Theorem we obtain the existence of smooth functions $\eta_1(t),\dots,\eta_k(t)$, defined in an open neighborhood of $\bar{t}$, such that

$$\frac{\partial}{\partial x_i} F(\eta_1(t),\dots,\eta_k(t),0,\dots,0,t) \equiv 0, \; i = 1,\dots,k. \tag{10.3.14}$$

Put $$\mu_j(t) = \frac{\partial}{\partial x_j} F(\eta_1(t),\dots,\eta_k(t),0,\dots,0,t), \; j = k+1,\dots,n. \tag{10.3.15}$$

Then, $\mu_j(t)$ is the Lagrange-parameter at the critical point $(\eta_1(t),\dots,\eta_k(t),0,\dots,0,t)$ for the function $F(\cdot,t)$ corresponding to the (active) linear inequality constraint "$x_j \geq 0$".

From (10.3.11) we see that $\mu_j(t) \neq 0$, $j \in \{k+1,\dots,n\}\setminus\{\ell\}$ for t in a neighborhood of $\bar{t}$.

However, $\mu_\ell(\bar{t}) = 0$. We contend that $\frac{d}{dt}\mu_\ell(\bar{t}) \neq 0$, i.e. at $\bar{t}$, $\mu_\ell(t)$ meets 0 "transversally". (Note that this is an analogous statement as in Theorem 10.2.2 (iii)).

From (10.3.14), (10.3.15) we obtain:

$$\frac{d\mu_\ell}{dt}(\bar{t}) = F_{\ell t} - \begin{pmatrix} F_{\ell 1} \\ \vdots \\ F_{\ell k} \end{pmatrix}^T \cdot \begin{pmatrix} F_{11} & \cdots & F_{1k} \\ \vdots & & \vdots \\ F_{k1} & \cdots & F_{kk} \end{pmatrix}^{-1} \cdot \begin{pmatrix} F_{1t} \\ \vdots \\ F_{kt} \end{pmatrix}. \tag{10.3.16}$$

The righthandside of (10.3.16) does not vanish; this follows from (10.3.10) and (10.3.12) (cf. also Example 7.3.3).
Next, we use a splitting argument in order to reduce the Case II to the special case "n = 1" which already has been treated in detail. From (10.3.10) and the Implicit Function Theorem we obtain the existence of smooth functions $\xi_i(x_\ell,t)$, $i = 1,\dots,k$, defined in an open neighborhood of $(\bar{x}_\ell,\bar{t})$ such that

$$\frac{\partial}{\partial x_i} F(\xi_1(x_\ell,t),\dots,\xi_k(x_\ell,t),0,\dots,0,x_\ell,0,\dots,0,t) \equiv 0,\ i = 1,\dots,k. \tag{10.3.17}$$

Note that $\xi_i(0,t) = \eta_i(t)$, η_i as in (10.3.14).
We proceed by considering the following local coordinate transformation, which -in particular- sends $(\bar{x},\bar{t})$ onto $(0,0)$:

$$y_i = x_i - \xi_i(x_\ell,t), i = 1,\dots,k \ ;\ y_j = x_j,\ j = k+1,\dots,n \ ;\ u = t-\bar{t},$$

$$\text{shortly written } (y,u) = \Phi(x,t). \tag{10.3.18}$$

$$\text{Put } G(y,u) = F\circ\Phi^{-1}(y,u). \tag{10.3.19}$$

Then, from the very construction we have:

$$\frac{\partial}{\partial y_i} G(0,\dots,0,y_\ell,0,\dots,0,u) \equiv 0,\ i = 1,\dots,k, \tag{10.3.20}$$

$$\frac{\partial}{\partial y_j} G(0,\dots,0,0) = F_j(\bar{x},\bar{t}),\ j = k+1,\dots,n. \tag{10.3.21}$$

A short calculation, using (10.3.17), (10.3.18), (10.3.19), shows:

$$\frac{\partial^2}{\partial y_\ell \partial u} G(0,0) = \frac{d}{dt}\mu_\ell(\bar{t}), \tag{10.3.22}$$

where $\mu_\ell(t)$ is defined in (10.3.15), $\frac{d}{dt}\mu_\ell(\bar{t})$ being calculated in (10.3.16).
Furthermore,

$$\frac{\partial^2}{\partial y_i \partial y_j} G(0,0) = F_{ij},\ 1 \le i,j \le k, \quad \frac{\partial^2}{\partial y_i \partial y_j} G(0,0) = 0,\ 1 \le i \le k, \tag{10.3.23}$$

$$\frac{\partial^2}{\partial y_\ell \partial y_\ell} G(0,0) = F_{\ell\ell} - \begin{pmatrix} F_{1\ell} \\ \vdots \\ F_{k\ell} \end{pmatrix}^T \cdot \begin{pmatrix} F_{11} & \cdots & F_{1k} \\ \vdots & & \vdots \\ F_{k1} & \cdots & F_{kk} \end{pmatrix}^{-1} \cdot \begin{pmatrix} F_{1\ell} \\ \vdots \\ F_{k\ell} \end{pmatrix}. \quad (10.3.24)$$

Note that the righthand side of (10.3.24) does not vanish in view of (10.3.13).

Put $\alpha = \mathrm{Index}(F_{ij})_{i,j=\{1,\dots,k\}\cup\{\ell\}} - \mathrm{Index}(F_{ij})_{i,j\in\{1,\dots,k\}}$. (10.3.25)

From Example 2.5.1 it follows, using the fact that $F_j = 0$, $j \in \{1,\dots,k\} \cup \{\ell\}$ and the Formulas (10.3.23), (10.3.24):

$\alpha = 0$ (resp. 1) iff $\frac{\partial^2 G}{\partial y_\ell \partial y_\ell}(0,0) > 0$ (resp. < 0).

We proceed with a splitting construction w.r.t. the function G and write:

$$G(y,u) = \underbrace{G(0,\dots,0,y_\ell,0,\dots,0,u)}_{\Psi(y_\ell,u)} + \underbrace{G(y,u)-G(0,\dots,0,y_\ell,0,\dots,0,u)}_{H(y,u)} \quad (10.3.26)$$

Since $H(0,\dots,0,y_\ell,0,\dots,0,u) \equiv 0$, we may write

$$H(y,u) = \int_0^1 \frac{d}{d\tau} H(\tau y_1,\dots,\tau y_{\ell-1},y_\ell,\tau y_{\ell+1},\dots,\tau y_n,u)d\tau = \sum_{\substack{i=1 \\ i\neq\ell}}^{n} y_i a_i(y,u). \quad (10.3.27)$$

From (10.3.20) it follows that $a_i(0,\dots,0,y_\ell,0,\dots,0,u) \equiv 0$, $i = 1,\dots,k$. Consequently, we may repeat the above "integral trick" for a_i, $i = 1,\dots,k$, and obtain, after collecting terms suitably:

$$H(y,u) = \sum_{i,j=1}^{k} y_i y_j b_{ij}(y,u) + \sum_{\substack{j=k+1 \\ j\neq\ell}}^{n} y_j b_j(y,u), \quad (10.3.28)$$

where we may assume without loss of generality that $b_{ij} = b_{ji}$.
From (10.3.11), (10.3.21), we have $b_j(0,0) \neq 0$ and from (10.3.23) it follows that the matrix $(b_{ij}(0,0))_{i,j=1,\dots,k}$ is nonsingular. Now we apply the Morse Lemma (Theorem 2.7.2) in a parametric form on the first k coordinates $y_1,\dots,y_k$, thereby treating $(y_{k+1},\dots,y_n,u)$ as the additional parameters (compare the Splitting Theorem, Theorem 6.2.1).
Let us denote the obtained new local coordinates again by (y,u) and write $\tilde{H}$, $\tilde{b}_j$ for the functions H, b_j in these new coordinates, then:

$$\tilde{H}(y,u) = \sum_{i=1}^{k} \pm y_i^2 + \sum_{\substack{j=k+1\\ j\neq \ell}}^{n} y_j \tilde{b}_j(y,u). \tag{10.3.29}$$

Finally we put $z_i = y_i$, $i = 1,\dots,k$, $z_\ell = y_\ell$, $z_j = y_j|\tilde{b}_j(y,u)|$, $j = k+1,\dots,n$, $j \neq \ell$, and $v = u$ which gives us a new local coordinate transformation. Let us denote by $\tilde{G}$ our original function G in these coordinates (z,v), then we have:

$$\tilde{G}(z,v) = \sum_{i=1}^{k} \pm z_i^2 + \sum_{\substack{j=k+1\\ j\neq \ell}}^{n} \pm z_j + \Psi(z_\ell,v). \tag{10.3.30}$$

Formula (10.3.30) actually serves as the required "splitting".
From (10.3.21), (10.3.22), (10.3.24) we obtain:

$$\left.\begin{aligned} &\frac{\partial}{\partial z_\ell}\Psi(0,0) = 0, \quad \frac{\partial^2}{\partial z_\ell \partial v}\Psi(0,0) = \frac{\partial^2}{\partial y_\ell \partial u} G(0,0) \neq 0, \\ &\frac{\partial^2}{\partial z_\ell \partial z_\ell}\Psi(0,0) = \frac{\partial^2}{\partial y_\ell \partial y_\ell} G(0,0) \neq 0. \end{aligned}\right\} \tag{10.3.31}$$

So, our original set $\Sigma^*(F)$ -in a neighborhood of $(\bar{x},\bar{t})$- transforms locally into a set of the form $\{\underbrace{(0,\dots,0)}_{n-1}\} \times \Sigma^*(\Psi)$, where $\Sigma^*(\Psi)$ is one of the types in Fig. 10.3.2, according to the Formula (10.3.7), by substitution of Ψ, z_ℓ, v for F, x, t in (10.3.7).
This completes our discussion about Condition B^*.

The Condition B^* can be replaced by an equivalent transversality condition on the 2-jet-extension of F; however, we will omit the details and merely state the resulting theorem. To this aim let the subset F^* of $C^\infty(\mathbb{R}^n \times \mathbb{R},\mathbb{R})$ be defined as follows:

$$F^* = \{F \in C^\infty(\mathbb{R}^n \times \mathbb{R},\mathbb{R}) \mid F \text{ satisfies Condition } B^*\}.$$

<u>Theorem 10.3.1.</u> The set F^* is C^k-dense in $C^\infty(\mathbb{R}^n \times \mathbb{R},\mathbb{R})$ for all k; moreover, F^* is C^k-open for all $k \geq 3$. □

<u>Remark 10.3.2.</u> For $F \in C^\infty(\mathbb{R}^n \times \mathbb{R},\mathbb{R})$ let $\Sigma_+(F)$ denote the (+)-Kuhn-Tucker set with respect to the constant feasible set $\mathbb{H}^n$ (compare Formula (10.1.12) with $p = n$, $q = 0$).

If $F \in F^*$, then $\Sigma_+(F)$ is a piecewise smooth one-dimensional manifold. To see this, note that $\Sigma_+(F)$ consists of those points of $\Sigma^*(F)$ (cf. (10.3.1)) at which the partial derivatives of F w.r.t. the active inequalities "$x_i \geq 0$, $i = 1,\ldots,n$" are nonnegative; a moment of reflection shows that nonsmoothness of $\Sigma_+(F)$ can only occur at those points corresponding to Case II in Condition B^*; in a neighborhood of those points the set $\Sigma_+(F)$ has the form as depicted in Fig. 10.3.3, where Fig. 10.3.3 corresponds to Fig. 10.3.2.

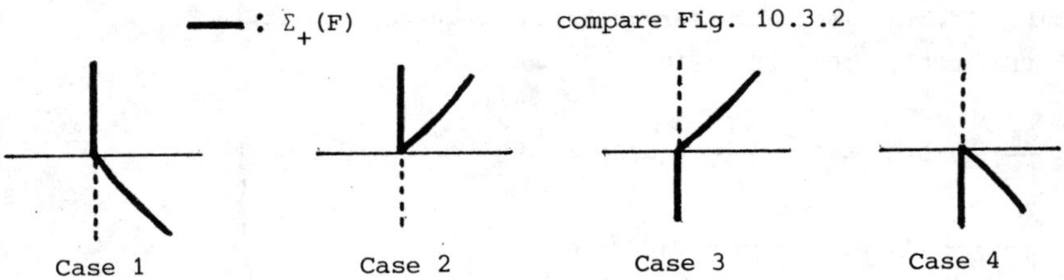

Fig. 10.3.3

10.4. One-parameter families of constraint-sets.

A compact and intrinsic study of one parameter families of sets defined by (in)equality constraints, is presented in [40]. In this section we will further clarify several aspects on this subject. For missing details we refer to [40].

Let $h_i, g_j \in C^\infty(\mathbb{R}^n \times \mathbb{R}, \mathbb{R})$, $i \in I$ and $j \in J$. The index sets I, J are fixed and finite sets, where $I = \{1,\ldots,m\}$ with $m < n$. The variables in $\mathbb{R}^n \times \mathbb{R}$ will be denoted by (x,t) as usual, where $t \in \mathbb{R}$ stands for the parameter. For each value of the parameter t, the functions h_i, g_j define the feasible set $M(t)$,

$$M(t) = \{x \in \mathbb{R}^n \mid h_i(x,t) = 0, i \in I, g_j(x,t) \geq 0, j \in J\}. \qquad (10.4.1)$$

We may identify the set $M(t) \subset \mathbb{R}^n$ with the set $M(t) \times \{t\} \subset \mathbb{R}^n \times \mathbb{R}$. From this point of view the set $M(t)$ becomes a "t-section" of the unfolded set M:

$$M = \{(x,t) \in \mathbb{R}^n \times \mathbb{R} \mid h_i(x,t) = 0, i \in I, g_j(x,t) \geq 0, j \in J\}. \tag{10.4.2}$$

In other words, $M(t)$ becomes a t-level of the function $\Pi_{|M}$, where Π is the projection on the parameter space (see Fig. 10.4.1),

$$\Pi(x,t) = t \tag{10.4.3}$$

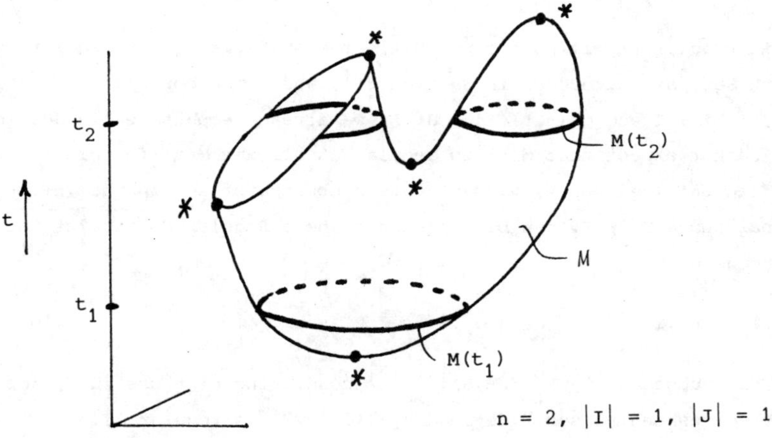

Fig. 10.4.1

In accordance with the latter viewpoint, local coordinate transformations $(y,u) = \Phi(x,t)$ around $(\bar{x},\bar{t})$ are assumed to have the following form:

$$\Phi(x,t) = (\phi(x,t), \psi(t)) \quad , \quad \psi'(\bar{t}) > 0. \tag{10.4.4}$$

From (10.4.4) we learn two facts:

(i) t-hyperplanes are mapped to u-hyperplanes,

(ii) the orientation of t (increasing/decreasing) is preserved.

Coordinate transformations of the type (10.4.4) will be called <u>canonical</u>. In fact, a typical example is already given in (10.1.9) where it has to be assumed that $t \in \mathbb{R}$. Note that a canonical coordinate transformation makes the following diagram (locally) commutative:

$$\begin{array}{ccc} \mathbb{R}^n \times \mathbb{R} & \xrightarrow{\Phi} & \mathbb{R}^n \times \mathbb{R} \\ \Pi \downarrow & & \downarrow \Pi \\ \mathbb{R} & \xrightarrow{\psi} & \mathbb{R} \end{array} \qquad \Pi \circ \Phi = \psi \circ \Pi \ .$$

Recall (Section 10.1) that $M(\bar{t})$ is called <u>regular</u> at an $\bar{x} \in M(\bar{t})$ if the following set of vectors is linearly independent:

$$\{D_x h_i, i \in I, D_x g_j, j \in J_0(\bar{x},\bar{t})\}_{|(\bar{x},\bar{t})} \ . \tag{10.4.5}$$

If $M(\bar{t})$ is regular at all points $\bar{x} \in M(\bar{t})$, then $M(\bar{t})$ is also a Regular Constraint Set, as introduced in Section 3.1, where the constraint functions are $h_i(\cdot,\bar{t})$, $i \in I$ and $g_j(\cdot,\bar{t})$, $j \in J$. As we already emphasized in Section 10.1, we cannot expect that $M(t)$ is regular at all $x \in M(t)$ for all $t \in \mathbb{R}$. However, the unfolded set $\mathcal{M}$, viewed at as a constraint set in the $(n+1)$-dimensional space $\mathbb{R}^n \times \mathbb{R}$, can be expected to be a Regular Constraint Set. In fact, denote

$$(H,G) = (h_i, i \in I, g_j, j \in J). \tag{10.4.6}$$

Let $\mathcal{R}$ be the subset of $C^\infty(\mathbb{R}^n \times \mathbb{R},\mathbb{R})^{|I|+|J|}$ consisting of those (H,G) for which the corresponding set $\mathcal{M}$, defined by (10.4.2), is regular; i.e. $(H,G) \in \mathcal{R}$ iff for all $(\bar{x},\bar{t}) \in \mathcal{M}$ the following set of vectors <u>(in $\mathbb{R}^n \times \mathbb{R}$)</u> is linearly independent:

$$\{Dh_i, i \in I, Dg_j, j \in J_0(\bar{x},\bar{t})\}_{|(\bar{x},\bar{t})} .$$

The next theorem is essentially contained in Theorem 7.1.5, Example 6.1.2 and Example 6.1.3.

<u>Theorem 10.4.1</u>. The set $\mathcal{R}$ is C^1-open and dense in $C^\infty(\mathbb{R}^n \times \mathbb{R},\mathbb{R})^{|I|+|J|}$. □

<u>Corollary 10.4.1</u>. Let $(H,G) \in \mathcal{R}$. Then, for each $(\bar{x},\bar{t}) \in \mathcal{M}$ the set of active constraints does not exceed $n+1$, i.e. $|I|+|J_0(\bar{x},\bar{t})| \leq n+1$. □

<u>Corollary 10.4.2</u>. Let $(H,G) \in \mathcal{R}$ and $(\bar{x},\bar{t}) \in \mathcal{M}$. Then:

$$\mathrm{rank}\{D_x h_i, i \in I, D_x g_j, j \in J_0(\bar{x},\bar{t})\}_{|(\bar{x},\bar{t})} \geq |I|+|J_0(\bar{x},\bar{t})| - 1. \qquad \square$$

<u>Until the part of this section where the discussion on the Mangasarian-Fromovitz constraint qualifiaction begins, we will assume that (H,G) belongs to $\mathcal{R}$.</u>

The following lemma relates the non-regularity of $M(\bar{t})$ with the appearance of critical points of $\Pi_{|M}$ (compare those points in Fig. 10.4.1 marked with a *).

<u>Lemma 10.4.1</u>. Let $\bar{x}$ belong to $M(\bar{t})$. Then we have:

$M(\bar{t})$ fails to be regular at $\bar{x}$ iff $(\bar{x},\bar{t})$ is a critical point for $\Pi_{|M}$.

<u>Proof</u>. Let $\bar{x}$ belong to $M(\bar{t})$ and suppose that $M(\bar{t})$ fails to be regular at $\bar{x}$. Then, there exist real numbers λ_i, $i \in I$, μ_j, $j \in J_0(\bar{x},\bar{t})$, not all vanishing, such that $\sum_{i\in I} \lambda_i D_x h_i + \sum_{j\in J_0(\bar{x},\bar{t})} \mu_j D_x g_j{}_{|(\bar{x},\bar{t})} = 0$.
Since $(H,G) \in \mathcal{R}$ it follows that $\sum_{i\in I} \lambda_i D_t h_i + \sum_{j\in J_0(\bar{x},\bar{t})} \mu_j D_t g_j{}_{|(\bar{x},\bar{t})} = \gamma \neq 0$.
Put $\bar{\lambda}_i = \lambda_i/\gamma$ and $\bar{\mu}_j = \mu_j/\gamma$. It then follows:

$$\begin{pmatrix} 0 \\ \vdots \\ 0 \\ 1 \end{pmatrix} = \sum_{i\in I} \bar{\lambda}_i \begin{pmatrix} D_x^T h_i \\ \\ D_t h_i \end{pmatrix} + \sum_{j\in J_0(\bar{x},\bar{t})} \bar{\mu}_j \begin{pmatrix} D_x^T g_j \\ \\ D_t g_j \end{pmatrix}_{|(\bar{x},\bar{t})} \tag{10.4.7}$$

But, (10.4.7) implies that $(\bar{x},\bar{t})$ is a critical point for $\Pi_{|M}$. On the other hand, if $(\bar{x},\bar{t})$ is a critical point for $\Pi_{|M}$, then (10.4.7) holds with unique Lagrange parameters $\bar{\lambda}_i$, $\bar{\mu}_j$; the latter numbers do not vanish simultaneously in view of the last row in (10.4.7). But then, it follows from the first n rows in (10.4.7) that the set $\{D_x h_i, i \in I, D_x g_j, j \in J_0(\bar{x},\bar{t})\}_{|(\bar{x},\bar{t})}$ is linearly independent, and hence, the set $M(\bar{t})$ is not regular at $\bar{x}$. □

The idea of the proof of the next theorem is essentially contained in the proof of Theorem 6.3.1, Step 3.

<u>Theorem 10.4.2</u>. (<u>Diffeomorphy-Theorem</u>).
Let t_1, t_2 be parameters with $t_1 < t_2$. Suppose that the set $M \cap (\mathbb{R}^n \times [t_1,t_2])$ is compact and that $M(t)$ is regular for all $x \in M(t)$ and $t \in [t_1,t_2]$.
Then, the set $M(t)$ is C^∞-<u>diffeomorphic</u> with $M(t_1)$ for all $t \in [t_1,t_2]$. □

The compactedness assumption in Theorem 10.4.2 cannot just be omitted as is shown in Fig. 10.4.2. In fact, if $M \cap (\mathbb{R}^n \times [t_1,t_2])$ is unbounded, then an appropriate "condition at infinity" has to be assumed (compare also the Palais-Smale "Condition C").

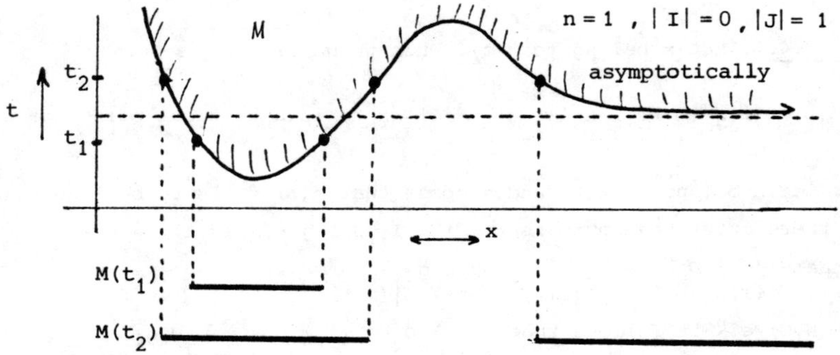

Fig.10.4.2

From Theorem 10.4.2 we learn that the structure of the feasible set M(t) can only change at points where M(t) fails to be regular. In view of Lemma 10.4.1 this gives rise to the introduction of the following set RR (compare also Lemma 3.2.6):

$$RR = \{(H,G) \in R \mid \text{all critical points of } \Pi_{|M} \text{ are nondegenerate}\}. \quad (10.4.8)$$

Theorem 10.4.3. The set RR is C^2-open and dense in $C^\infty(\mathbb{R}^n \times \mathbb{R},\mathbb{R})^{|I|+|J|}$.

Sketch of the proof. We restrict ourselves to the case with equality-constraints only (i.e. $J = \emptyset$).

Subcase 1: $|I| = 1$, $|J| = 0$.
In this subcase we are dealing with one equality constraint, say h(x,t). We firstly contend:

$$\circledast \qquad (h) \in RR \text{ iff at every } (x,t) \text{ with } \begin{cases} h = 0 \\ D_x h = 0 \end{cases} \text{ we have } \begin{cases} D_t h \neq 0 \\ D_x^2 h \text{ nonsingular.} \end{cases}$$

To see the validity of ⊛, suppose that (h) $\in$ RR. Let $(\bar{x},\bar{t})$ be a point at which both h and $D_x h$ vanish. Since $RR \subset R$ we see that $D_t h \neq 0$ since $Dh(\bar{x},\bar{t}) \neq 0$ and $D_x h(\bar{x},\bar{t}) = 0$. Next, note that $(\bar{x},\bar{t})$ is a critical point for $\Pi_{|M}$. In fact, there exists a real number $\bar{\lambda}$ with $(0,\ldots,0,1) = \bar{\lambda} Dh(\bar{x},\bar{t})$. Hence, $\bar{\lambda} = D_t h(\bar{x},\bar{t})^{-1}$. The corresponding Lagrange function (cf. Lemma 3.2.6) will be the function $\Pi-\bar{\lambda}h$ $(= t-\bar{\lambda}h(x,t))$, and the tangentspace $T_{(\bar{x},\bar{t})}M$ equals $\mathbb{R}^n \times \{0\}$. It follows that the critical point $(\bar{x},\bar{t})$ for $\Pi_{|M}$ is nondegenerate iff $D^2(\Pi-\bar{\lambda}h)_{|T_{(\bar{x},\bar{t})}M}$ is nondegenerate (cf. Lemma 3.2.6), which is equivalent with: $\bar{\lambda} D_x^2 h(\bar{x},\bar{t})$ is nonsingular.
Hence, $D_x^2 h(\bar{x},\bar{t})$ is nonsingular, since $\bar{\lambda}$ in this particular case happens to be nonzero. This proves one direction of ⊛. The converse direction is now easily shown, and will be omitted.

Next, we introduce the following reduced 1-jet extension $\tilde{j}^1 h$ of h:

$$\tilde{j}^1 h: (x,t) \mapsto (h, D_x h). \tag{10.4.9}$$

From Theorem 7.4.1 and Remark 7.4.3 it follows that the set of those h for which $\tilde{j}^1 h \pitchfork (\{0_1\} \times \{0_n\})$ is C^2-open and dense in $C^\infty(\mathbb{R}^n \times \mathbb{R},\mathbb{R})$. In order to finish Subcase 1, it suffices to show that
$RR = \{h \in C^\infty(\mathbb{R}^n \times \mathbb{R},\mathbb{R}) \mid \tilde{j}^1 h \pitchfork (\{0_1\} \times \{0_n\})$. To see the latter, note that $\tilde{j}^1 h \pitchfork (\{0_1\} \times \{0_n\})$ iff at each $(\bar{x},\bar{t})$ with vanishing h and $D_x h$, the following matrix is nonsingular:

$$\begin{pmatrix} D_x^T h & D_x^2 h \\ D_t h & D_t D_x h \end{pmatrix}_{|(\bar{x},\bar{t})} . \tag{10.4.10}$$

Since $D_x h(\bar{x},\bar{t}) = 0$, the determinant of the matrix in (10.4.10) equals $\pm D_t h \cdot \det(D_x^2 h)_{|(\bar{x},\bar{t})}$. This establishes Subcase 1 in view of the equivalence ⊛.

<u>Subcase 2</u>. $I = \{1,\ldots,m\}$, $m > 1$ and $J = \emptyset$.
Put $H = (h_1,\ldots,h_m)$. We start with $H \in R$. Since R is C^1-open (Theorem 10.4.1) we can perturb H slightly without leaving the set R, and it suffices to perturb H step by step in local coordinates. From Corollary 10.4.2 we know that $\text{rank}\{D_x h_i, i = 1,\ldots,m\} \geq m-1$ at all points $(\bar{x},\bar{t})$ with $\bar{x} \in M(\bar{t})$.

So, choose a point $\bar{x} \in M(\bar{t})$ and suppose (without loss of generality) that $\text{rank}\{D_x h_1,\dots,D_x h_{m-1}\}|_{(\bar{x},\bar{t})} = m-1$. The set $M(t)$ is the zero set of $h_m(\cdot,t)$ on the common zero set of $h_1(\cdot,t),\dots,h_{m-1}(\cdot,t)$. Now we choose local <u>canonical</u> C^∞-coordinates with respect to the zero set of the functions h_i, $i = 1,\dots,m-1$ (as suggested in (10.1.9)), and in this way we have locally reduced Subcase 2 to Subcase 1. A subsequent approximation of the function h_m in these new local coordinates (as in Subcase 1) clarifies the desired local approximation of $h_1,\dots,h_m$. □

Now we can describe -in the generic and compact case- the change of the structure of $M(t)$ if t passes a parameter value $\bar{t}$ at which $M(\bar{t})$ fails to be regular. The main idea is closely connected with the idea in Morse Theory (see Chapter 1-5): it consists of a <u>local part</u>, namely an analysis of the behaviour of $M(t)$ in a neighborhood of a point where $M(t)$ fails to be regular ("normal forms"); and, on the other hand, it consists of a <u>global part</u>, based on Theorem 10.4.2.

<u>Theorem 10.4.4</u>. Let (H,G) belong to $\mathcal{RR}$, and suppose that $M(\bar{t})$ fails to be regular at $\bar{x} \in M(\bar{t})$.
Then, there exist real numbers $\bar{\lambda}_i$, $i \in I$, $\bar{\mu}_j$, $j \in J_0(\bar{x},\bar{t})$, unique up to a common multiple, such that (cf. also (10.4.7) and Corollary 10.4.2):

$$\left.\begin{aligned} &\sum_{i\in I} \bar{\lambda}_i D_x h_i + \sum_{j\in J_0(\bar{x},\bar{t})} \bar{\mu}_j D_x g_j \Big|_{(\bar{x},\bar{t})} = 0 \\ &\bar{\mu}_j \neq 0 \ , \ j \in J_0(\bar{x},\bar{t}) \end{aligned}\right\} \qquad (10.4.11)$$

Moreover, there exist local <u>canonical</u> C^∞-coordinates such that in the new coordinates the set $M(t)$ takes the form ((0,0) corresponding with $(\bar{x},\bar{t})$):

<u>Type 1</u>: $J_0(\bar{x},\bar{t}) = \emptyset$.

$$t = -\sum_{i=1}^{k} x_i^2 + \sum_{j=k+1}^{n-m+1} x_j^2 \qquad (10.4.12)$$

<u>Type 2</u>: $J_0(\bar{x},\bar{t}) \neq \emptyset$; the numbers $\bar{\mu}_j$ in (10.4.11) do <u>not</u> have the same sign.

$$\left.\begin{aligned} &\pm t \geq -\sum_{i=1}^{k} x_i^2 + \sum_{j=k+1}^{c} x_j^2 - \sum_{\ell_1=c+1}^{c+d} x_{\ell_1} + \sum_{\ell_2=c+d+1}^{e} x_{\ell_2} \\ &\text{where } x_\ell \geq 0,\ \ell = c+1,\dots,e \ \text{ and } \ d \neq 0 \end{aligned}\right\} \qquad (10.4.13)$$

<u>Type 3</u>: $J_0(\bar{x},\bar{t}) \neq \emptyset$; the numbers $\bar{\mu}_j$ in (10.4.11) have the same sign.

$$\left.\begin{aligned} & t \geq -\sum_{i=1}^{k} x_i^2 + \sum_{j=k+1}^{c} x_j^2 + \sum_{\ell=c+1}^{e} x_\ell, \\ & \text{where } x_\ell \geq 0,\ \ell = c+1,\dots,e. \end{aligned}\right\} \tag{10.4.14}$$

<u>Type 4</u>: As Type 3, but t in (10.4.14) replaced by -t. □

For a complete proof of Theorem 10.4.4, and for an intrinsic characterization of the number of positive (negative) squares resp. linear terms in (10.4.12), (10.4.14), we refer to [40]. Here, we clarify the appearance of the inequalities in (10.4.13) and (10.4.14). So, we assume that $J_0(\bar{x},\bar{t}) \neq \emptyset$. The starting point is (10.4.11). From Corollary 10.4.2 and the fact that none of the $\bar{\mu}_j$ vanishes, it follows that for every $\tilde{j} \in J_0(\bar{x},\bar{t})$ the following set is linearly independent:

$$\{D_x h_i, i \in I, D_x g_j, j \in J_0(\bar{x},\bar{t})\setminus\{\tilde{j}\}\}_{|(\bar{x},\bar{t})} \tag{10.4.15}$$

Choose $\tilde{j} \in J_0(\bar{x},\bar{t})$ and consider the set

$$\tilde{M}(t) = \{x \in \mathbb{R}^n \mid h_i(x,t) = 0, i \in I, g_j(x,t) \geq 0, j \in J_0(\bar{x},\bar{t})\setminus\{\tilde{j}\}\}. \tag{10.4.16}$$

From the linear independence in (10.4.15) we see that $\tilde{M}(\bar{t})$ is regular at the point $\bar{x}$. Using suitable coordinates (compare (10.1.9)), the set $\tilde{M}(t)$ transforms locally into the <u>constant set</u>

$$\mathbb{R}^q \times \mathbb{H}^p, \text{ with } q = n-m-|J_0(\bar{x},\bar{t})| + 1 \text{ and } p = |J_0(\bar{x},\bar{t})| - 1.$$

The number of squares in (10.4.13), (10.4.14) is equal to q, whereas the number of linear terms x_ℓ is precisely p.
After this local coordinate transformation we can assume that the set M(t) is described as follows (with $g(0,0) = 0$):

$$\{(x_1,\dots,x_n) \mid g(x,t) \geq 0\ ,\ x_\ell \geq 0 \text{ for } \ell = q+1,\dots,n\}, \tag{10.4.17}$$

where the point <u>$(\bar{x},\bar{t})$</u> is transformed to the <u>origin</u> in $\mathbb{R}^n \times \mathbb{R}$. Note that the conditions for *RR* (cf. (10.4.8)) are invariant under canonical smooth coordinate transformations. It then follows that $\frac{\partial}{\partial t} g(0,0) \neq 0$ and (use

Exercise 3.2.3) that the origin $0 \in \mathbb{R}^n$ is a <u>nondegenerate</u> critical point for $g(\cdot,0)_{|\mathbb{R}^q \times \mathbb{H}^p}$.

In particular we have (exercise):

$$\left.\begin{array}{l}\frac{\partial g}{\partial x_\ell}(0) < 0 \text{ for } \ell = q+1,\ldots,n \text{ iff all } \bar{\mu}_j \text{ in (10.4.11) have} \\ \text{the same sign.}\end{array}\right\} \quad (10.4.18)$$

Since $0 \in \mathbb{R}^n$ is a nondegenerate critical point for $g(\cdot,0)_{|\mathbb{R}^q \times \mathbb{H}^p}$, we obtain (use the Implicit Function Theorem) unique locally defined smooth functions $\tilde{x}_1(t),\ldots,\tilde{x}_q(t)$ such that

$$\frac{\partial}{\partial x_i} g(\tilde{x}_1(t),\ldots,\tilde{x}_q(t),0,\ldots,0,t) \equiv 0 \ , \ i = 1,\ldots,q.$$

In fact, the point $(\tilde{x}_1(t),\ldots,\tilde{x}_q(t),0,\ldots,0)$ is the only critical point for $g(\cdot,t)_{|\mathbb{R}^q \times \mathbb{H}^p}$ in a neighborhood of the origin for $t \approx 0$. Consider the <u>canonical</u> local coordinates defined by:

$$\left\{\begin{array}{l} y_i = x_i - \tilde{x}_i(t) \ , \ i = 1,\ldots,q \\ y_j = x_j \ , \ j = q+1,\ldots,n \\ u = t \end{array}\right. \quad (10.4.19)$$

In the new coordinates (10.4.19) the critical point is shifted to the origin as t varies. So, having performed the coordinate transformation (10.4.19), and denoting g as well as the variables x, t in these coordinates again by g, x, t, we get:

$$\frac{\partial}{\partial x_i} g(0,t) = 0 \ , \ i = 1,\ldots,q. \quad (10.4.20)$$

Since $g(0,0) = 0$, we obtain $g(x,t) = tF_1(x,t) + \sum_{i=1}^{n} x_i F_{1i}(x,t)$ (cf. Theorem 2.7.2, Step 1).

Put $F_1(x,t) = F_1(0,t) + F_2(x,t)$. Then, $F_2(0,t) = 0$ and we obtain $F_2(x,t) = \sum_{i=1}^{n} x_i F_{2i}(x,t)$. Substitution yields:

$$g(x,t) = tF_3(t) + \sum_{i=1}^{n} x_i F_{3i}(x,t).$$

From (10.4.20) we see that $F_{3i}(0,t) = 0$, $i = 1,\ldots,q$. Hence,

$$F_{3i}(x,t) = \sum_{j=1}^{n} x_j F_{4ij}(x,t) \ , \ i = 1,\dots,q.$$

Substitution with a subsequent collection of terms yields:

$$g(x,t) = tF_3(t) + \sum_{i,j=1}^{q} x_i x_j F_{5ij}(x,t) + \sum_{j=q+1}^{n} x_j F_{5j}(x,t). \qquad (10.4.21)$$

Now, with (10.4.21) in mind, the formulas (10.4.13), (10.4.14) become transparent.

Note that we have Type 3 (resp. Type 4) iff $\frac{\partial g}{\partial t}(0) > 0$ (resp. < 0) and $\frac{\partial g}{\partial x_j}(0) < 0$, $j = q+1,\dots,n$ (compare also (10.4.18)).

So far the discussion on the proof of Theorem 10.4.4.

From Theorem 10.4.4 we see that, in a neighborhood of a point where $M(t)$ fails to be regular, the set $M(t)$ behaves as:

- either the level set of a function (without constraints) in a neighborhood of a nondegenerate critical point (Type 1),
- or the lower level set of a function (subject to inequality constraints) in a neighborhood of a nondegenerate critical point which might be a (+)-Kuhn-Tucker point (Type 3, Type 4) or not (Type 2).

So, for a study of the change of the structure of $M(t)$ we can use the ideas as developed in Chapter 3, as far as points of Type 2-4 are concerned (cf. Theorem 3.3.7 and Theorem 3.3.8). For a description of such a change in case of Type 1 we need some preparation. To this aim let D^k, resp. S^k stand for a homeomorphic image of $\{x \in \mathbb{R}^k \mid \|x\| \le 1\}$, resp. $\{y \in \mathbb{R}^{k+1} \mid \|y\| = 1\}$. For convenience, we put $S^{-1} = \emptyset$. Considering the product set $S^k \times D^\ell$ as a topological manifold with boundary ∂, we have $\partial(S^k \times D^\ell) = S^k \times S^{\ell-1}$. In particular, $\partial(S^k \times D^0) = \emptyset$. Let M be a $(k+\ell)$-dimensional topological manifold with boundary ∂M, and let $S^k \times D^\ell$ be embedded in $M\backslash\partial M$. Then, we may delete $S^k \times D^\ell$ from M and put $D^{k+1} \times S^{\ell-1}$ in its place by sending $\partial(D^{k+1})$ resp. $S^{\ell-1}$ homeomorphically onto S^k, resp. $\partial(D^\ell)$. In this way we obtain a manifold $\tilde{M}$. If N, $\tilde{N}$ are manifolds, homeomorphic to M, $\tilde{M}$, we say that $\tilde{N}$ is obtained from N by deleting $S^k \times D^\ell$ and implanting $D^{k+1} \times S^{\ell-1}$. Now, the next theorem becomes transparent, and we will omit its proof.

<u>Theorem 10.4.5</u>. Suppose that $(H,G) \in \mathcal{RR}$. Let $t_1, t_2 \in \mathbb{R}$ be parameters with $t_1 < t_2$, and suppose that $M \cap (\mathbb{R}^n \times [t_1,t_2])$ is compact.
Suppose that $M(t)$ is regular for all $x \in M(t)$ and $t \in [t_1,t_2]$, <u>except</u> for the point $\bar{x} \in M(\bar{t})$, where $t_1 < \bar{t} < t_2$. Then exactly one of the following four alternatives holds:

<u>Type 1</u>. $M(t_2)$ is obtained from $M(t_1)$ by deleting $S^{\beta-1} \times D^{\alpha-\beta}$ and implanting $D^{\beta} \times S^{\alpha-\beta-1}$, where, corresponding to (10.4.12), $\beta = k$ and $\alpha = n-m+1$.

<u>Type 2</u>. $M(t_2)$ is homotopy-equivalent to $M(t_1)$.

<u>Type 3</u>. $M(t_2)$ is homotopy-equivalent to $M(t_1)$ with a k-cell attached.

<u>Type 4</u>. $M(t_1)$ is homotopy-equivalent to $M(t_2)$ with a k-cell attached.

(Types 3 and 4: the number k corresponds to (10.4.14)) □

In Figure 10.4.3 the effect of the appearance of points of Type 1-4 is depicted, where the parameter t increases from the left to the right.

In the situation of Theorem 10.4.5, Type 3 and Type 4, the topological structure of $M(t)$ changes since a cell of specific dimension is attached (or "cut off") as t increases thereby passing the critical value $\bar{t}$ (compare also Chapter 5).
Concerning Type 1 the topological structure before and after passing the critical value $\bar{t}$ may ramain the same in the special case that $\partial(S^{\beta-1} \times D^{\alpha-\beta})$ is "symmetric" in the sense that $\partial(\cdot) = S^{\gamma} \times S^{\gamma}$, i.e. $2\beta = \alpha$. This global phenomenon is depicted in Fig. 10.4.4 (Type 1, $\beta = 1$, $\alpha = 2$).

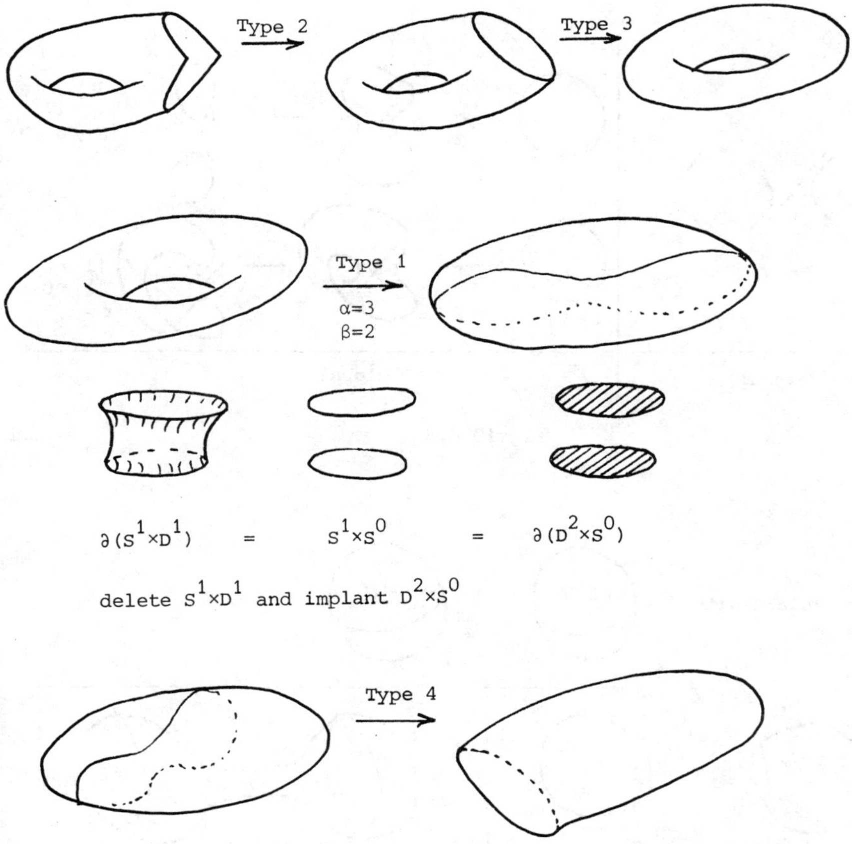

Fig.10.4.3

One may pose the question: when does the number of connected components of $M(t)$ change? For compact sets $M(t)$ a classification of all possible cases is presented in [40]. For an example, concerning Type 1, see Fig. 10.4.5 (compare with Fig. 10.4.3).

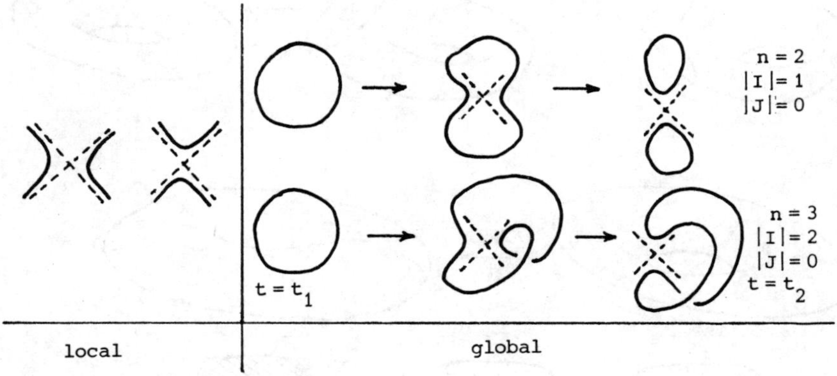

Fig.10.4.4

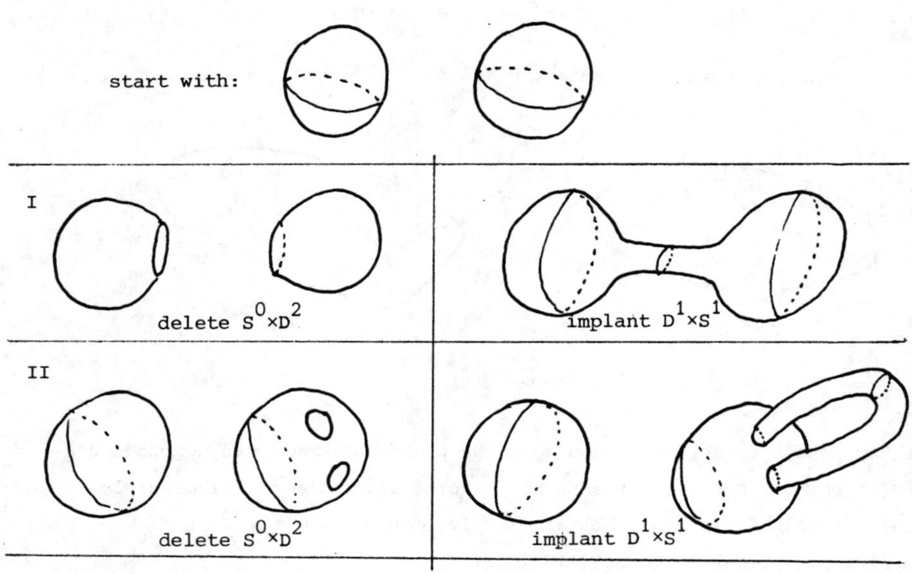

Fig.10.4.5

We conclude this section with a further discussion on the effect of passing points of Type 2. In fact, from Theorem 10.4.5 we know that (in the compact case) the homotopy type of $M(t)$ does not change when passing a critical value of t corresponding to Type 2. On the other hand, in the situation of Fig. 10.4.3, Type 2, even the homeomorphy type does not change (the sets $M(t)$ remain homeomorphic to each other when passing the critical value $\bar{t}$). This is indeed true in the general (compact) case, and it has a deep background: in fact, it relates the socalled Mangasarian-Fromovitz constraint qualification with the homeomorphy-stability of the feasible set ([21]). In order to get an understanding of this, we introduce the Mangasarian-Fromovitz constraint qualification and discuss a "dual" equivalent condition which is obtained by means of a "theorem on alternatives".

Definition 10.4.1. Let $\bar{x}$ belong to $M(t)$.
The Mangasarian-Fromovitz constraint qualification (shortly: MFCQ) is said to hold at $\bar{x}$ if the following two conditions are satisfied:

(i) $D_x h_i(\bar{x},\bar{t})$, $i \in I$, are linearly independent

(ii) There exists a vector $\bar{\xi} \in \mathbb{R}^n$ solving the system

$$\left.\begin{array}{l} D_x h_i(\bar{x},\bar{t})\xi = 0 \,, \quad i \in I, \\ D_x g_j(\bar{x},\bar{t})\xi > 0, \quad j \in J_0(\bar{x},\bar{t}). \end{array}\right\} \qquad (10.4.22)$$

Exercise 10.4.1. If $M(\bar{t})$ is regular at $\bar{x} \in M(\bar{t})$, then MFCQ holds at $\bar{x}$.

The Mangasarian-Fromovitz constraint qualification can be seen as a "positive" linear independence condition (the word positive is related with the active inequality constraints):

Theorem 10.4.6. For $\bar{x} \in M(\bar{t})$ the statements MFCQ 1 and MFCQ 2 are equivalent:

MFCQ 1. MFCQ holds at $\bar{x}$.

MFCQ 2. There do not exist real numbers λ_i, $i \in I$, $\mu_j \geq 0, j \in J_0(\bar{x},\bar{t})$, such that

$$\left.\begin{aligned} &\sum_{i\in I} \lambda_i D_x h_i + \sum_{j\in J_0(\bar{x},\bar{t})} \mu_j D_x g_j \big|_{(\bar{x},\bar{t})} = 0, \\ &\sum_{i\in I} |\lambda_i| + \sum_{j\in J_0(\bar{x},\bar{t})} \mu_j > 0. \end{aligned}\right\} \qquad (10.4.23)$$

□

The proof of Theorem 10.4.6 follows from a "theorem on alternatives". The first such theorem is the famous "Farkas' Lemma".

Lemma 10.4.2. (Farkas' Lemma).
Let $a_1,\ldots,a_s \in \mathbb{R}^n$ and $b \in \mathbb{R}^n$. Then, exactly one of the alternatives I, II holds:

I. $b = \sum_{i=1}^{s} \lambda_i a_i$ with $\lambda_i \geq 0$, $i = 1,\ldots,s$.

II. There exists a vector $\xi \in \mathbb{R}^n$ such that
$\xi^T b > 0$ and $\xi^T a_i \leq 0$, $i = 1,\ldots,s$.

Proof. Obviously, the validity of II contradicts I.
Recall that $\|\cdot\|$ is the Euclidean distance. Let K be a closed, convex nonempty subset of $\mathbb{R}^n$ and choose $\bar{x} \in \mathbb{R}^n$ with $\bar{x} \notin K$. Then, there exists a unique $\bar{y} \in K$ with $\|\bar{x}-\bar{y}\| = \inf_{y\in K} \|\bar{x}-y\|$: In fact, the existence is trivial; if there are two different minimizers $\bar{y}_1,\bar{y}_2 \in K$, then all points of the line-segment $[y_1,y_2] \subset K$ are minimizers and hence, the line-segment $[y_1,y_2]$ is contained in the sphere $\{y \in \mathbb{R}^n | \ \|\bar{x}-y\| = \|\bar{x}-y_1\|\}$, which is a contradiction.

Now, let $a_1,\ldots,a_s \in \mathbb{R}^n$ and put $K = \{\sum_{i=1}^{s} \lambda_i a_i$, all $\lambda_i \geq 0\}$, i.e. K is the nonnegative convex cone generated by the vertices a_i, $i = 1,\ldots,s$. The set K is also closed: each converging sequence $(x^k) \subset K$ has its limit in K since each x^k has some minimal nonnegative representation (by linearly independent elements from $\{a_1,\ldots,a_s\}$), and since the set $\{a_1,\ldots,a_s\}$ is finite.

Obviously, either $b \in K$ (= Alternative I), or $b \notin K$. If $b \notin K$, take the unique point $\bar{y} \in K$ with $\|b-\bar{y}\| = \inf_{y\in K} \|b-y\|$. See Fig. 10.4.6.

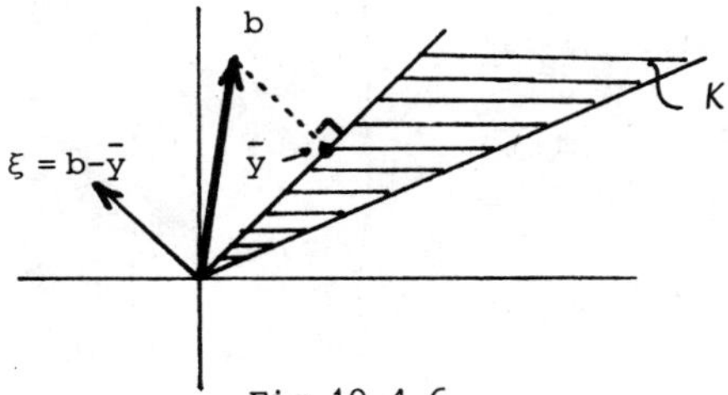

Fig.10.4.6

Put $\xi = b-\bar{y}$ and note that $\xi^T\bar{y} = 0$. A short calculation, using the fact that $\|\cdot\|$ is induced by the standard inner product $\langle x,y\rangle = x^Ty$, shows the validity of Alternative II. □

<u>Theorem 10.4.7</u>. Consider the following system of linear (in)equalities, where the vectors a_i, b_j, c_k belong to $\mathbb{R}^n$:

$$\left.\begin{aligned} &\xi^T a_i < 0 \ , \ i = 1,\ldots,m_a \ , \ m_a \geq 1 \\ &\xi^T b_j \leq 0 \ , \ j = 1,\ldots,m_b \ , \\ &\xi^T c_k = 0 \ , \ k = 1,\ldots,m_c \ . \end{aligned}\right\} \qquad (10.4.24)$$

Then, exactly one of the alternatives I, II holds:

I. The system (10.4.24) is solvable.

II. There exist real numbers $u_i \geq 0$, $i = 1,\ldots,m_a$, $v_j \geq 0$, $j = 1,\ldots,m_b$ and $w_k \in \mathbb{R}$, $k = 1,\ldots,m_c$, such that:

$$\left.\begin{aligned} &\sum_{i=1}^{m_a} u_i a_i + \sum_{j=1}^{m_b} v_j b_j + \sum_{k=1}^{m_c} w_k c_k = 0, \\ &\sum_{i=1}^{m_a} u_i > 0 \ . \end{aligned}\right\} \qquad (10.4.25)$$

<u>Proof</u>. Firstly, we note that the system (10.4.24) (in $\mathbb{R}^n$) is solvable iff the following system (in $\mathbb{R}^{n+1}$) is solvable:

$$\left.\begin{array}{ll} \xi_{n+1} > 0 & \\ \xi^T a_i + \xi_{n+1} \le 0 \ , & i = 1,\dots,m_a \ , \\ \xi^T b_j \le 0 \ , & j = 1,\dots,m_b \ , \\ \left.\begin{array}{l} \xi^T c_k \le 0 \\ \xi^T(-c_k) \le 0 \end{array}\right\} & k = 1,\dots,m_c \ . \end{array}\right\} \tag{10.4.26}$$

Indeed, if (ξ,ξ_{n+1}) solves (10.4.26), then ξ solves (10.4.24). On the other hand, if ξ solves (10.4.24) then, $(\xi,\tilde{\xi}_{n+1})$, with $\tilde{\xi}_{n+1} = -\max_{1\le i\le m_a} \xi^T a_i$, solves (10.4.26).

Now, we apply Lemma 10.4.2 in the space $\mathbb{R}^{n+1}$, with $b = (0,\dots,0,1)^T$, etc. It follows that (10.4.26) is not solvable iff there exist real numbers $u_i \ge 0$, $v_j \ge 0$, $w_k^+ \ge 0$ and $w_k^- \ge 0$ such that:

$$\begin{pmatrix} 0 \\ \vdots \\ 0 \\ \hline 0 \end{pmatrix} = \sum_{i=1}^{m_a} u_i \begin{pmatrix} a_i \\ \hline 1 \end{pmatrix} + \sum_{j=1}^{m_b} v_j \begin{pmatrix} b_j \\ \hline 0 \end{pmatrix} + \sum_{k=1}^{m_c} w_k^+ \begin{pmatrix} c_k \\ \hline 0 \end{pmatrix} + \sum_{k=1}^{m_c} w_k^- \begin{pmatrix} -c_k \\ \hline 0 \end{pmatrix}. \tag{10.4.27}$$

Put $w_k = w_k^+ - w_k^-$. The first n equations in (10.4.27) give the first equation in (10.4.25), whereas the last equation in (10.4.27) yields the second one in (10.4.25). □

<u>Proof of Theorem 10.4.6</u>. Suppose that MFCQ1 holds. Now, suppose that there exist numbers λ_i, $i \in I$ and μ_j, $j \in J_0(\bar{x},\bar{t})$ such that (10.4.23) is satisfied. Then, in view of Definition 10.4.1(i) not all μ_j vanish; but then, in virtue of Theorem 10.4.7, (ii) in Definition 10.4.1 is violated (the role of a_i, resp. c_k being played by $-D_x^T g_j(\bar{x},\bar{t})$, resp. $D_x^T h_i(\bar{x},\bar{t})$). Now, suppose that MFCQ 1 does not hold. If (i) in Definition 10.4.1 is not satisfied, then (10.4.23) obviously holds with $\mu_j = 0$, $j \in J_0(\bar{x},\bar{t})$. So, now suppose that (i) in Definition 10.4.1 is satisfied, but (ii) is violated. Then, again in virtue of Theorem 10.4.7, it follows that (10.4.23) holds. This completes the proof. □

Let $\bar{x} \in M(\bar{t})$ and suppose that $M(\bar{t})$ is <u>regular</u> at $\bar{x}$. If $\bar{x}$ is, in addition, a local minimum for $f(\cdot,\bar{t})_{|M(\bar{t})}$, then $\bar{x}$ is necessarily a (+)-Kuhn-Tucker point (cf. Definition 3.2.3 and Lemma 3.2.4).

Now, we extend the concept of a (+)-Kuhn-Tucker point to the case that $M(\bar{t})$ is possibly not regular at $\bar{x}$. In order to make this extension unambiguous, we delete the (+)-symbol in the next definition.

<u>Definition 10.4.2</u>. Let $\bar{x}$ belong to $M(\bar{t})$. The point $\bar{x}$ is called a Kuhn-Tucker point for $f(\cdot,\bar{t})_{|M(\bar{t})}$ if there exist real numbers $\bar{\lambda}_i$, $i \in I$ and $\bar{\mu}_j \geq 0$, $j \in J_0(\bar{x},\bar{t})$, such that

$$D_x f = \sum_{i \in I} \bar{\lambda}_i D_x h_i + \sum_{j \in J_0(\bar{x},\bar{t})} \bar{\mu}_j D_x g_j{}_{|(\bar{x},\bar{t})} . \tag{10.4.28}$$

<u>Remark 10.4.1</u>. Note that the numbers $\bar{\lambda}_i$, $\bar{\mu}_j$ in (10.4.28) need <u>not</u> be unique.

<u>Theorem 10.4.8</u>. Let $\bar{x} \in M(\bar{t})$ be a local minimum for $f(\cdot,\bar{t})_{|M(\bar{t})}$. Then, there exist real numbers $\bar{\lambda}_i$, $i \in I$, $\bar{\lambda} \geq 0$, $\bar{\mu}_j \geq 0$, $j \in J_0(\bar{x},\bar{t})$,

$$\bar{\lambda} D_x f = \sum_{i \in I} \bar{\lambda}_i D_x h_i + \sum_{j \in J_0(\bar{x},\bar{t})} \bar{\mu}_j D_x g_j{}_{|(\bar{x},\bar{t})} . \tag{10.4.29}$$

If, moreover, the Mangasarian-Fromovitz constraint qualification holds at $\bar{x}$, then $\bar{\lambda}$ in (10.4.29) must be unequal to zero, and hence, $\bar{x}$ is a Kuhn-Tucker point for $f(\cdot,\bar{t})_{|M(\bar{t})}$.

<u>Proof</u>. If $D_x h_i(\bar{x},\bar{t})$, $i \in I$, are linearly dependent, then (10.4.29) can be satisfied with $\bar{\lambda} = 0$ and $\bar{\mu}_j = 0$, $j \in J_0(\bar{x},\bar{t})$. Now, suppose that $D_x h(\bar{x},\bar{t})$, $i \in I$, are linearly independent, and that $\bar{x} \in M(\bar{t})$ is a local minimum for $f(\cdot,\bar{t})_{|M(\bar{t})}$. Then, the following system is not solvable at $(\bar{x},\bar{t})$:

$$\left.\begin{aligned} &D_x f\xi < 0 \\ &D_x h_i \xi = 0 , \ i \in I \\ &D_x g_j \xi > 0, \ \ j \in J_0(\bar{x},\bar{t}) . \end{aligned}\right\} \tag{10.4.30}$$

In fact, if (10.4.30) would have a solution $\tilde{\xi}$, then, using the fact that $D_x h_i(\bar{x},\bar{t})$, $i \in I$, are linearly independent, there exists a C^1-curve $u \mapsto x(u)$ with the properties (cf. also Lemma 3.2.1):

$$x(0) = \bar{x}\ ,\ h_i(x(u)) \equiv 0\ ,\ i \in I \text{ and } \frac{dx}{du}(0) = \tilde{\xi}.$$

Obviously, for $u \in (0,\varepsilon)$, ε sufficiently small, we have $g_j(x(u)) > 0$, $j \in J$ (for $j \in J_0(\bar{x},\bar{t})$ this follows from (10.4.30)). Hence, for $u \in [0,\varepsilon)$, the point $x(u)$ lies in $M(\bar{t})$. But, $\frac{d}{du} f(x(u))_{|u=0} = Df(\bar{x})\tilde{\xi} < 0$, and, consequently, $f(x(u)) < f(x(0))$ for all $u > 0$ and u sufficiently small. This, however contradicts the fact that $\bar{x}$ is a local minimum for $f(\cdot,t)_{|M(\bar{t})}$.

So, (10.4.30) is <u>not</u> solvable, and application of Theorem 10.4.7 gives us (10.4.29). In particular, not all $\bar{\lambda}$, $\bar{\mu}_j$, $j \in J_0(\bar{x},\bar{t})$ vanish.

Next, suppose in addition, that MFCQ holds and that $\bar{\lambda} = 0$. Then, Condition (i) in Definition 10.4.1 implies that at least one of the numbers $\bar{\mu}_j$, $j \in J_0(\bar{x},\bar{t})$ does not vanish. Finally, we multiply (10.4.29) from the right with a solution vector $\bar{\xi}$ of the system (10.4.22) and we get a contradiction. □

<u>Remark 10.4.2</u>. A point $\bar{x} \in M(\bar{t})$ which satisfies (10.4.29) with $\bar{\lambda} \geq 0$ and $\bar{\mu}_j \geq 0$, $j \in J_0(\bar{x},\bar{t})$, is also called a Fritz-John point. So, a local minimum is always a Fritz-John point; but, in order to be also a Kuhn-Tucker point, some additional condition has to be imposed, such as MFCQ for example. In fact, deleting the parameter $\bar{t}$, consider the following example in $\mathbb{R}^2$, with data $f(x_1,x_2) = x_1$, $g_1(x) = x_2-x_1^2$, $g_2(x) = 2x_1^2-x_2$ and $g_3(x) = x_1x_2$. The feasible set $M[g_1,g_2,g_3]$ is depicted in Fig. 10.4.7. Obviously, the origin is a local minimum for $f_{|M}$, but $\frac{\partial f}{\partial x_1}(0) = 1$, whereas $\frac{\partial}{\partial x_1} g_i(0) = 0$, $i = 1,2,3$. Consequently, the origin cannot be a Kuhn-Tucker point for $f_{|M}$. For a condition which is both necessary and sufficient for a local minimum to be a Kuhn-Tucker point we refer to [27]. □

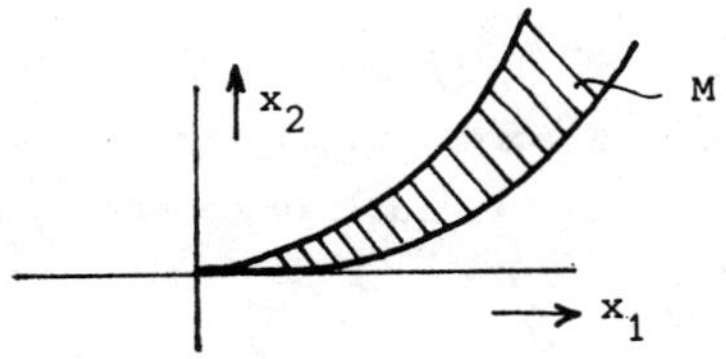

Fig.10.4.7

We recall that the Lagrange parameters in (10.4.28) need not be unique. However, if $\bar{x} \in M(\bar{t})$ is a Kuhn-Tucker point, then the set $\Lambda = \{(\lambda_i, i \in I, \mu_j, j \in J_0(\bar{x},\bar{t}) \,|\, \lambda_i, \mu_j$ satisfy (10.4.28)$\}$ is <u>compact</u> iff MFCQ holds at $\bar{x}$. This is Gauvin's result [14], and it is easily proved by using Theorem 10.4.7. Of course, Λ is also a convex set (even a polyhedron). Now, we state the theorem that we had in mind (compare with Theorem 10.4.2, the Diffeomorphy-Theorem).

<u>Theorem 10.4.9</u>. (<u>Homeomorphy-Theorem</u>).
Let t_1, t_2 be parameters with $t_1 < t_2$. Suppose that the set $M \cap (\mathbb{R}^n \times [t_1,t_2])$ is compact and that for all $t \in [t_1,t_2]$ the set $M(t)$ satisfies the Mangasarian-Fromovitz constraint qualification at all points $x \in M(t)$. Then, $M(t)$ is <u>homeomorphic</u> with $M(t_1)$ for all $t \in [t_1,t_2]$. □

We will not give the proof of Theorem 10.4.9 here. In fact, since the set $M \cap (\mathbb{R}^n \times [t_1,t_2])$ is assumed to be compact, the proof can be deduced from the following basic stability theorem in [21] (the proof of the latter theorem is rather complicated since no stratification can be used; further, tools from algebraic topology are needed in order to show that specific sets are <u>not</u> homeomorphic).

<u>Theorem 10.4.10</u>. (<u>Stability Theorem</u>) ([21]).
Let h_i, $g_j \in C^\infty(\mathbb{R}^n,\mathbb{R})$, $i \in I$, $j \in J$ (fixed index sets) and $|I|+|J| < \infty$. Denote $M[h,g] = \{x \in \mathbb{R}^n \,|\, h_i(x) = 0, i \in I, g_j(x) \geq 0, j \in J\}$.
Suppose that $M[h,g]$ is <u>compact</u>.
Then, there exists a C^1-neighborhood $\mathcal{O}$ of $(h_i, i \in I, g_j, j \in J)$ in $C^\infty(\mathbb{R}^n,\mathbb{R})^{|I|+|J|}$ with the property that $M[\tilde{h},\tilde{g}]$ is <u>homeomorphic</u> with $M[h,g]$

for all $(\tilde{h}_i, i \in I, \tilde{g}_j, j \in J) \in \mathcal{O}$, <u>iff</u> MFCQ is satisfied at all points of $M[h,g]$. □

<u>Remark 10.4.3</u>. It suffices in Theorem 10.4.10 to take all equality constraints (h_i) of class C^2 and all inequality constraints (g_j) of class C^1 (cf. [21])

<u>Remark 10.4.4</u>. In the situation of Theorem 10.4.4., the Mangasarian-Fromovitz constraint qualification fails to hold at the points of Type 1, 3 and 4. However, MFCQ holds at a point of Type 2 (use Theorem 10.4.6). So, in Theorem 10.4.5, <u>Type 2</u>, we may replace "homotopy-equivalent" by "homeomorphic".

10.5. Final remarks.

A different approach in studying one-parameter families of optimization problems is proposed in the interesting work of M. Kojima and R. Hirabayashi [52]. The underlying idea was developed by M. Kojima in [51], where he introduced a special type of <u>piecewise</u> differentiable mapping (the word "piecewise" refers to Lagrange parameters corresponding to <u>inequality constraints</u>).

Let again $f, h_i, g_j \in C^\infty(\mathbb{R}^n \times \mathbb{R}, \mathbb{R})$, $i \in I$, $j \in J$ and $|I|+|J| < \infty$. For each value of the parameter $t \in \mathbb{R}$ let the feasible set $M(t)$ be defined as in (10.4.1). For $\alpha \in \mathbb{R}$ we define (and this construction makes the subsequent mapping $\mathcal{T}$ piecewise smooth):

$$\alpha^+ = \max\{0,\alpha\}, \quad \alpha^- = \min\{0,\alpha\}. \tag{10.5.1}$$

In our terminology, the mapping introduced by M. Kojima becomes:

$$\mathcal{T}: \begin{pmatrix} x \\ \lambda \\ \mu \\ \hline t \end{pmatrix} \longmapsto \begin{pmatrix} D_x^T\left(f - \sum\limits_{i\in I} \lambda_i h_i - \sum\limits_{j\in J} \mu_j^+ g_j\right)\Big|_{(x,t)} \\ h_i(x,t), \quad i \in I \\ \mu_j^- + g_j(x,t), \quad j \in J \end{pmatrix} \tag{10.5.2}$$

The mapping T in (10.5.2) is piecewise smooth, due to the stratification of the μ-space into its orthants.
Now, suppose that $T(\bar{x},\bar{\lambda},\bar{\mu},\bar{t}) = 0$. Then, it follows that $\bar{x} \in M(\bar{t})$ and, moreover, $\bar{x}$ is a Kuhn-Tucker point for $f(\cdot,\bar{t})_{|M(\bar{t})}$ (compare Definition 10.4.2). It is easily verified that also the converse holds. So, the projection of the set $T^{-1}(0)$ to the (x,t)-space is precisely the Kuhn-Tucker set Σ_{KT} (in analogy with (10.1.12)). The concept of a regular value of the piecewise smooth mapping T can be introduced in a natural way subject to the stratification of the μ-space (cf. [52]). In particular, if zero is a regular value of T, then the set $T^{-1}(0)$ is a piecewise smooth curve in (x,λ,μ,t)-space. If, in addition, the Mangasarian-Fromovitz qualification holds at all points in the closure of the Kuhn-Tucker set Σ_{KT}, then Σ_{KT} turns out to be a one-dimensional topological manifold without boundary ([52]).
Points at which MFCQ fails to hold will cause difficulties in the above approach. In [47], [48] these difficulties are studied, and there a complete (intrinsic) study of critical sets depending on one parameter is presented, including all possible changes of the indices LI, LCI, QI, QCI; the approach in [47], [48] is based on the ideas as developed in the present chapter.

In this chapter we considered problems with a finite number of constraints and depending on external parameters. However, parametric aspects also arise when the index set of the inequality constraints becomes infinite, or even a continuum (for example, an interval, square etc.). This type of problems in intimately related with Chebyshev approximation (cf. Chapter 4). In order to get an idea of the structure of these problems, let Y be an index set and G a mapping from the product $\mathbb{R}^n \times Y$ to $\mathbb{R}$. The mapping G defines a feasible set M as follows:

$$M = \{x \in \mathbb{R}^n | G(x,y) \geq 0 \quad \text{for all} \quad y \in Y\}. \tag{10.5.3}$$

If Y is a finite set, then M in (10.5.3) is defined by means of a finite number of inequality constraints, say $G(\cdot,y_i)$; $i = 1,\dots,s$. In the infinite case, the following observation is crucial. Let $\tilde{x} \in M$ and define the active index set $Y_0(\tilde{x})$ as follows:

$$Y_0(\tilde{x}) = \{y \in Y | G(\tilde{x},y) = 0\} . \tag{10.5.4}$$

‖ Then, <u>each point of</u> $Y_0(\bar{x})$ <u>is a (global) minimum for</u> $G(\tilde{x},\cdot)_{|Y}$.

Now, we treat the variable x as a parameter and study the set of global minima of $G(x,\cdot)_{|Y}$ as x varies. The corresponding marginal value then decides whether x is a feasible point (i.e. $x \in M$) or not. So, in this way, we enter in a natural way into parametric problems. For general information on these socalled "semi-infinite" problems see[29]. For a specific study, using tools from singularity theory, we refer to [44], [45], [46].

REFERENCES

[1] Abraham, R., Robbin, J.: Transversal mappings and flows. Benjamin, New York (1967).

[2] Andronov, A.A., Leontovich, E.L., Gordon, I.I., Maier, A.G.: Qualitative theory of second-order dynamical systems. John Wiley & Sons (1973).

[3] Aubin, J.P., Ekeland, I.: Applied nonlinear analysis. Wiley-Interscience Publ. (1984).

[4] Bank, B., Guddat, J., Klatte, D., Kummer, B., Tammer, K.: Nonlinear parametric optimization. Akademie-Verlag, Berlin (1982).

[5] Braess, D.: Ueber die Einzugsbereiche der Nullstellen von Polynomen beim Newton-Verfahren. Numer. Math. 29, pp. 123-132 (1977).

[6] Branin, F.H.: A widely convergent method for finding multiple solutions of simultaneous non-linear equations. IBM J. Res. Develop., pp. 504-522 (1972).

[7] Bröcker, Th.: Differentiable germs and catastrophes; translated by L. Lander. London Math. Soc. Lecture Notes 17, Cambridge Univ. Press (1975).

[8] Coddington, E.A. and Levinson, N.: Theory of ordinary differential equations. McGraw-Hill; New York, Toronto, London (1955).

[9] Diener, I.: On the global convergence of path-following methods to determine all solutions to a system of nonlinear equations. Göttingen, NAM-Bericht No. 47 (1985).

[10] Diener, I.: Trajectory nets connecting all critical points of a smooth function. Göttingen, NAM-Bericht No. 51 (1986).

[11] Fujiwara, O.: A note on differentiability of global optimal values. Mathematics of Operations Research, Vol. 10, No. 4, pp. 612-618 (1985).

[12] Gantmacher, F.R.: Matrizenrechnung, Berlin (1970).

[13] Garcia, C.B., Gould, F.J.: Relations between several path-following algorithms and local and global Newton-methods. SIAM Review, Vol. 22, No. 3, pp. 263-274 (1980).

[14] Gauvin, J.: A necessary and sufficient regularity condition to have bounded multipliers in nonconvex programming. SIAM J. Control and Optimization, 17, pp. 321-338 (1979).

[15] Giblin, P.J.: Graphs, surfaces and homology. John Wiley & Sons (1977).

[16] Gibson, C.G., Wirthmüller, K., du Plessis, A.A., Looijenga, E.J.N.: Topological stability of smooth mappings. Lect. Notes Math., Vol. 552, Springer Verlag (1976).

[17] Gibson, C.G.: Singular points of smooth mappings. Research Notes in Mathematics, Vol. 25, Pitman (1979).

[18] Golubitsky, M., Guillemin, V.: Stable mappings and their singularities. Graduate Texts in Mathematics, Springer Verlag (1973).

[19] Gomulka, J.: Remarks on Branin's method for solving non-linear equations. In: Towards global optimization (L.C.W. Dixon and G.P. Szegö, eds.), Acad. Press (1976).

[20] Grobman, D.: Homeomorphisms of systems of differential equations. Dokl. Akad. Nauk SSSR Vol. 128, pp. 880-881 (1959).

[21] Guddat, J., Jongen, H.Th., Rueckmann, J.: On stability and stationary points in nonlinear optimization. Twente University of Technology, Memorandum Nr. 526 (1985); to appear in J. Australian Math. Soc., Series B.

[22] Guillemin, V., Pollack, A.: Differential topology. Prentice Hall (1974).

[23] Halmos, P.R.: Measure theory. Van Nostrand-Reinhold, New York (1950).

[24] Harary, F.: Graph theory. Addison-Wesley Publ. (1969).

[25] Hartman, P.: A lemma in the theory of structural stability of differential equations. Proceedings American Math. Soc., Vol. 11, pp. 610-620 (1960).

[26] Hestenes, M.R.: Conjugate direction methods in optimization. Applications of Mathematics, Vol. 12, Springer Verlag (1980).

[27] Hettich, R., Jongen, H.Th.: On first and second order conditions for local optima for optimization problems in finite dimensions. Methods of Operations Research, Vol. 23, pp. 82-97 (1977).

[28] Hettich, R., Jongen, H.Th.: On the local continuity of the Chebyshev operator. J. Approximation Theory, Vol. 33, No. 4, pp. 296-307 (1981).

[29] Hettich, R., Zencke, P.: Numerische Methoden der Approximation und semi-infiniten Optimierung. Teubner Studienbücher, Stuttgart (1982).

[30] Hirsch, M.W., Smale, S.: Differential equations, dynamical systems and linear algebra. Academic Press (1974).

[31] Hirsch, M.W.: Differential topology. Graduate Texts in Mathematics, Vol. 33, Springer Verlag (1976).

[32] Hirsch, M.W., Smale, S.: Algorithms for solving f(x) = 0. Comm. Pure Appl. Math. 32, pp. 281-312 (1979).

[33] Hocking, J.G., Young, G.S.: Topology. Addison-Verlag Publ. Co. Inc. (1961).

[34] Jongen, H.Th.: On nonconvex optimization. Dissertation, Twente University of Technology (1977).

[35] Jongen, H.Th.: Zur Geometrie endlichdimensionaler nichtkonvexer Optimierungsaufgaben.Int. Ser. Num. Math., Vol. 36, pp. 111-136 (1977).

[36] Jongen, H.Th., Jonker, P., Twilt, F.: On Newton-flows in optimization. Methods of Operations Research, Vol. 31, pp. 345-359 (1979).

[37] Jongen, H.Th., Jonker, P., Twilt, F.: The continuous Newton-method for meromorphic functions. In: Geometric Approaches to Differential Equations (R. Martini, ed.), Lect. Notes in Math., Vol. 810, Springer Verlag, pp. 181-239 (1980).

[38] Jongen, H.Th.: Optimalitätskriterien und lokale Stetigkeit des Tschebyscheff-Operators. Int. Series Num. Math., Vol. 55, Birkhäuser Verlag, pp. 121-130 (1980).

[39] Jongen, H.Th., Sprekels, J.: The index-k-stabilizing differential equation. OR-Spektrum 2, pp. 223-225 (1981).

[40] Jongen, H.Th., Jonker, P., Twilt, F.: On one-parameter families of sets defined by (in)equality constraints. Nieuw Archief v. Wiskunde (3), XXX, pp. 307-322 (1982).

[41] Jongen, H.Th., Jonker, P., Twilt, F.: On index-sequence realization in parametric optimization. Seminarberichte Humboldt Universität zu Berlin, Vol. 50, pp. 159-166 (1983).

[42] Jongen, H.Th., Jonker, P., Twilt, F.: The continuous, desingularized Newton-method for meromorphic functions. Twente University of Technology, Memorandum Nr. 501 (1985).

[43] Jongen, H.Th., Jonker, P., Twilt, F.: Parametric optimization: the Kuhn-Tucker set. Twente University of Technology, Memorandum Nr. 549 (1986); to appear in proceedings of the conference "Parametric Optimization and Related Topics" (Plaue, GDR, 1985).

[44] Jongen, H.Th., Zwier, G.: On the local structure of the feasible set in semi-infinite optimization. Int. Series Num. Math., Vol. 72, pp. 185-202 (1985).

[45] Jongen, H.Th., Zwier, G.: Structural analysis in semi-infinite optimization. Proceedings Third Franco-German Conference in Optimization; INRIA (C. Lemarechal, ed.), pp. 56-67 (1985).

[46] Jongen, H.Th., Zwier, G.: On regular semi-infinite optimization. In: Infinite Programming (E.J. Anderson, A.B. Philpott, eds.). Lecture Notes in Economics and Math. Systems, Vol. 259, Springer Verlag, pp. 53-64 (1985).

[47] Jongen, H.Th., Jonker, P., Twilt, F.: One-parameter families of optimization problems: equality constraints. Journal of Optimization Theory and Applications 48, pp. 141-161 (1986).

[48] Jongen, H.Th., Jonker, P., Twilt, F.: Critical sets in parametric optimization. Mathematical Programming 35, pp. 1-21 (1986).

[49] Keller, H.B.: Global homotopies and Newton-methods. In: Recent advances in numerical analysis (De Boor, C., Golub, G.H., eds.), Acad. Press, pp. 73-94 (1978).

[50] Kelley, J.K.: General topology. Van Nostrand-Reinhold (1969).

[51] Kojima, M.: Strongly stable stationary solutions in nonlinear programs. In: Analysis and Computation of Fixed Points (S.M. Robinson, ed.), Acad. Press, pp. 93-138 (1980).

[52] Kojima, M., Hirabayashi, R.: Continuous deformations of nonlinear programs. Mathematical Programming Study 21, pp. 150-198 (1984).

[53] Lang, S.: Introduction to differentiable manifolds. Interscience Publishers (1962).

[54] Lang, S.: Algebra. Addison Wesley Publ. (1965).

[55] Lefschetz, S.: Differential equations: geometric theory. Interscience Publ. (1962).

[56] Ljusternik, L.A., Sobolew, W.I.: Elemente der Funktionalanalysis. Akademie-Verlag, Berlin (1968).

[57] Lu, Y.-C.: Singularity theory and an introduction to catastrophe theory. Universitext, Springer Verlag (1976).

[58] Markushevich, A.I.: Theory of functions of a complex variable, Vol. I, Prentice Hall (1965).

[59] Markushevich, A.I.: Theory of functions of a complex variable, Vol. II, Prentice Hall (1965).

[60] Mather, J.N.: Stability of C^∞-mappings: V, transversality. Advances in Mathematics 4, pp. 301-336 (1970).

[61] Mather, J.N.: Stratifications and mappings. In: Dynamical Systems (Ed.: M.M. Peixoto), Academic Press, pp. 195-232 (1973).

[62] Milnor, J.: Lectures on the h-cobordism theorem. Mathematical Notes 1, Princeton University Press (1965).

[63] Narasimhan, R.: Analysis on real and complex manifolds. North-Holland Publ. Co., Amsterdam (1968).

[64] Peixoto, M.C.: Structural stability on two-dimensional manifolds. Topology 1, pp. 101-120 (1962).

[65] Palis, J., Smale, S.: Structural stability theorems. In: Global Analysis; Proc. A.M.S. Symp. in Pure Math. XIV, pp. 223-232 (1970).

[66] Palis, J., Takens, F.: Stability of parametrized families of gradient vector fields. Annals of Mathematics, Vol. 118, pp. 383-421 (1983).

[67] Schecter, S.: Structure of the first-order solution set for a class of nonlinear programs with parameters. Mathematical Programming 34, pp. 84-110 (1986).

[68] Shub, M., Williams, B.: The Newton graph of a complex polynomial. Preprint (1985).

[69] Smale, S.: On gradient dynamical systems. Annals of Mathematics, Vol. 74, No. 1, pp. 199-206 (1961).

[70] Smale, S.: An infinite dimensional version of Sard's theorem. Ann. J. Math., 87, pp. 861-866 (1965).

[71] Smale, S.: A convergent process of price adjustment and global Newton-methods. J. Math. Economics, Vol. 3, pp. 107-120 (1976).

[72] Smale, S.: On the efficiency of algorithms of analysis. Bull. Am. Math. Soc., Vol. 13, No. 2, pp. 87-121 (1985).

[73] Spanier, E.H.: Algebraic Topology. McGraw-Hill; New York, Toronto, London (1966).

[74] Sternberg, S.: Lectures on differential geometry. Prentice Hall, Inc. (1964).

[75] Twilt, F.: Newton-flows for meromorphic functions. Dissertation, Twente University of Technology (1981).

[76] Whitney, H.: A function not constant on a connected set of critical points. Duke Mathematical Journal, Vol. 1, pp. 514-517 (1935).

INDEX

Note: References to page 1-264 are contained in the first volume:

NONLINEAR OPTIMIZATION IN $\mathbb{R}^n$, I. MORSE THEORY, CHEBYSHEV APPROXIMATION. Vol. 29, Methoden und Verfahren der mathematischen Physik, Peter Lang Verlag (1983).

GLOSSARY OF SYMBOLS

Symbol	Meaning	Page
$<\cdot>$	standard inner product	
$<<\cdot>>$	arbitrary inner product	
$\|\cdot\|$	Euclidean norm	
$\|\cdot\|_{\cdot}$	norm	267
$\|\|\|\cdot\|\|\|$	norm induced by $<<\cdot>>$	
$\lvert\cdot\rvert_{\cdot}$	$(C^k$-) semi-norm	281
$\bar{\pitchfork}$		325, 329
$\bar{\pitchfork}\ M$		339
$\tilde{A}$	adjoint matrix of A	442
C(F)		479
$C^k(\cdot,\cdot)$		3,25
$C^2(B)$		267
Crit$(\cdot)$		441, 455
Ess$(\cdot)$		444, 455
Ext$(\cdot)$		444, 455
$f_{\mid X}$	restriction of f to X	
G(n,k)		351
G(n,k,r)		353
$\text{grad}_{\mathcal{R}}f$		405
$\text{grad}_{\mathcal{R}}f_{\mid M}$		411
J_0	activity map	94
$\mathcal{J}(n,m,\ell)$		363
$J_p(n,m,\ell)$		369
$j^{\ell}f$		363
$\widetilde{j^{\ell}}f$		368
$j_p^{\ell}f$		369
LI		114
LCI		114
MFCQ		543
MGB		93
M[h,g]		93

Symbol	Meaning	Page
ND1, ND2		116
$\mathcal{N}(\cdot)$		442, 455
$\tilde{\mathcal{N}}(\cdot)$		479
$\bar{\mathcal{N}}(\cdot)$		480
$\bar{\bar{\mathcal{N}}}(\cdot)$		487
N(F)		479
N_xM	normal space to M at x	328
P(F)		479
QI		113
QCI		113
$\mathcal{R}(\cdot)$	Riemannian metric	405
$\mathcal{R}$		487
$\tilde{\mathcal{R}}$		489
$\mathcal{RR}$		534
Reg$(\cdot)$		444, 455
(S,Σ)		359
stratum*		359
$\Sigma(F)$, $\Sigma_d(F)$		497, 498
$\Sigma_+(F)$		502
TF$(\cdot)$	trajectory-map	445
T_xX	tangentspace to X at x	
$W_{\bar{x}}^s$		388
$W_{\bar{x}}^u$		391
$V_{\phi,f}^k$		280

Die Reihe „Methoden und Verfahren der Mathematischen Physik" dient der möglichst raschen Veröffentlichung von Beiträgen aus Gebieten der angewandten Mathematik und der mathematischen Physik, die insbesondere neuere Entwicklungen auf diesen Gebieten berücksichtigen. Es sollen Vorträge von Tagungen, Übersichtsberichte über neuere Entwicklungen, Projektbeschreibungen, Vorlesungs- und Seminarausarbeitungen veröffentlicht werden. Die Beiträge sollen sowohl auf die zugrundeliegenden naturwissenschaftlichen bzw. technischen Probleme selbst eingehen, wie auch den Methoden zu ihrer Behandlung bis hin zur verfahrensmäßigen Gewinnung der in der Praxis tatsächlich interessierenden Lösung gewidmet sein. Auf diese Weise wird angestrebt, den Mathematiker enger an Anwendungsmöglichkeiten seines Faches heranzuführen, den Physiker und Ingenieur hingegen näher mit mathematischen Methoden, die er für seine Arbeit benötigt, vertraut zu machen.
Manuskripte von Beiträgen, die möglichst in englischer Sprache abgefaßt sein sollen, können an einen der unten genannten Herausgeber eingesandt werden. Die Manuskripte werden photomechanisch vervielfältigt und müssen daher mit großer Sorgfalt angefertigt werden. Hinweise für ihre Anfertigung sind bei den Herausgebern erhältlich.

The series „Methoden und Verfahren der Mathematischen Physik" serves as a quick publication of contributions to applied mathematics and mathematical physics, and gives special regard to new developments in these fields. To be published are proceedings of conferences, project descriptions, synopses of new developments, and lecture and seminar works. The contributions should go into the underlying scientific and technical problems themselves, as well as the methods used in their treatment, and the procedural attainment of those solutions which are interesting for practical applications. In this way, efforts are made not only to aquaint the mathematician more closely with applications of his field, but at the same time to familiarize the physicist and the engineer with mathematical methods which are very helpful in their work.
Manuscripts should be written in English if possible, and can be sent to one of the editors listed below. Because the manuscripts are duplicated photomechanically, they should be prepared with great care. Suggestions for their preparation are obtainable from the editors.

Prof. Dr. Bruno Brosowski
Johann Wolfgang Goethe-Universität
Fachbereich Mathematik
Robert Mayer-Straße 6-10
D-6000 Frankfurt 1

Prof. Dr. Erich Martensen
Universität Karlsruhe
Mathematisches Institut II
Englerstraße 2
D-7500 Karlsruhe 1

METHODEN UND VERFAHREN DER MATHEMATISCHEN PHYSIK

Band 1-16 ist erschienen im Bibliographischen Institut, Mannheim.

Band 17 Bruno Brosowski und Erich Martensen: Proceedings of the Conference Oberwolfach, December 4-10, 1977. 1978.

Band 18 Rainer Kress und Norbert Weck: Free and Mixed Boundary Value Problems. Proceedings of a Conference held in Oberwolfach, April 16-22, 1978. 1979.

Band 19 Helmut Neunzert und Donald Pack: Mathematical Problems in the Kinetic Theory of Gases. Proceedings of a Conference Oberwolfach, May 20-26, 1979. 1980.

Band 20 Rainer Kress und Joachim Wick: Mathematical Problems of Plasmaphysics. Proceedings of a Conference held in Oberwolfach, September 16-22, 1979. 1980.

Band 21 Bruno Brosowski und Erich Martensen: Methods and Techniques of Mathematical Physics. Proceedings of a Conference held in Oberwolfach, February 3-9, 1980. 1981.

Band 22 Bruno Brosowski: Parametric Semi-Infinite Optimization. 1982.

Band 23 Ernst Stephan und Wolfgang Wendland: Mathematische Grundlagen der finiten Element-Methoden. Ausarbeitung zum Mittelseminar im WS 1980/81. 1982.

Band 24 Erhard Meister, Karl Nickel und Jan Polášek: Mathematical Methods in Fluid Mechanics. Proceedings of a Conference Oberwolfach, November 29 - December 5, 1981. 1983.

Band 25 Rudolf Gorenflo und Karl-Heinz Hoffmann: Applied nonlinear Functionalanalysis. Variational Methods and Ill-posed Problems. 1983.

Band 26 Bruno Brosowski und Erich Martensen (Eds.): Dynamical Problems in Mathematical Physics. 1983.

Band 27 Bruno Brosowski und Erich Martensen (Eds.): Approximation and Optizimation in Mathematical Physics. 1983.

Band 28 Erhard Meister: Integraltransformationen mit Anwendungen auf Probleme der mathematischen Physik. 1983.

Band 29 H. Th. Jongen; P. Jonker; F. Twilt: Nonlinear Optimization in $\mathbb{R}^n$. I. Morse Theory, Chebyshev Approximation. 1983.

Band 30 D.S. Gilliam; J.R. Schulenberger: The Propagation of Electromagnetic Waves Through, Along, and Over a Three-Dimensional Conducting Half Space. EM Waves Over a Conducting Earth. 1986.

Band 31 Johannes Jahn: Mathematical Vector Optimization in Partially Ordered Linear Spaces. 1986.

Band 32 H.Th. Jongen/P. Jonker/F. Twilt: Nonlinear Optimization in $\mathbb{R}^n$. II. Transversality, Flows, Parametric Aspects. 1986.